Discovery EDUCATION | MATH TECHBOOK

ALGEBRA 1 WITH HONORS

FLORIDA INTERACTIVE STUDENT EDITION

Log in to Math Techbook at
www.DiscoveryEducation.com

ISBN 13: 978-1-68220-450-4

Library of Congress Cataloging-in-Publication Data available upon request from the Library of Congress.

Printed in the United States of America.

1 2 3 4 5 6 7 8 9 10

Discovery EDUCATION™

800-323-9084

One Discovery Place, Silver Spring, Maryland 20910

Table of Contents

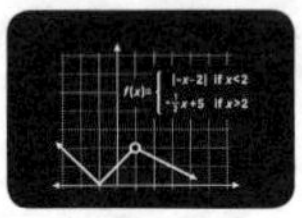

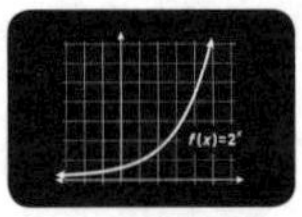

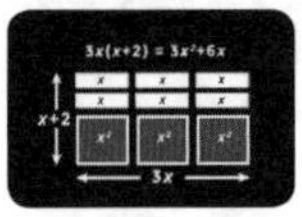

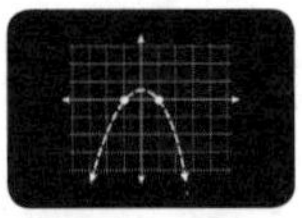

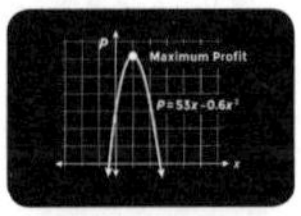
Maximum Profit

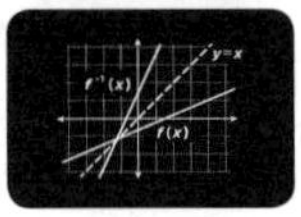

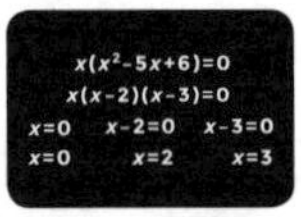
x(x²-5x+6)=0
x(x-2)(x-3)=0
x=0 x-2=0 x-3=0
x=0 x=2 x=3

How many minutes to fill?

Letter to the Student

Dear Student,

You are about to experience mathematics like you never have before! In this class, you'll be using Math Techbook™—a comprehensive, digital math instructional program developed by the educators and designers at Discovery Education. Math Techbook is full of interactives, videos, digital tools, and game-like activities to help you learn mathematical concepts and apply them to the real world. You won't just study the concepts; you'll develop real-world skills and tackle relevant problems that are worth solving. There are multiple pathways for you to work at your own pace and ability. You'll even be able to monitor your progress in real time using the Student Learning Dashboard.

As a print resource to use in conjunction with Math Techbook, this *Interactive Student Edition (ISE)* allows you to practice the skills you've learned in class and online. You can use this resource to record questions, find connections to the digital content, and develop your own mathematical understanding. Math Techbook is divided into units, and each unit is divided into concepts. Each concept is then divided into three tabs: Discover, Practice, and Apply. This print resource is organized by concept and includes the following:

- OVERVIEW: Lesson Objectives, Essential Questions, key vocabulary, and short descriptions of the online investigations will help you make connections to the mathematical content.
- DISCOVER: This page includes QR codes that connect to the online Discover investigations and provides a place for you to take notes and record important information.
- CHECKS FOR UNDERSTANDING: There is a Check for Understanding page for each investigation within Discover. The exercises in each Check for Understanding are your chance to show what you've learned. The final question in each set asks you to justify your reasoning and communicate mathematically.
- SUMMARY: The Summary is a written description of what you've learned in each investigation and encourages you to make connections.
- PRACTICE EXERCISES: These exercises provide additional review material to sharpen your skills. You can find more problems in the online Practice section, with Coach exercises that provide feedback and support and Play exercises that allow you to earn badges while practicing what you've learned.
- APPLY: The Apply problem uses real-world examples to put your knowledge to the test. A list of criteria and a rubric will give you direction on what's required.

Within each section of this resource, you'll find QR codes that take you to the corresponding online section of Math Techbook for that concept. For instance, the QR codes on the Discover page provide a direct link to each online investigation, and the QR codes on the Apply page connect you to related videos and additional problems. Once inside Math Techbook, you'll have access to activities that develop conceptual understanding in Discover, practice exercises that develop procedural fluency in Practice and real-world problems worth solving in Apply.

We hope this program offers you the chance to dive deeper into math and have fun, too! Best of luck for a fantastic year!

Sincerely,

The Discovery Education Math Team

Letter to the Parent/Guardian

Dear Parent/Guardian,

This year, your student will be using Math Techbook™, a comprehensive, digital program developed by the educators and designers at Discovery Education. Math Techbook is an innovative program that offers engaging, real-world problems to help your student master key mathematical concepts and procedures. In class, students experience dynamic content, interactives, videos, digital tools, and game-like activities that increase their motivation to learn math.

As a print resource to use in conjunction with Math Techbook, this *Interactive Student Edition (ISE)* offers a way for students to practice the skills they've learned in class and online. Students are encouraged to use this resource to capture important ideas, seek connections, and develop their own mathematical understanding. Math Techbook is divided into units, and each unit is divided into concepts. Each concept is then divided into three tabs: Discover, Practice, and Apply. This print resource is organized by concept and includes the following:

- OVERVIEW: Students learn mathematics through inquiry and active participation. In the overview for each concept, the Essential Questions and short descriptions of the in-class investigations provide a roadmap for what students are expected to learn and help them make connections to the mathematical content.
- DISCOVER: The Discover page is a place for students to record important information that they learn while completing the activities within the online Math Techbook. QR codes link directly to the online investigations.
- CHECKS FOR UNDERSTANDING: There is a Check for Understanding page for each investigation within Discover. The exercises in each Check for Understanding give students an opportunity to demonstrate what they've learned. The final question in each set allows them to justify their reasoning and communicate mathematically.
- SUMMARY: Math Techbook is aligned with math standards that require deep exploration of critical topics and encourage students to make connections. The Summary provides a reference for what students have learned in each concept.
- PRACTICE EXERCISES: Students need additional review material to practice what they've learned. In addition to the Coach and Play items found in the online Practice section, the Practice Exercises in this print resource allow students to sharpen their skills and show what they've learned.
- APPLY: Students experience learning when they apply their mathematical thinking to problems worth solving, rooted in real-world contexts that are relevant to their lives. Online, students have access to several Apply problems. In this print resource, one of these in-depth tasks is provided with a rubric and evaluation criteria.

Within each section, QR codes will take you and your student to the corresponding section of Math Techbook. Once inside, your student will have access to the investigations that develop conceptual understanding in Discover, practice exercises that develop procedural fluency in Practice, and real-world problems worth solving in Apply.

While this print resource can be used at home to complete assignments when the Internet is not available, it is only through Math Techbook that students develop the type of deep conceptual understanding that fosters mathematical proficiency. QR codes are provided throughout the *ISE* to redirect students to Math Techbook's digital resources.

To use the QR codes, you will need a QR reader. Readers are available for phones, tablets, laptops, desktops, and virtually any device in between. Most use the device's camera, but there are some that scan documents that are on your screen. Download a free QR reader in the App Store or Google Play. To access Math Techbook resources, follow these steps:

1. Open the QR code reader on your device.
2. Hold your device so the QR code is visible within your device's screen. One of two things will happen:
   - The device may automatically scan the code; or,
   - The device will scan the code when you press a button, similar to taking a picture.
3. Once scanned, the QR code will direct you to a page or resource on the web.
4. For resources in Math Techbook, you may need to log in with your username and password the first time you access a QR code. After that, you won't need to log in again, unless you sign out or remain inactive for too long.

Scan this QR code to access a video that provides a deeper introduction to Math Techbook:

We encourage you to support your student in using the interactive materials provided in Math Techbook, as well as the exercises and material in this print resource at home. Together, may you and your student enjoy a fantastic year of mathematics!

Sincerely,

The Discovery Education Math Team

Math Tools

Within Math Techbook, you'll have access to the following tools:

Calculator: Solve addition, subtraction, multiplication, and division problems with large numbers or decimals.

Scientific Calculator: Solve complex problems involving multiple operations, exponents, and functions.

Graphing Calculator: Display equations and functions on a coordinate plane.

Whiteboard: Draw lines and shapes, insert images and grids, display text and graphics together.

Dynamic Geometry Tool: Create more advanced shapes, polygons, and conics, transform them, and discover the equations behind them.

Unit Converter: Easily change units of measurement for length, mass, time, temperature, electric current, data storage, and angles from one to another.

Construction Tool: Practice your freehand drawing skills with a ruler and compass for a digital pencil.

Matrix Solver: Solve complex problems involving one or multiple matrices by changing the size and inputting values.

Glossary: View the definitions, key context, animations, and related videos for key mathematical terms.

Interactives: Investigate mathematical problems with game-like activities.

Board Builder: Use images and text to create presentations with this handy tool.

UNIT 1: Foundations of Algebra

1.1 Analyze Expressions and Equations

photo: Getty Images

Lesson Objectives

- Interpret variables and quantities in terms of the context.
- Create expressions and equations in two variables.
- Describe real-world situations using multiple representations.

Essential Question

- What can the structure of an algebraic expression reveal about the mathematical or real-world situation it models?

photo: Getty Images

Investigations

Can You Model Motion?

How could you use numbers and variables to describe a roller coaster ride?

Frames

Express yourself! Create a mirror and describe it using expressions.

How Do the Models Match?

Different expressions can convey the same thing. What does your expression convey?

Which Wireless Plan Should You Choose?

Choose the best plan for your cell phone using algebraic expressions and equations.

Key Vocabulary

square number, square root, term (in an expression), variable

Discover

As you complete Engage and the investigations, record the most important ideas you've learned.

Engage	Investigation 1
Investigation 2	Investigation 3

Name ______________________________ Date ____________

Check for Understanding

Analyze Expressions and Equations: Investigation 1

Brian creates boxes for Food Fads, Inc. He must follow the standards required by the Food Fads company.

- All boxes must maintain the same length-height and length-width relationship.
- The surface area of a box cannot exceed 70,000 cm^2.

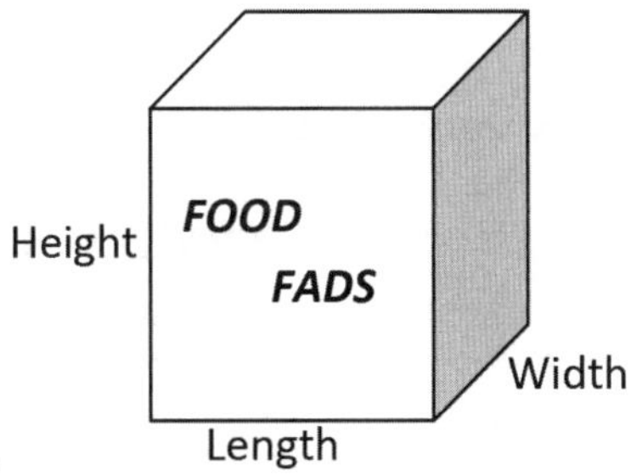

The expression for the surface area of a box is $2LW + 2HW + 2HL$.

Brian wants to take only one measurement to check the surface area criteria, so he rewrites the expression as $2L(L - 20) + 2(L + 50)(L - 20) + 2(L + 50)L$.

1. Compare Brian's expression with the original expression for surface area. What does the factor $(L + 50)$ mean in this scenario? *Select all that apply.*

A. The width of a box is 50 centimeters more than its height.

B. The height of a box is 50 centimeters more than its length.

C. The height of a box is half a meter more than its length.

D. The length of a box is half a meter less than its width.

2. What does the term $2(L + 50)(L - 20)$ represent in this scenario? *Select all that apply.*

A. the sum of the areas of the right and left sides of a box

B. the area of the back of a box

C. two times the area of the top of a box

D. twice the product of the height and width of a box

3. The potato chip box has a length of 45 centimeters. What is the surface area of the box?

____

4. Using the same length-height and length-width relationships, explain how you can rewrite the expression so that Brian will only need to measure the height of a box.

Name ______________________________ Date ______________

Check for Understanding

Analyze Expressions and Equations: Investigation 2

During a strength training workout, Jacoby does 10 sets of push-ups. He does 30 push-ups during the first set, and he does 3 fewer push-ups in each set after the first.

1. Complete the following table of values to match the scenario described above.

x	1	2	3	4	5					10
value	30	27								3

2. Which of the following expressions could be used to represent this scenario?

A. $10x + 3$ B. $10x - 3$ C. $30x + 3$

D. $33 - 3x$ E. $3x + 10$ F. $3x + 30$

3. Show how tiles could be used to represent the expression $2x + 5$.

4. What is the benefit of being able to represent situations in various ways?

Name ______________________________ Date ______________

Check for Understanding

Analyze Expressions and Equations: Investigation 3

On a long-distance biking trip, Annike started biking at 7 a.m., and her average speed was 11 miles per hour. Celia started at 8 a.m., and her average speed was 14 miles per hour.

1. Write an expression that represents the total number of miles completed by each biker n hours after 7 a.m.

   Annike: ______________________ Celia: ______________________

2. By 10 a.m, who had completed more miles?

   A. Annike

   B. Celia

   C. Neither; they had both biked the same number of miles by 10 a.m.

3. At what time will the two bikers have completed the same number of miles?

   ____

4. Briefly explain the method that you used to answer question 3, and then describe a different method that you could have used to find the answer.

Summary

Before you attempt the Practice Exercises, review what you've learned.

Numerical expressions contain numbers and operational symbols. Numerical equations contain two numerical expressions with an equal sign between them. They are used to represent numerical relationships.

EXAMPLE

Grace works 7 hours this week and earns $12 per hour.

- The expression $7 \cdot 12$ represents her earnings for the week.
- The equation $7 \cdot 12 = 84$ indicates that her earnings are equal to $84.

This numerical representation describes Grace's earnings for this week only. To generalize the situation, you can use algebraic expressions and equations. Algebraic expressions contain numbers, operational symbols, and variables.

A **variable** is a letter that represents a quantity that can change.

An algebraic equation contains two algebraic expressions with an equal sign between them.

EXAMPLE

Let h equal the number of hours Grace works in a week and let e equal Grace's earnings in a week.

- The expressions $12h$ and e both represent her weekly earnings.
- Since the expressions represent the same quantity, the expressions can be set equal to each other, forming the equation $12h = e$.

This algebraic equation generalizes the relationship between the number of hours Grace works and her earnings, which makes it a useful tool that can be applied to any given week.

To translate a verbal expression to an algebraic expression, look for keywords and phrases that can be represented symbolically. Here are some common phrases for each operation:

$n + 1$	$x - 3$	$2y$	$\frac{a}{3}$
• 1 more than n	• 3 less than x	• 2 multiplied by y	• a divided by 3
• n plus 1	• x minus 3	• 2 times y	• the ratio of a and 3
• the sum of n and 1	• the difference of x and 3	• the product of 2 and y	• the quotient of a and 3

You may need to translate parts of a verbal expression separately and combine the results.

My Notes

Summary *(continued)*

EXAMPLE

Write an algebraic expression for the quotient of a number squared and six less than the number.

- "the quotient" indicates the result of division
- "a number squared" indicates a power, where the base is the number and the exponent is 2
- "six less than the number" indicates the number minus 6

Since "a number" could be any number, use a variable. Let n = any number. Then the given verbal expression translates to $\frac{n^2}{n-6}$.

To evaluate an algebraic expression, substitute a value for the variable and simplify. You can generate a table by evaluating an expression for several values of the variable.

EXAMPLE

Make a table of values for $\frac{n^2}{n-6}$.

n	$\frac{n^2}{n-6}$
0	$\frac{0^2}{0-6} = \frac{0}{-6} = 0$
1	$\frac{1^2}{1-6} = \frac{1}{-5} = -\frac{1}{5}$
3	$\frac{3^2}{3-6} = \frac{9}{-3} = -3$
9	$\frac{9^2}{9-6} = \frac{81}{3} = 27$

If you are given a table, you may need to look for patterns in the table in order to write an expression. If the values of the variable are equally spaced, subtraction is a good way to look for a pattern.

My Notes

Summary *(continued)*

EXAMPLE

Let t represent the number of hours that an air-conditioner technician spends on a service call. Use the table of values shown to write an expression that represents the cost of the call.

t	Cost of Service Call
1	\$75
2	\$110
3	\$145
4	\$180

To look for a pattern in the table, first notice that the values of t are equally spaced with a difference of 1. Now subtract to compare the values of the cost of the service call.

$$110 - 75 = 35$$

$$145 - 110 = 35$$

$$180 - 145 = 35$$

The constant difference is 35, which represents the hourly rate. So, part of the expression will be $35t$.

Because the cost of the call for 1 hour is not \$35, there must also be a fixed rate that is part of the cost. Subtract 35 from the rate for 1 hour.

$$75 - 35 = 40$$

The fixed cost is \$40. The expression is $35t + 40$.

Test this expression for the values of t.

$$35(1) + 40 = 75$$

$$35(2) + 40 = 110$$

$$35(3) + 40 = 145$$

$$35(4) + 40 = 180$$

The values match the table. The expression $35t + 40$ is correct.

My Notes

Practice Exercises

Review what you've learned using these practice problems. For practice problems with feedback, try the Coach and Play items in the Practice section online.

1. An online shopping company has T-shirts for sale. The sales tax rate is 7.25% on each shirt ordered, and there is a flat fee of $4.75 shipping for the entire order. There is no limit on the number of T-shirts one person can purchase.

   Consider the expression that models the total cost of any online purchase. What is the value of the constant term and what does it represent?

   A. 0.0725 represents the sales tax

   B. 0.725 represents the sales tax

   C. 4.75 represents the shipping charge

   D. 7.25 represents the shipping charge

   E. 7.25 represents the sales tax

   F. 47.5 represents the sales tax

2. Every weekday after tennis practice, Will buys either a carton of milk or a bottle of power drink.

   Suppose that m is the number of cartons of milk that Will purchased during the last month. His total cost for beverages in dollars last month can be represented by the algebraic expression $2m + 3(20 - m)$.

   Which of the following statements are true? *Select all that apply.*

   A. The cost of the power drink is $3 per bottle.

   B. Will purchased 20 more bottles of power drink than cartons of milk.

   C. Will purchased 20 more cartons of milk than bottles of power drink.

   D. Will purchased a total of 20 beverages during the last month.

   E. The total amount Will spent on power drinks was $60.

   F. An equivalent expression for the total cost of Will's beverages during this month is $60 - m$ dollars.

3. The area of a triangle is $A = \frac{1}{2}bh$. The area of a particular triangle is $6xy$ and its base measures $4y$. What is the height of the triangle?

   A. $2xy$

   B. $3x$

   C. $12x$

   D. $24x$

4. Choose the expression that represents the following:

   the sum of the square of a number x and a number y, minus the product of three times the sum of x and 4.

   A. $2x + y - 3(x + 4)$

   B. $x^2 + y - 3(x + 4)$

   C. $(x + y)^2 - 3(x + 4)$

   D. $x^2 + y - 3x + 4$

   E. $2x + y - 3x + 4$

5. The perimeter of a rectangle is represented by the expression $2(3x + 1) + 2x$.

   Which statements explain correctly what each part of the expression represents? *Select all that apply.*

   A. One side of the rectangle measures $3x + 1$.

   B. One side of the rectangle measures $3x$.

   C. One side of the rectangle measures $2x$.

   D. One side of the rectangle measures x.

   E. 2 is the coefficient in each term that represents 2 lengths and 2 widths in a rectangle.

   F. x is the coefficient in each term that represents the number of sides in a rectangle.

Practice Exercises *(continued)*

6. A rectangular photo measuring 8 inches by 10 inches is surrounded by a rectangular mat. The mat that surrounds the frame is w inches wide on each side of the photo.

Which expression represents the area of the mat and photo in terms of w, the width of the mat?

A. $(2w + 8)(2w + 10)$ **B.** $4w^2 + 36w$

C. $4w^2 + 18w$ **D.** $4w^2 + 80$

7. Janice makes 3 identical charm bracelets for her friends. Each bracelet is made with 2 charms and a $4 chain. If x represents the cost of each charm, which models represent the cost of the bracelets?

Select one tile model and one algebraic model to represent the total cost of the bracelets that Janice makes.

A.

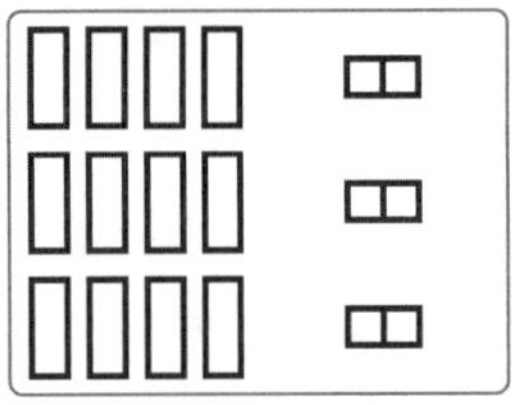

B.

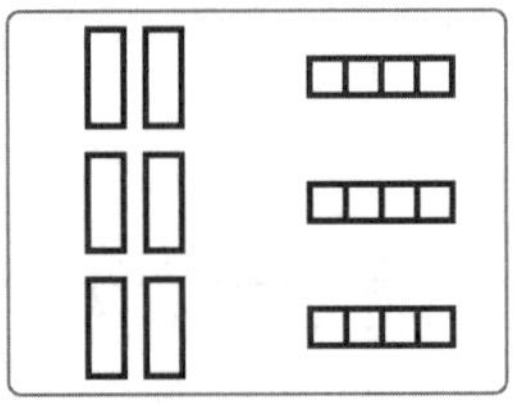

C.

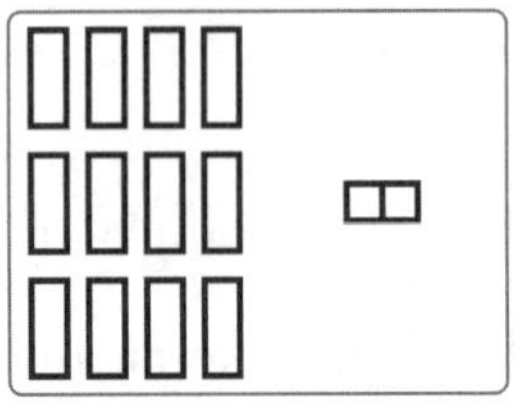

D.

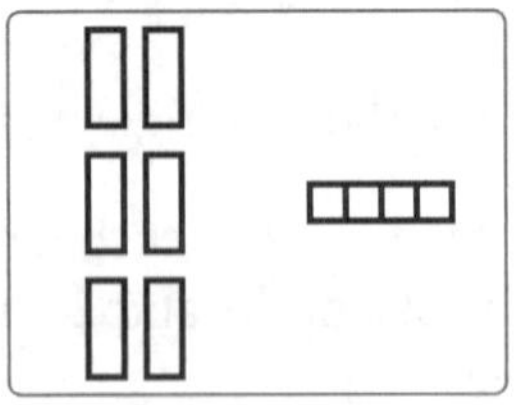

E. $3(4x) + 2$ **F.** $3(2x) + 4$

G. $3(2x + 4)$ **H.** $3(4x + 2)$

8. Taylor works at a garden shop that sells flower pots. One kind of flower pot has a rim height of 6 centimeters and a planting height of 22 centimeters. Taylor begins to stack the flower pots as shown.

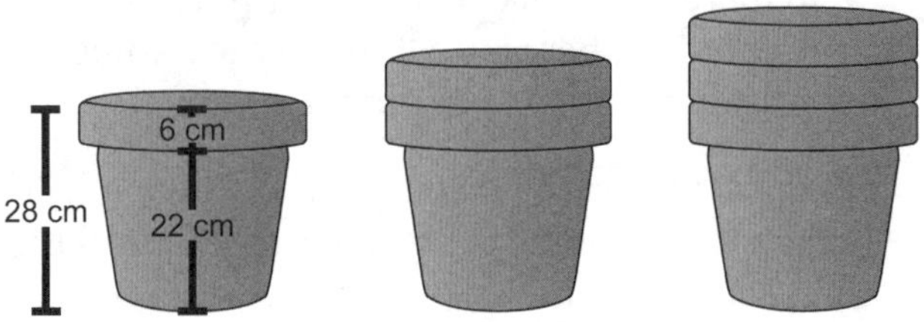

Complete the table so that it shows the height of each stack for the given number of flower pots. Then write a simplified algebraic expression to represent the height of a stack of n flower pots.

Number of Flower Pots, n	1	2	3	4	5
Height of Stack in cm	28	___	___	___	___

A simplified algebraic expression that represents the height of a stack of n flower pots is ________________ cm.

9. The internal temperature of a new refrigerator is 82 degrees. For each hour h the refrigerator is plugged in, the temperature decreases by 3 degrees. What expression can represent the temperature of the refrigerator after h hours?

The temperature of the refrigerator after h hours can be written as ________________.

10. The EasyCall cell phone company offers a plan with a flat $30 monthly fee for unlimited phone calls, but charges for data usage at a monthly rate of $10 per gigabyte (GB) used. An additional 35% in taxes is added to each month's bill. The monthly cost of the plan can be calculated using this equation:

$y = 1.35(10x + 30)$.

What is the monthly cost of the plan for someone who makes 5 hours of phone calls each month and uses 3GB of data?

____

Apply

How Can You Model Your Initials with Algebraic Expressions?

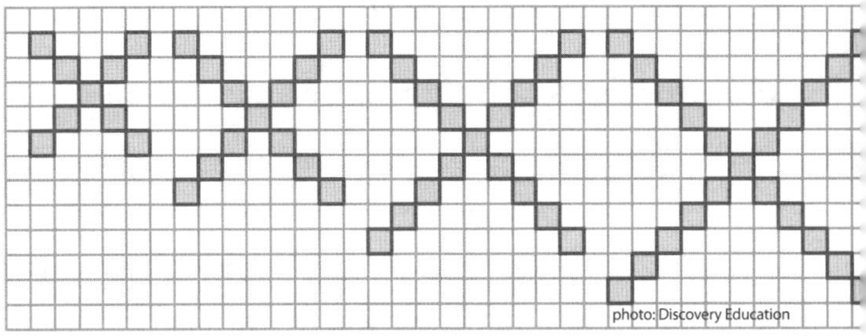
photo: Discovery Education

A sequence is an ordered list of numbers, such as 1, 3, 5, 7, ... Each number in a sequence is a term. If you can determine a pattern in the terms of a sequence, you can write an algebraic expression that you can use to find any term in the sequence based on the term number.

Watch this video to see one example of a sequence involving blocks that form a pyramid.

An Algebraic Expression for a Pyramid: View this video segment to see the formula for the number of blocks in each row of a pyramid.

Draw the first letter of your name on graph paper. Draw the letter by shading complete squares on the grid. Then draw at least 3 larger versions of the letter.

An example with the letter X is shown below.

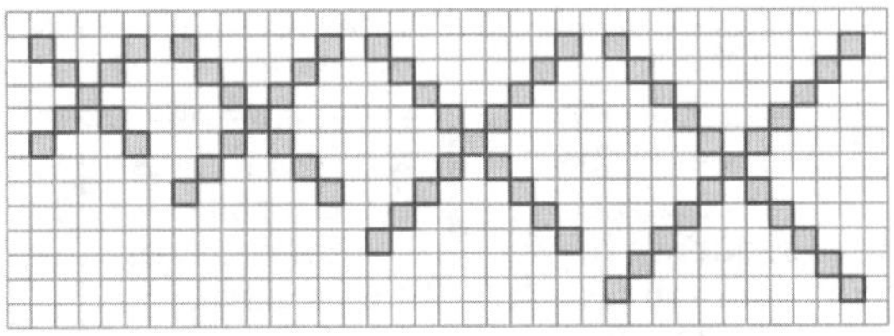

The number of squares in each X forms a sequence as shown in the table below:

Term Number	1	2	3	4
Term of Sequence	9	13	17	21

Draw the larger versions of your initial so that there is consistent increase in the number of shaded squares that are used.

Model a Pattern

Use the diagram you created with your initial to generate a sequence. The terms of the sequence should be the numbers of squares in the letters you drew.

Describe your sequence with a table of values as well as an algebraic expression.

How many squares would be needed to draw the 150th letter in the sequence for your initial? Explain how you determined your answer.

Show what you've learned by completing the other performance tasks in the online Apply section.

Apply *(continued)*

Your answer to Apply will be assessed on the following criteria:

1. Creating a pattern based on the initial of your first name and writing a sequence based on the pattern
2. Writing an algebraic expression that can be used to find any term of the sequence
3. Explaining how you determined the algebraic expression
4. Using your expression to make a prediction

Criteria \ Scale	4 Exceeds Criteria	3 Meets Criteria	2 Progressing to Criteria	1 Below Expectations	0 No Expectation
Creating a Pattern and Sequence	Clearly displays the first four terms (at least) of a pictorial pattern and correctly translates to a numerical sequence.	Clearly displays the first three terms of a pictorial pattern and correctly translates to a numerical sequence.	Clearly displays the first four terms of a pictorial pattern but incorrectly translates to a numerical sequence.	Creates an incorrect pictorial pattern and incorrectly translates to a numerical sequence.	Does not attempt the task.
Writing an Expression	Writes a correct algebraic expression to represent the terms of the sequence including testing each of the known terms in the expression.	Writes a correct algebraic expression to represent the terms of the sequence but does not test all known terms.	Writes an algebraic expression that works for the first few terms but not the remainder of the sequence.	Writes an algebraic expression unrelated to the numerical sequence.	Does not attempt the task.
Explaining Reasoning	Fully explains the method used for developing the correct algebraic representation of the sequence.	Mostly explains the method used for developing the correct algebraic representation of the sequence.	Explains the method behind developing an incorrect sequence that holds true for a few terms of the sequence, but not all terms.	Provides an incomplete or unclear explanation.	Does not attempt the task.
Making Predictions	Correctly uses the expression to make a prediction, performing all calculations correctly.	Correctly uses an (incorrect) expression to make a prediction, performing all calculations correctly.	Correctly uses the expression to make a prediction, but performs a minor computation error.	Correctly uses the expression to make a prediction, but performs a major computation error.	Does not attempt the task.

UNIT 1: Foundations of Algebra

1.2 Reason with Expressions and Equations

photo: Discovery Education

Lesson Objectives

- Interpret complex expressions and equations in terms of the context.
- Create complex expressions and equations in one variable.

Essential Question

- How can you use real-world patterns to create models in the form of algebraic expressions?

photo: Getty Images

Investigations

So, You Want to See a Trick?

Trick the trickster! Watch for patterns, and then use expressions and algebra to predict the outcome.

Target Heart Rates

What is your target heart rate, and what does it have to do with age? Use algebra to find out.

Pumpkin Launch!

Flying pumpkins? Analyze an equation to find out how far a launched pumpkin is likely to fly.

Algebraic Card Tricks

Complex tricks have more variables, but they still have patterns. Model the patterns with expressions.

Exploring Diagonals in a Polygon

Geometric patterns can be modeled with variables. Investigate string art to find out how.

Key Vocabulary

diagonal, heptagon, hexagon, pentagon, polygon, term (in an expression), variable

Discover

As you complete Engage and the investigations, record the most important ideas you've learned.

Engage	
Investigation 1	**Investigation 2**
Investigation 3	**Investigation 4**

Name ______________________________ Date ____________

Check for Understanding

Reason with Expressions and Equations: Investigation 1

A Tae Kwon Do academy offers two different membership options.

- **Regular:** \$80 per month, plus a one-time cost for the uniform
- **Promotional:** \$95 per month, cost of the uniform is waived

The sales manager of the academy created the following table for customers to compare the cost for 1, 2, 3, or 4 months.

Months	Regular Membership	Promotional Membership
1	\$230	\$95
2	\$310	\$190
3	\$390	\$285
4	\$470	\$380

1. Create an expression for the total cost of the regular membership for *n* months.

____

2. Create an expression for the total cost of the promotional membership for *n* months.

____

3. How much would each membership cost for 24 months?

Regular: ________________ Promotional: ________________

4. Which of these conclusions about the total cost of the two memberships is true?

A. The total cost for a promotional membership is less at first, but becomes greater than a regular membership over time.

B. The total cost of a promotional membership will always be less than the cost of a regular membership.

C. After a certain month, the total promotional and regular membership costs will be equal going forward.

5. What is the benefit of using an expression to calculate the cost for 24 months instead of using some other method?

Name ______________________________ Date ______________

Check for Understanding

Reason with Expressions and Equations: Investigation 2

An object is launched from the top of a building. The initial upward velocity is 19.5 meters per second. The acceleration due to gravity on the object is –9.8 meters per second squared. An equation that models the height h, in meters, of the object at time t seconds after launch is $h = -4.9t^2 + 19.5t + 25$.

1. What does the constant term in the equation mean in this context?

A. the distance the object travels

B. the height of the building

C. the maximum height the object will reach

D. how long the object will be in the air

2. How high above the ground is the object 1.5 seconds after it is launched?

$h =$ ____________________

The length of a box can be calculated from its volume using the formula $L = \frac{V}{WH}$, where L is the length, V is the volume, W is the width, and H is the height.

3. If the volume and width of a box remain constant, use the equation to explain the effect on the length of the box as the height increases. Include a specific example in your explanation.

Name ______________________ Date ____________

Check for Understanding

Reason with Expressions and Equations: Investigation 3

1. Match the verbal expression to its correct algebraic expression.

A. three times the difference of p and q	**I.** $3p - 3q$
B. the product of three and p minus the product of three and q	**II.** $qp - 3$
C. the product of q and three less than p	**III.** $3(p - q)$
D. three less than the product of q and p	**IV.** $q(p - 3)$

A local store is selling bananas for \$0.44 per pound and apples for \$0.98 per pound. You pick up b pounds of bananas and a pounds of apples. At the checkout, you give the cashier \$10. The tax on food is 6%.

2. Which equation describes the change, C, you should receive? *Select all that apply.*

A. $C = 10 - 1.06(0.44b + 0.98a)$ **B.** $C = 10 - 1.06(0.44b) - 1.06(0.98a)$

C. $C = 1.06(10 - 0.44b + 0.98a)$ **D.** $C = 1.06(0.44b + 0.98a) - 10$

3. If you buy 2 pounds of bananas and 5 pounds of apples, how much change should you receive?

$C =$ ______________________

You are asked to pick up some fruit, bread, and vegetables on your way home from school. You are given \$20 and told you can keep the change for gas money.

4. If tax on food is 6%, which option will leave you with the most change?

A. 2 pounds of grapes at \$1.95 per pound, 2 loaves of multi-grain bread at \$2.55 per loaf (buy one get one free), and 10 ears of corn at 10 for \$1.00

B. 1 pineapple at \$3.47 each, 2 packages of hamburger buns at \$1.50 each, and 1 package of frozen mixed vegetables at \$1.15 each

C. 1 large cantaloupe at \$2.00 each, 2 loaves of whole wheat bread at \$1.35 each, and 1 package of mixed greens salad at 2 for \$5.00

Name ______________________________ Date ______________

Check for Understanding

Reason with Expressions and Equations: Investigation 4

Observe the process used below to derive an expression for {3, 6, 11, 18, 27, 38}.

Create a table. The first term in the sequence is 3, so when $n = 1$, the corresponding number is 3. Likewise, the second term in the sequence is 6, so when $n = 2$, the corresponding number is 6. This process continues until $n = 6$.

n	1	2	3	4	5	6
Sequence	3	6	11	18	27	38

Because the terms of the sequence do not increase by the same amount, the expression is not linear. A reasonable guess is that it might contain an n^2 term.

Use the table above for questions 1–3.

1. How does each value in the sequence compare to the value of n^2?

____

2. What is the value in the sequence when $n = 47$?

____

3. Which is an expression for the nth term of the sequence?

A. $2n^2$ **B.** $n^2 - 2$ **C.** $n^2 + 2$ **D.** $\frac{n^2}{2}$

4. What are the 1st, 2nd, and 15th terms for the sequence {__, __, 12, 20, 30, 42, 56, 72, . . .}? What is an expression for the nth term of the sequence?

1st term: __________ 2nd term: __________ 15th term: __________

Expression: __________

Summary

Before you attempt the Practice Exercises, review what you've learned.

An algebraic expression is a mathematical phrase that includes one or more variables. To evaluate an algebraic expression, first substitute a given number for each variable. Then simplify the resulting numerical expression using the order of operations.

EXAMPLE: Create and Interpret an Expression

The blueprint represents a portion of a house. The side length of the square bedroom has not yet been decided on.

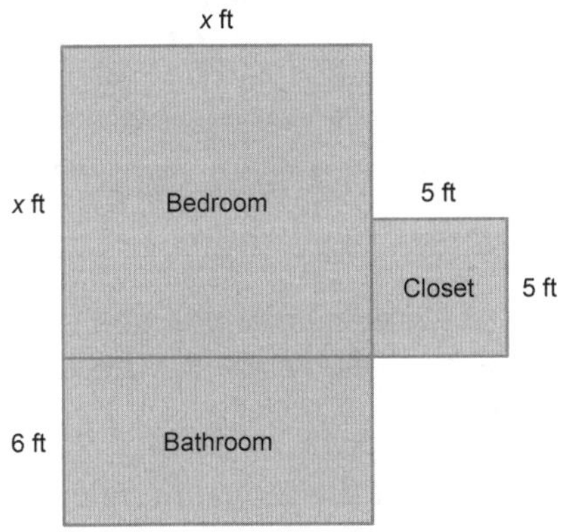

Write an expression that represents the total area in square feet of this portion of the house. Interpret each part of the expression in the context of this situation.

SOLUTION:

The bedroom is a square with a side length of x feet, so x^2 represents the area of the bedroom in square feet.

The bathroom is a rectangle with a length of x feet and a width of 6 feet, so the product $6x$ represents the area of the bathroom in square feet.

The closet is a square with a side length of 5 feet, so 5^2 or 25 represents the area of the closest in square feet.

The expression $x^2 + 6x + 25$ represents the total area in square feet, where each term is the area of one of the rooms.

An equation consists of two expressions that are set equal to each other. All equations include an equal sign.

EXAMPLE: Create and Interpret an Equation

Max bought 4 gallons of wall paint on sale, for a total cost of $86.24 before tax. The regular price of the paint is $23.96 per gallon. Write an equation that models this situation. Interpret each part of the equation in terms of the context. What would the total discount be if Max bought 6 gallons of paint at the sale price?

SOLUTION:

Let d represent the discount in dollars on each gallon.

The difference $\$23.96 - d$ represents the sale price per gallon of paint because it shows the regular price per gallon, $23.96, minus the discount per gallon, d.

Max bought 4 gallons of paint, so the product $4(\$23.96 - d)$ gives the sale price for 4 gallons.

The total cost of Max's paint was $86.24, so set the sale price for 4 gallons equal to the total cost.

$$4(\$23.96 - d) = \$86.24$$

So, the equation that models the situation is $4(\$23.96 - d) = \$86.24$, where each side of the equation is equal to the total cost of the paint.

My Notes

Summary *(continued)*

To find the amount of the discount, solve the equation for *d*.

$$4(\$23.96 - d) = \$86.24$$

$$\frac{4(23.96 - d)}{4} = \frac{86.24}{4}$$

$$\begin{aligned} 23.96 - d &= 21.56 \\ -23.96 \quad & \quad -23.96 \end{aligned}$$

$$-d = -2.40$$

$$d = 2.40$$

So, the amount of the discount is \$2.40 per gallon.

If Max bought 6 gallons of paint, his total discount would be 6(\$2.40) = \$14.40.

EXAMPLE: Create and Evaluate an Expression

A regular polygon can be divided into triangles by drawing diagonals from one vertex to each of the other nonadjacent vertices. Use the number of triangles in a polygon to calculate the measure of one interior angle of a regular polygon. What is the measure of the interior angle of a regular decagon?

My Notes

SOLUTION:

Start by determining the number of triangles in the first few regular polygons. Organize the number of vertices and number of triangles in a table.

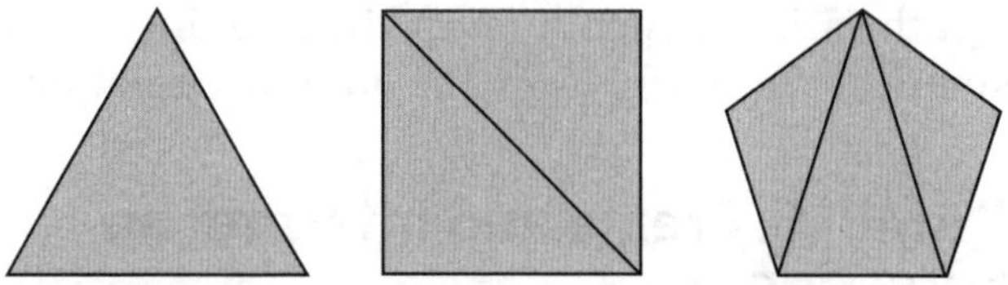

Polygon	Number of Vertices	Number of Triangles
Triangle	3	1
Square	4	2
Pentagon	5	3
Hexagon	6	4
Heptagon	7	5

The number of triangles in a polygon is 2 less than the number of vertices, or $n - 2$, where n is the number of vertices.

The total measure of the interior angles of a polygon is equal to the sum of all the angle measures in the triangles inside it. The sum of the angle measures of a triangle is always 180°. So, the sum of the interior angle measures must be 180° times the number of triangles, or $180°(n - 2)$.

The measure of one interior angle is the total measure of the interior angles divided by the number of angles. The number of vertices and the number of angles in a polygon are the same, so the number of angles can be represented by n. The measure of one interior angle can be modeled by the expression $\frac{180°(n-2)}{n}$.

To find the measure of the interior angle of a regular decagon, substitute 10 for the number of angles and vertices in the polygon.

$$\frac{180°(n-2)}{n} = \frac{180°(10-2)}{10} = \frac{180°(8)}{10} = \frac{1{,}440°}{10} = 144°$$

So, the measure of the interior angle of a regular decagon is 144°.

Practice Exercises

Review what you've learned using these practice problems. For practice problems with feedback, try the Coach and Play items in the Practice section online.

1. Andre is at the grocery store and purchases 5 pounds of bananas at x dollars per pound. He also purchases 6 pounds of apples at y dollars per pound. There is a 10% sales tax on groceries.

Which expression can be used to find the total cost of Andre's purchases?

A. $5x + 6y$

B. $(5x + 6y)(0.10)$

C. $(5x + 6y) + (5x + 6y)(0.10)$

D. $(5x + 6y) + (5x + 6y)(0.90)$

E. $(5x + 6y) + (5x + 6y)$

2. The length, l, of a rectangle is 5 times its width. The perimeter of the rectangle is 96 centimeters.

Which equation can be used to correctly calculate the length of the rectangle?

A. $2l + 2(l + 5) = 96$

B. $2l + 2(l - 5) = 96$

C. $2l + 2\left(\frac{l}{5}\right) = 96$

D. $2l + 10l = 96$

3. Holly is purchasing new uniform pants and shirts for school this year. She buys x pairs of pants at \$34.98 each. She buys 3 shirts to go with each pair of pants and 4 additional shirts. Each shirt costs \$17.49.

Write an expression for the amount Holly spends on clothes.

Simplify the expression completely.

The amount that Holly spends on clothes can be calculated using this expression:

__________ x + __________

4. The students in Drama Club need to arrange 275 available chairs in the cafeteria for the audience of their latest performance.

They place most of the chairs into a rectangular arrangement. There are twice as many rows as there are chairs within a single row. When they finish, there are enough chairs left over to add three more rows.

If c represents the number of chairs in each row, which equation describes the total number of chairs?

A. $c^2 + 3c = 275$

B. $3c^2 + 2c = 275$

C. $c^2 - 3c = 275$

D. $2c^2 + 3c = 275$

5. An online ticket broker sells concert tickets. The cost for an adult ticket is \$55 and the cost for a student ticket is \$20. The ticket broker charges a convenience fee of \$3 for each ticket purchased.

Suppose that Jamie purchases a adult tickets and s student tickets.

Which of these equations represents the total cost of her purchase?

Select all that apply.

A. $C = 55a + 20s + 3(a + s)$

B. $C = 55(a + 3) + 20(s + 3)$

C. $C = 55a + 20s + 3$

D. $C = 58a + 23s$

E. $C = 75(a + s) + 3(a + s)$

Practice Exercises *(continued)*

6. At a math competition, each team wrote 2 questions for each of the other teams. All of the teams then solved the questions that were written for them and judges scored all of the questions. The judges scored a total of 40 questions.

Let n be the number of teams at the competition.

Using the letters of the terms below, define each of the other unknowns in terms of n and complete the equation that can be used to find n.

Letters may be used more than once.

A. $(n - 2)$

B. 2

C. n^2

D. $(n - 1)$

E. n

F. $(2n - 1)$

Number of teams for which each team wrote questions: __________

Number of questions each team wrote:

____

Total number of questions written, in terms of n: __________

Equation that can be used to find n: $40 =$ __________

7. A landscaping and tree maintenance company charges a flat fee of \$30 plus \$20 per hour and \$5 per tree.

Which expression represents the company's total charges based on x hours and t trees?

A. $(30 + 20 + 5t)x$

B. $30 + 20x + 5t$

C. $30y + 20x + 5t$

D. $30 + 25x$

E. $30 + 20 + 5t$

8. Last year, 300 people attended the school play. The cost of one ticket was \$8. This year, the Drama Club wants to raise prices to increase revenue. It estimates that for each \$1 increase in the ticket price, 8 fewer people will attend this year's play.

The expected revenue from this year's play for x number of \$1 ticket price increases is modeled by: Revenue $= (300 - 8x)(8 + x)$.

Which expression represents the expected number of people relative to the number of price increase, x?

A. x

B. $8x$

C. $(300 - 8x)$

D. $(8 + x)$

E. $(x + 8)(300 - 8x)$

9. A motel charges its guests \$79.95 for each night's stay. But on Friday and Saturday nights, the rate increases to \$99.95. A family is planning a vacation, and they want to stay overnight on both Friday and Saturday.

If they have budgeted \$450 for a motel room, how many nights in all could they stay at the motel? Choose the equation below that represents the situation. Let n be the total number of nights.

A. $2(99.95) + n(79.95) = 450$

B. $n(99.95) + n(79.95) = 450$

C. $n(99.95) + (n - 2)(79.95) = 450$

D. $2(99.95) + (n - 2)(79.95) = 450$

10. A shoe store is having a 30%-off sale.

Complete the equation to represent the sales price, P, for a purchase if n is the original price.

$P =$ ________ − (________)

Apply

How Much Should You Charge to Make the Most Profit?

Have you ever been to a street fair? How much money do you think the sellers make?

photo: Paul Fuqua

Street Vendor Profits: View this video segment to see how much should you charge to make the most profit.

A group of students at your school decides to hold a fundraiser for a charity. To raise money, you plan to sell an item at a street fair where renting a booth costs $40.

How should you price your items to make the most profit?

Investigate Costs

First, decide on an item to sell. Research how much it will cost to purchase 125 items from a supplier, including shipping. What other costs will you have?

Investigate Pricing

Next, you'll need to know how much people are willing to pay for your item. How can you find out? Use the information you gather to decide on a low price, a medium price, and a high price.

Investigate Profit

Write an expression for the profit in dollars you will make for buying 125 items from your supplier and selling n of them at a price of p dollars. Explain what each part of your expression represents.

Based on advice from sellers who have worked at other street fairs, you estimate that you could sell 125 items at your low price, 115 items at your medium price, and 100 items at your high price.

Choose Your Price

Which of the three selling prices should you choose for your item? Explain your reasoning.

Show what you've learned by completing the other performance tasks in the online Apply section.

Apply *(continued)*

Your answer to Apply will be assessed on the following criteria:

1. Gathering data to determine how much people will pay for your item and determining low, medium, and high selling prices based on your data
2. Modeling profit with an algebraic expression and explaining what each part of the expression represents
3. Calculating the net profit for the low, medium, and high selling prices you established in #1
4. Deciding on a selling price and justifying your decision with both mathematical reasoning and any other supporting evidence

Scale / Criteria	4	3	2	1	0
	Exceeds Criteria	Meets Criteria	Progressing to Criteria	Below Expectations	No Expectation
Pricing Analysis	Clearly demonstrates research of purchasing a quantity of a certain item, including potential low, medium, and high selling prices.	Demonstrates adequate research of purchasing a quantity of a certain item, including potential low, medium, and high selling prices.	Demonstrates adequate research about purchasing an item but does not include reasoning behind the choices for low, medium, or high selling prices.	Provides low, medium, and high selling prices for an item without evidence of research.	Does not attempt the task.
Modeling Profit	Provides correct expression for profit, with detailed reasoning as to the different parts of the expression.	Provides correct expression for profit, but makes a minor error in explaining the parts of the expression.	Provides an expression for profit, but makes a major error in explaining the parts of the expression.	Provides an expression for profit, but does not explain the parts of the expression.	Does not attempt the task.
Calculating Net Profit	Correctly determines the net profit for all three selling prices, clearly showing and labeling all work.	Determines the net profit for the three selling prices, making a minor computation error.	Determines the net profit for the three selling prices, making more than one minor computation error.	Does not calculate the net profit for all three requested prices or makes major computation errors.	Does not attempt the task.
Selecting a Selling Price	Identifies the selling price that results in the greatest net profit, including an explanation that clearly demonstrates reasoning.	Identifies the selling price that results in the greatest net profit, including an explanation that mostly demonstrates reasoning.	Identifies a selling price that does not result in the greatest net profit or does not include an explanation of reasoning.	Identifies a selling price that does not result in the greatest net profit and does not include an explanation of reasoning.	Does not attempt the task.

UNIT 1: Foundations of Algebra

1.3 Apply and Evaluate Expressions and Equations

photo: Getty Images

Lesson Objectives

- Create expressions and equations to describe real-world situations.
- Evaluate algebraic expressions and equations and explain solutions in context.

Essential Question

- How do you translate complex real-world relationships to algebraic expressions and equations to solve finance problems?

photo: Getty Images

Investigations

Flight Arrangements

Plan a trip for a group. Use expressions to find the best deal.

The Trip Continues

It's more than airfare. Use expressions to consider other costs of the trip.

Money Makes Money!

Save some money for a rainy day. Use expressions to compare simple and compound interest.

Key Vocabulary

compound interest, simple interest, term (in an expression), variable

Discover

As you complete Engage and the investigations, record the most important ideas you've learned.

Engage	
Investigation 1	Investigation 2

Name ______________________________ Date ______________

Check for Understanding

Apply and Evaluate Expressions and Equations: Investigation 1

Freshman students at Mason High School are planning their first Homecoming Dance and decided to hire the party planner Blowout Bashes to coordinate the event.

Blowout Bashes

Entertainment (***E***): DJ @ $50.00 per half hour, ($t$), plus a flat fee of $100.00 for equipment set up

Decorations (***D***): $10.00 per 100 sq. ft. of space, (***s***), plus a flat fee of $100.00 for set up

Refreshments (***R***): $10.00 per person, ($p$), plus $50.00 per wait staff (1 staff person/50 people required)

1. Write an equation to represent the cost for each service provided by Blowout Bashes.

A. Entertainment equation: ______________________

B. Decorations equation: ______________________

C. Refreshments equation: ______________________

2. The entertainment is scheduled for 7:00–11:00 p.m. Calculate the total entertainment cost.

A. $300.00 **B.** $400.00 **C.** $500.00 **D.** $600.00

3. Two rooms are required to accommodate the event. One room is for music and photography, and the other is for serving and eating food. Calculate the cost for each room to determine the total decorating cost.

Room 1:
50 ft × 75 ft
Area = ______________________
Cost equation: ______________________
Cost: $ ______________________
Total Decorating Cost: ______________________

Room 2:
50 ft × 20 ft
Area = ______________________
Cost equation: ______________________
Cost: $ ______________________

4. Last year, 375 students attended the Homecoming Dance. The committee expects at least 500 to attend this year's event. What is the expected cost for refreshments based on expected attendance?

Cost equation: ______________ Total Refreshment Cost: ______________

5. The committee wants to make a profit of at least $20,000 to invest toward their Junior Prom. Explain a minimum amount the committee can charge per ticket to meet their goal.

Name ______________________________ Date ______________

Check for Understanding

Apply and Evaluate Expressions and Equations: Investigation 2

The student committee made a profit of $20,000 from the Homecoming Dance. They now want to invest the money to earn enough to fund their Senior Prom. They are considering two different types of investments: one that earns simple interest and one that earns compound interest.

1. Complete the table to represent money earned on a 5% simple interest 3-year money market account.

Year	0	1	2	3
Account Balance $	20,000			

2. Complete the table representing money earned on a 3-year money market account at 5% interest compounded annually.

Year	0	1	2	3
Interest Earned				
Account Balance USD				

3. $20,000 was invested in a 3-year simple interest money market account. After 3 years, the account balance was $32,000. What was the interest rate for the account?

A. 20% **B.** 40% **C.** 12% **D.** 25%

4. $20,000 was invested in a 3-year money market account with interest compounded annually. After 3 years, the account balance was $59,719.70. What was the interest rate for the account?

A. 44% **B.** 26% **C.** 55% **D.** 33%

5. Examine the equations for calculating simple versus compound interest. Explain how you would know which equation calculates simple interest and which equation calculates compound interest.

Summary

Before you attempt the Practice Exercises, review what you've learned.

Algebraic Expressions

You can use algebraic expressions to model real-life situations. When you evaluate an algebraic expression, you substitute values for the variables and then simplify using the order of operations.

EXAMPLE: Write and Evaluate an Algebraic Expression

An art supply store is distributing coupons offering one tube of watercolor paint for $2.99. Tubes of watercolor paint regularly cost $4.49.

- Write an expression for the cost, before tax, of n tubes of watercolor paint for a customer who has a coupon.
- Then evaluate the expression for $n = 6$ and tell what the value of the expression represents.

SOLUTION:

First, write an expression.

Discounted tube: This tube will cost $2.99.

Regularly priced tubes: The other tubes will cost $4.49 each. There are n tubes in all, so there are $n - 1$ regularly priced tubes. Multiply the cost per tube by the number of tubes: $4.49($n - 1$).

Total: Add the cost of the discounted tube to the cost of the regularly priced tubes: $2.99 + $4.49($n - 1$).

Next, evaluate the expression for $n = 6$.

$2.99 + 4.49(n - 1)$

$2.99 + 4.49(6 - 1)$	Substitute 6 for n.
$2.99 + 4.49(5)$	Follow the order of operations; subtract within parentheses.
$2.99 + 22.45$	Multiply.
25.44	Add.

The cost, before tax, for 6 tubes of watercolor paint for a customer who has a coupon is $25.44.

Algebraic Equations

You can also use algebraic equations to model real-life situations. An algebraic equation may have one variable or more than one variable.

EXAMPLE: Write an Equation in One Variable

Three friends are in the same art class and want to split the cost of a large set of colored pencils equally. The pencil set is priced at $36.95, and the sales tax rate is 8%. Write an equation that can be used to determine each friend's cost, including sales tax. Be sure to define any variables you use.

SOLUTION:

Cost for each friend: Let c represent the cost in dollars for each friend. There are 3 friends.

Total cost: The total cost is equal to the price of the pencil set, $36.95, plus the sales tax on the pencil set, 8% of $36.95. Mathematically, the total cost in dollars is $36.95 + 0.08(36.95)$.

Now write the equation:
$3c = 36.95 + 0.08(36.95)$.

My Notes

Summary *(continued)*

EXAMPLE: Write and Evaluate an Equation with More than One Variable

An art store provides picture-framing services. The store generally adds a 2-inch border to each picture. The glass that fits in the frame will cover both the picture and the border.

- Write an equation that gives a, the area in square inches, of the glass needed for a square picture with a side length of s inches.
- Then find the value of a when $s = 12$ and tell what the resulting value of a represents.

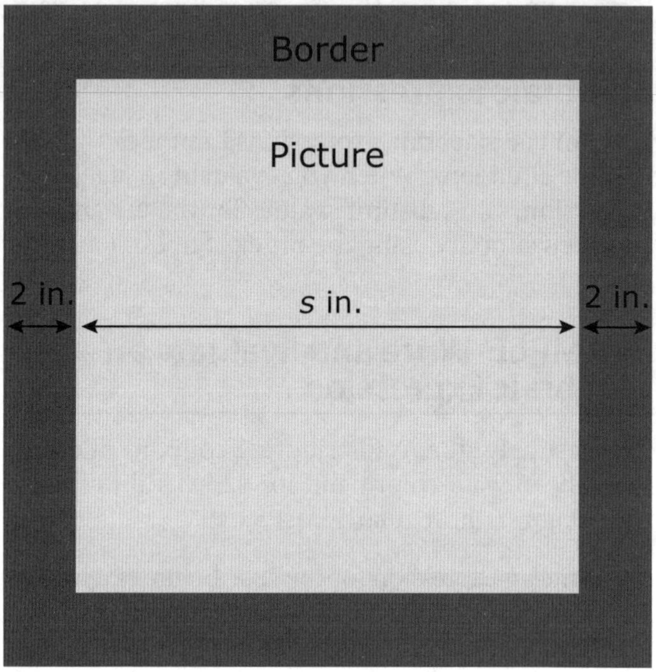

SOLUTION:

First, write an equation.

Think: The glass is square, so the area of the glass is equal to the side length of the glass times itself.

The side length of the glass in inches is equal to the side length of the picture plus 2 border widths: $s + 2 + 2$, or $s + 4$.

Area of glass: $(s + 4)(s + 4)$, or $(s + 4)^2$.

Now set the area of the glass a equal to the expression for the area: $a = (s + 4)^2$.

Next, evaluate the equation for $s = 12$.

$a = (12 + 4)^2$	Substitute 12 for s.
$a = 16^2$	Follow the order of operations. Add within parentheses.
$a = 256$	Evaluate the power.

The area of the glass for a square picture with a side length of 12 inches is 256 square inches.

My Notes

Practice Exercises

Review what you've learned using these practice problems. For practice problems with feedback, try the Coach and Play items in the Practice section online.

1. Amaya made a deposit in an account that earns annual interest. Her account balance, in dollars, after t years is represented by the expression $250 \cdot 1.03^t$.

Trevor made a deposit in an account that earns annual interest. His account balance, in dollars, after t years is represented by the expression $200 \cdot 1.06^t$.

Which statement describes the situation correctly?

A. Amaya deposited \$50 more than Trevor, and Trevor's account earns 3% less annual interest than Amaya's account.

B. Trevor deposited \$3 more than Amaya, and Amaya's account earns 50% more annual interest than Trevor's account.

C. Amaya deposited \$50 more than Trevor, and Trevor's account earns 3% more annual interest than Amaya's account.

D. Trevor deposited \$50 more than Amaya, and Amaya's account earns 3% less annual interest than Trevor's account.

2. Three friends are dining at a restaurant. When the bill arrives, they add a tip of 20% of the original bill, n, and then split the total amount 3 ways.

Which equations below show how much each friend will pay, P? *Select all that apply.*

A. $P = 0.2n + \left(\frac{n}{3}\right)$

B. $P = \left(\frac{n}{3}\right) + \left[\left(\frac{n}{3}\right)0.2\right]$

C. $P = \left(\frac{n}{3}\right)0.2n$

D. $P = \left(\frac{n}{3}\right)0.2$

E. $P = \frac{n + (0.2n)}{3}$

3. Nadine pays \$100 for three baseball cards. The cost of the JJ Vaughn card, v, was half the cost of a Derek Thomas card, t. The JJ Vaughn card was \$20 more than the cost of a Miguel Sosa card, s. Nadine uses the equation $t + v + s = 100$ to show how much she paid for the three cards.

Which equation represents the total cost of the three baseball cards in terms of the cost of a JJ Vaughn card, using only v?

A. $\frac{1}{2}v + v + v - 20 = 100$

B. $2v + v + v - 20 = 100$

C. $\frac{1}{2}v + v + v + 20 = 100$

D. $2v + v + v + 20 = 100$

4. Lamar buys \$30.00 of PexCorp stock that increases \$0.25 a day. Jay buys \$40.00 of DepCom stock that increases \$0.05 per day.

If days are represented by x, which of the following statements are true? *Select all that apply.*

A. Lamar's stock worth can be represented by the equation $y = 0.25x + 30$.

B. Jay's stock worth can be represented by the equation $y = 0.05x + 40$.

C. It will take 40 days for Lamar's stock to be worth \$40.00.

D. In 50 days, Lamar's stock and Jay's stock will both be worth \$42.50.

E. Jay's stock value increases at a faster rate than Lamar's.

5. A cable company charges a flat fee of \$50 per month plus \$8 per month for each movie channel, m, and \$4 per month for each sports package, s.

Create an equation to represent T, the amount of money a person pays for cable in 1 year (12 months).

$T = 12(\_\_\_\_ + \_\_\_\_ m + 4s)$

Practice Exercises *(continued)*

6. A hexagon has 6 sides. The length of each side of the hexagon is 4 centimeters less than the length of each side of a square. Suppose *s* is the length of one side of the square.

Complete the equation below so that it shows the perimeter of the square is the same as the perimeter of the hexagon. Then, solve the equation to find the length of each side of the square and each side of the hexagon.

$4s =$ ______

The length of the side of the square is ______ centimeters, and the length of the side of the hexagon is ______ centimeters.

7. A ferry carries cars, drivers, and passengers across a large lake. The total charge includes $18 for each car and its driver, plus a fare of *f* dollars for each passenger. When Jerry took his family across the lake in his car on the ferry, the total cost was $30.

Suppose *n* represents the number of passengers in Jerry's car. Model this situation by completing the equation.

$30 =$ ______

8. For a simple interest account, the amount of interest earned annually is represented by the expression $750(0.0325)t$.

Which statements define the meaning of each part of this expression accurately? *Select all that apply.*

A. $750 is the amount of interest earned.

B. $750 is the original principal.

C. 3.25% is the interest rate.

D. 0.0325% is the interest rate.

E. *t* is the total amount in the account, with principal and interest.

F. *t* is time in years.

9. A rectangular garden has a perimeter of 54 feet. Its length is 3 less than twice its width.

Using the letters of the options below, write and solve an equation for the garden's dimensions.

A. $w - 3$

B. $54 - w$

C. $2w + 3$

D. $2(w - 3)$

E. $27 - w$

F. $2w - 3$

G. $=$

H. 10

I. 11

J. 17

K. 19

L. 20

M. 27

N. 33

O. 54

The equation is ______.
The width is ______.
The length is ______.

10. The balance of a compound interest savings account is calculated using the expression $P\left(1 + \frac{r}{n}\right)^{nt}$, where *P* is the principal amount and *r* is the interest rate compounded *n* times in *t* years. This expression represents the function $A(t)$, which is used to determine the final balance of an account.

If Neelam has a compound interest savings account that compounds monthly, then what is $A(t)$ after 1 year at 3.5% interest with an initial principal of $1,000?

Apply

How Much Will It Cost to Drive across the Country?

How much will it cost to drive across the country?

Which is farther from your home, the Atlantic Ocean or the Pacific Ocean? How much would it cost you to drive there?

To help determine the estimated cost of the trip, write an algebraic expression and define the variables for each of the following:

- the cost of gasoline needed for the trip
- the number of days the trip will take, assuming you drive a reasonable number of hours per day
- the cost of meals and lodging (either in a hotel or a campground) for the trip

Use your expressions to estimate the total cost of the trip.

Show what you've learned by completing the other performance tasks in the online Apply section.

Apply *(continued)*

Your answer to Apply will be assessed on the following criteria:

1. Defining variables for each of the given items
2. Writing algebraic expressions for each of the given items
3. Researching reasonable values for each variable in your expressions
4. Using your expressions to estimate the total cost of the trip

Criteria \ Scale	4	3	2	1	0
	Exceeds Criteria	**Meets Criteria**	**Progressing to Criteria**	**Below Expectations**	**No Expectation**
Defining Variables	Clearly defines each variable used and states the unit of measurement for each.	Defines each measurement used, but omits one or two units of measurement.	Defines each variable used, but does not state any units of measurement, or defines some variables used.	Uses variables in expressions, but does not define them or defines only one or two variables.	Does not use or define variables.
Writing Expressions	Writes algebraic expressions that correctly model the cost of gasoline, the number of days the trip will take, and the cost of meals and lodging.	Writes algebraic expressions for the cost of gas, the number of days, and the cost of food and lodging, but makes a minor error in one of the expressions.	Attempts to write algebraic expressions for the cost of gas, the number of days, and the cost of food and lodging, but the expressions indicate incomplete understanding of the relationships between quantities.	Attempts to write at least one algebraic expression, but the expressions indicate little or no understanding of relationships between quantities.	Does not write any expressions.
Research	Uses research to determine a reasonable value for each variable and cites reliable sources.	Uses research to determine a reasonable value for each variable but omits one or two sources.	Appears to have used research to determine values for the variables but does not cite sources.	Researches only one or two values or appears to have guessed at the values of the variables.	Does not attempt research and does not give possible values for any variables.
Estimating the Total Cost	Calculates a reasonable estimate for the total cost and clearly shows or explains how the estimate was obtained.	Calculates a reasonable estimate for the total cost, but work or explanation contains a minor error.	Calculates a reasonable estimate without showing work or giving any explanation, or gives an unreasonable estimate, but work or explanation shows some understanding of the relationships and quantities involved.	Gives an unreasonable estimate, and work or explanation, if any, indicates little understanding of the relationships and quantities involved.	Does not attempt to estimate the total cost.

UNIT 2: Equations and Inequalities

2.1 Solve Equations and Inequalities

photo: Getty Images

Lesson Objectives

- Create equations and inequalities to represent real-world situations.
- Apply properties of equality to solve equations and justify the solution process.
- Apply properties of inequality to solve inequalities and graph their solutions.

Essential Questions

- How are algebraic properties used to justify solution methods for linear equations and inequalities?
- How can you discern between the different special case scenarios for equations and inequalities that result in no solution or the solution set of all real numbers?

photo: Getty Images

Investigations

Reasoning with Equations

How does the value of one variable in an equation affect another? Analyze to discover properties.

Solving Algebraic Equations

Solving equations can take several steps. Use properties and commonalities to shorten the distance.

Name That Property

There's more than one way to get there. Consider different solution paths, justifying each step.

Justifying Inequalities

What's different about solving an inequality? Not a lot, but there are some special cases.

Tickets to Ride

You know when a real-life situation has no solution. When does an inequality have no real solution?

Posters for the High School Musical

Compare the costs of printing posters. What's the better deal?

Key Vocabulary

additive inverse, identity, inequality, multiplicative inverse, solution, variable

Discover

As you complete Engage and the investigations, record the most important ideas you've learned.

Engage	**Investigation 1**
Investigation 2	**Investigation 3**
Investigation 4	**Investigation 5**

Name ______________________________ Date ______________

Check for Understanding

Solve Equations and Inequalities: Investigation 1

1. Justify the solution to $2(x - 5) = 56 + 6(x - 5)$ by matching a property to each step.

$2(x - 5) = 56 + 6(x - 5)$

Step 1: $-4(x - 5) = 56$

Step 2: $x - 5 = -14$

Step 3: $x = -9$

A. addition property of equality

B. subtraction property of equality

C. multiplication property of equality

D. division property of equality

E. distributive property

2. When Richard solved the original equation above, he used the distributive property in Step 1 to write an equivalent equation. What equation did Richard write?

Step 1: ____________________ = ____________________

3. Solve each equation.

A. $9(2x + 1) - 4(x - 6) = 6(2x + 1) + 5(x - 6)$

B. $11\left(\frac{x}{4}\right) - 4\left(\frac{x}{3}\right) + 12 = 2\left(\frac{x}{3}\right) - 9\left(\frac{x}{4}\right)$

4. When Lucy solved the equation $9(3x - 2) - 4(3x - 2) = 15 + 10(3x - 2)$, one of the equations in her solution was $(3x - 2) = 3 + 2(3x - 2)$. Identify the sequence of the properties of equality that Lucy could have used to obtain this step in her solution.

Name ______________________________ Date ____________

Check for Understanding

Solve Equations and Inequalities: Investigation 2

1. Which of these equations could be a correct first step in solving the equation $0.10(p - 3) = 1.1 - 0.25p$? *Select all that apply.*

A. $10(p - 3) = 110 - 25p$

B. $10(p - 3) = 11 - 25p$

C. $10(100p - 300) = 110 - 25p$

D. $0.10(p - 3) + 0.25p = 1.1$

E. $0.10p - 0.30 = 1.1 - 0.25p$

2. Determine whether each equation has one solution, no solution, or infinitely many solutions.

A. $9x - 20 = 9x + 16$ ____________________

B. $8(x - 4) - 3x = 4x + 20$ ____________________

C. $3x - 10 = 2(4x - 5) - 5x$ ____________________

D. $4x - 12 = -4x - 12$ ____________________

3. Solve the equation: $10x + 4 = 5(x + 2) - 3(2 - x)$.

4. Find a value for b so that the equation $4(2x - 3) - 3x = 13 + 5(x - b)$ has infinitely many solutions.

5. Janae needs to solve the equation $3x + \frac{5}{16} = \frac{3}{4} - \frac{1}{8}x - \frac{1}{2}$. What is an advantage of using the multiplication property of equality before combining like terms using other properties? In your explanation, tell the value by which you would multiply both sides.

Name ______________________________ Date ____________

Check for Understanding

Solve Equations and Inequalities: Investigation 3

1. If $p < q$ and $r < 0$, then which of these statements are true? *Select all that apply.*

A. $-8p < -8q$ **B.** $pr < qr$ **C.** $pr > qr$

D. $p + r < q + r$ **E.** $p - r > q - r$ **F.** $\frac{p}{r} > \frac{q}{r}$

2. Match each inequality with the number line that shows its solution. *Each number line may be used once, more than once, or not at all.*

A. $4x > -8$

B. $2x + 7 > 2x - 10$

C. $-6 > x - 4$

D. $-12 < -3(x + 2)$

E. $6(x - 5) - 4x > 2(x - 10)$

F. $-\frac{3}{4}x < -\frac{3}{2}$

I. 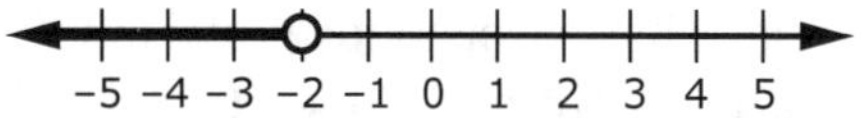

II. 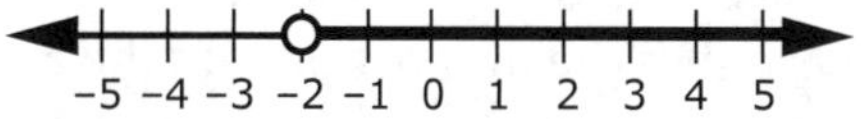

III. 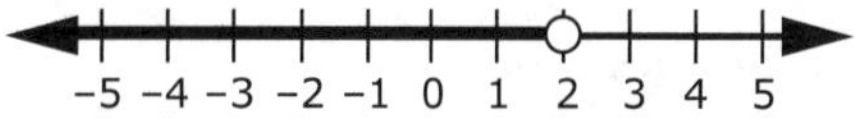

IV. 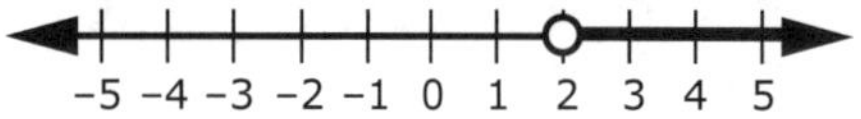

V. 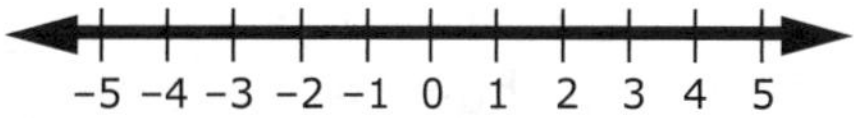

VI. 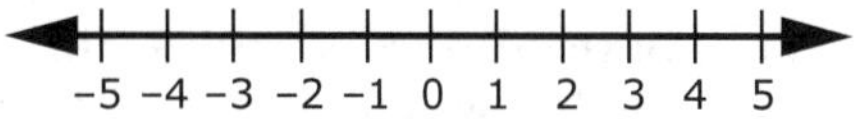

Jamal stops by the Gas & Go Mart to buy gas for his car. His car is very dirty so he will also have it washed. He sees the price of gas is \$3.58 per gallon and a car wash costs \$9.50.

3. Which of these inequalities represents the relationship between the amount of money Jamal has available to spend, m, and the gallons of gas that Jamal buys, g?

A. $m > 9.50 + 3.58g$ **B.** $m \leq 9.50 + 3.58g$

C. $m < 9.50 + 3.58g$ **D.** $m \geq 9.50 + 3.58g$

4. Jamal has at most \$30 to spend at the Gas & Go Mart. Solve the inequality from question 3 to find the amount of gas that Jamal can buy to the nearest tenth of a gallon.

5. Sammy and Lisa solve $3x < -30$ by dividing each side by 3. Lisa thinks the direction of the inequality sign must change, but Sammy disagrees. Who is correct? Explain why.

Name ______________________________ Date ______________

Check for Understanding

Solve Equations and Inequalities: Investigation 4

1. To enter a bowling tournament, David must score an average of at least 180 points on his 4 qualifying games. So far, David scored 164, 190, and 174. Which inequality describes *s*, a score on the next game that allows David to enter the tournament?

   A. $s \leq 4(180) - (164 + 190 + 174)$ **B.** $s \geq 4(180) - (164 + 190 + 174)$

   C. $s \geq 180 - \frac{(164 + 190 + 174)}{3}$ **D.** $s \geq 180 - \frac{(164 + 190 + 174)}{4}$

2. An elevator has a weight limit of 825 kilograms. A 93-kilogram construction worker takes a 30-kilogram wheelbarrow and some concrete mix to the top floor. If each bag of concrete weighs 42 kilograms, which best describes the number of bags the elevator can carry?

   A. no more than 16 **B.** at most 17 **C.** 19 or fewer

   D. at least 16 **E.** no less than 17 **F.** at most 20

Jade found a rule-of-thumb to make sure fish kept in an aquarium are not overcrowded: the surface area of the water in square inches should be at least 12 times the total combined lengths of the fish in inches. Jade's aquarium has a water surface area of 276 square inches. A list of her fish is shown in the table.

Type of Fish	Number	Length
Tiger Barb	1	3 in.
Gold Tetra	5	1 in.
Neon Tetra	2	1.75 in.
Cobra Gripper	3	1.5 in.

3. What is the total combined length of the fish in Jade's aquarium?

4. Jade will add *n* flagfish to her aquarium. Each is 2 inches long. Solve an inequality for *n*. Then, state the number of flagfish Jade can safely add to her aquarium.

   $n \leq$ ________________ Maximum number of flagfish to add: ________________

5. In solving an inequality, how are a contradiction and an identity the same and how are they different? In your response, include an example of each.

Name ______________________________ Date ______________

Check for Understanding

Solve Equations and Inequalities: Investigation 5

The cost to rent a storage unit is different for two storage companies. Stack 'n Store charges $150 per month plus a nonrefundable deposit of $175. Smart Storage charges $185 each month with no deposit. Complete the table showing the total cost for each company during the first six months and use it to answer the questions.

Months	Cost in $ at Stack 'n Store	Cost in $ at Smart Storage
1	325	185
2		
3		
4		
5		
6		

1. Lee plans to store clothes for four months. Which company offers the better deal?

2. After how many months is the total cost the same at both storage companies?

3. Write the equation for the total cost, C, of renting the unit m months at each company.

   Stack 'n Store: ________________

   Smart Storage: ________________

Your class will sell school T-shirts for $10 each to make money for a class trip. Graphs of the equations for revenue earned, R, and cost of producing the T-shirts, C, are shown. The class will break even when the revenue equals the cost to produce the T-shirts.

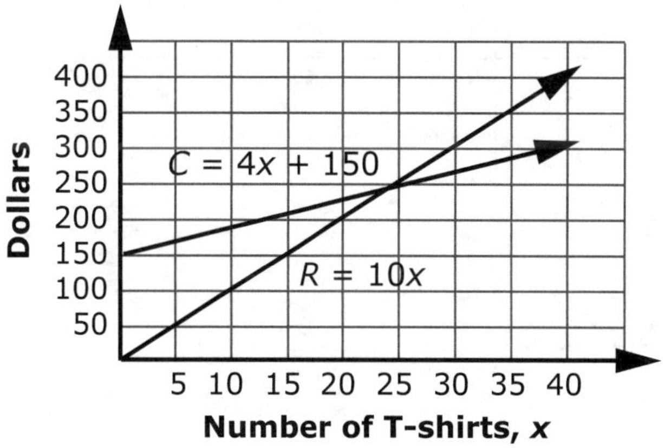

4. Which of these describes the number of T-shirts the class must sell to break even?

   A. $x = 25$

   B. $x > 25$

   C. $x < 25$

   D. $x > 250$

   E. $x < 250$

   F. $x = 250$

5. Suppose you do not have graphs of the revenue and cost equations. Explain how you can find the number of T-shirts the class must sell to make a profit. Show your solution.

Summary

Before you attempt the Practice Exercises, review what you've learned.

Properties of Equality

Linear equations can be solved using inverse operations. You can use the properties of equality to write equivalent equations.

A **linear equation** in one variable can be written in the form $ax + b = c$, where a, b, and c are real numbers and $a \neq 0$.

Properties of Equality		
Addition property of equality	If you add the same number to both sides of an equation, the equation remains true.	If $a = b$, $a + c = b + c$.
Subtraction property of equality	If you subtract the same number from both sides of an equation, the equation remains true.	If $a = b$, $a - c = b - c$.
Multiplication property of equality	If you multiply both sides of an equation by the same number, the equation remains true.	If $a = b$, $a \cdot c = b \cdot c$.
Division property of equality	If you divide both sides of an equation by the same nonzero number, the equation remains true.	If $a = b$ and $c \neq 0$, $\frac{a}{c} = \frac{b}{c}$.

My Notes

Solving Linear Equations

To solve an equation:

- First, simplify each side of the equation by combining like terms. You can also simplify expressions that are within grouping symbols.
- Then, identify which operations were done to the variable.
- Next, undo the operations in the reverse order by applying the properties of equality.

EXAMPLE

SOLVE:

$-3.5(2x - 16) + 6 = 3x + 1 - 1.5x$

SOLUTION:

$-3.5(2x - 16) + 6 = 3x + 1 - 1.5x$

$-7x - 56 + 6 = 3x + 1 - 1.5x$ Distributive property

$-7x - 50 = 1.5x + 1$ Combine like terms.

$\underline{-1.5x \quad} = \underline{-1.5x \quad}$

$-8.5x - 50 = 1$ Subtraction property of equality

$\underline{+50} = \underline{+50}$

$-8.5x = 51$ Addition property of equality

$\frac{-8.5x}{-8.5} = \frac{51}{-8.5}$

$x = -6$ Division property of equality

You can check the solution to an equation by substituting the solution value for the variable in the original equation. Substitute –6 for x.

$-3.5[2(-6) + 16] + 6 \stackrel{?}{=} 3(-6) + 1 - 1.5(-6)$

$-3.5(-12 + 16) + 6 \stackrel{?}{=} -18 + 1 + 9$

$-3.5(4) + 6 \stackrel{?}{=} -6$

$-8 = -8$

Summary *(continued)*

Identities and Contradictions

Sometimes when you are solving an equation, the variable will "disappear." The final statement has only constants on each side of the equation. If this resulting statement is true, the equation is an identity. If it is false, the equation is a contradiction.

> An identity equation is true for all values of the variable.
>
> A contradiction equation is not true for any value of the variable.

EXAMPLE: Solving an Identity Equation

SOLVE:

$8\left(-\frac{1}{2}v - \frac{1}{8}\right) = -\frac{1}{4}(16v + 12) + 2$

SOLUTION:

$$8\left(-\frac{1}{2}v - \frac{1}{8}\right) = -\frac{1}{4}(16v + 12) + 2$$
$$-4v - 1 = -4v - 3 + 2$$
$$-4v - 1 = -4v - 1$$
$$-1 = -1$$

Because the resulting statement is true, the equation is an identity. You can substitute any value for v, and the equation will always be true. This means all real numbers are solutions of the equation.

EXAMPLE: Solving a Contradiction Equation

SOLVE:

$6.5y - 18 + 3.5y = 2(5y - 18)$

SOLUTION:

$$6.5y - 18 + 3.5y = 2(5y - 18)$$
$$6.5y - 18 + 3.5y = 10y - 36$$
$$10y - 18 = 10y - 36$$
$$-18 \neq -36$$

Because the resulting statement is false, the equation is a contradiction. You can substitute any value for y, and the equation will never be true. This means the equation has no solutions.

So, a linear equation in one variable can have no, one, or infinitely many solutions.

Solving and Graphing Linear Inequalities

Linear inequalities are also solved using inverse operations. You use the properties of inequality to write equivalent inequalities. The properties of inequality are similar to the properties of equality. However, if you multiply or divide both sides of an inequality by a negative number, you must reverse the inequality symbol.

Solutions to linear inequalities are often restricted when real-world problems are modeled.

My Notes

Summary *(continued)*

Properties of Inequality		
Addition property of inequality	If you add the same number to both sides of an inequality, the inequality remains true.	If $a < b$, $a + c < b + c$.
Subtraction property of inequality	If you subtract the same number from both sides of an inequality, the inequality remains true.	If $a < b$, $a - c < b - c$.
Multiplication property of inequality	If you multiply both sides of an inequality by the same positive number, the inequality remains true. If you multiply both sides of an inequality by the same negative number, you must reverse the symbol to keep the inequality true.	If $a < b$, $a \cdot c < b \cdot c$ when $c > 0$, and $a \cdot c > b \cdot c$ when $c < 0$.
Division property of inequality	If you divide both sides of an inequality by the same positive number, the inequality remains true. If you divide both sides of an inequality by the same negative number, you must reverse the symbol to keep the inequality true.	If $a < b$, $\frac{a}{c} < \frac{b}{c}$ when $c > 0$, and $\frac{a}{c} > \frac{b}{c}$ when $c < 0$.

My Notes

EXAMPLE: Solving a Real-World Problem

Jake is saving money for a tablet computer that costs $375.99 plus 6% sales tax. He has $50 saved so far. If he saves $25 each month, when will he have enough money for the tablet?

SOLUTION:

Write and solve an inequality. Let t be the number of months until Jake has enough money.

$$25t + 50 \geq 375.99 \cdot 1.06$$

$$25t + 50 \geq 398.55$$

$$25t \geq 348.55$$

$$t \geq 13.94$$

If we consider only whole months, then Jake will have enough money after 14 months. So, the solution set is all whole numbers greater than or equal to 14.

Points of Intersection and Solutions to the Equation

The x-coordinates of the points where the graphs of the equations $y = f(x)$ and $y = g(x)$ intersect are the solutions to the equation $f(x) = g(x)$. You can determine these coordinates using tables and graphs.

Practice Exercises

Review what you've learned using these practice problems. For practice problems with feedback, try the Coach and Play items in the Practice section online.

1. Find values for a and b so that $2(3x + 5) + ax = 5(4x - 6)$ is an identity equation.

A. $a =$ ________

B. $b =$ ________

2. Inez solves the equation $2m - 6 = 4m + 10$ but needs to give the reasons for each step.

Help her complete her solution by giving the correct reason for each step from the choices below.

A. addition property of equality

B. division property of equality

C. multiplication property of equality

D. subtraction property of equality

E. given

F. simplify

G. substitution property

$2m - 6 = 4m + 10$ ________

$2m = 4m + 16$ ________

$-2m = 16$ ________

$m = \frac{16}{-2}$ ________

$m = -8$ ________

3. Solve the equation.

$4\left(\frac{x}{5}\right) - (2x + 9) = 5(2x + 9) - 2\left(\frac{x}{5}\right)$

The solution is $x =$ ________.

4. A shipping company is hired to deliver 125 large mirrors to a furniture store for $15 each. The shipping company agrees to pay $45 for each mirror, n, that is damaged during transport and expects the cost for the driver and fuel to be $140.

If the shipping company hopes to earn at least $1,250, which inequality can be used to represent this situation?

What is the maximum number of mirrors, n, that can be damaged?

Select both an inequality and the maximum number of mirrors.

A. $(125 \cdot 15) - 45n + 140 \geq 1{,}250$

B. $(125 \cdot 15) - 45n - 140 \leq 1{,}250$

C. $(125 \cdot 15) - 45n - 140 \geq 1{,}250$

D. 10 mirrors

E. 11 mirrors

F. 17 mirrors

5. The average high temperature during the week from Monday through Friday was less than 81°. The daily high temperatures for Monday through Thursday were 82°, 79°, 81°, and 76°. What might the high temperature have been on Friday?

Write and solve an inequality for the high temperature, t, on Friday.

____

Practice Exercises *(continued)*

6. Solve the inequality.

$5 \leq 1 - 2n$

Which graph models the solution set?

A. 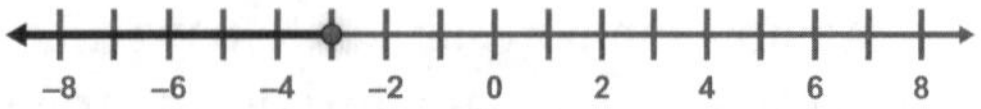

B. 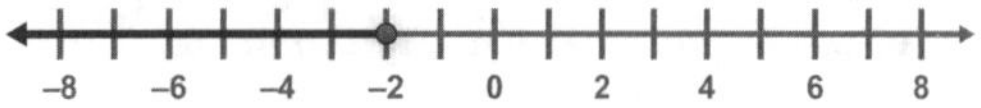

C. 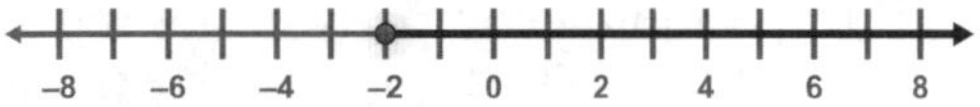

D. 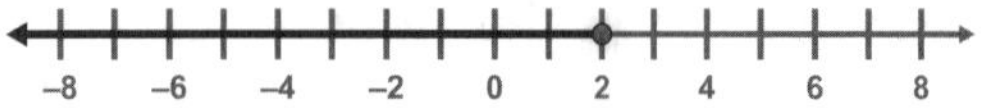

E. 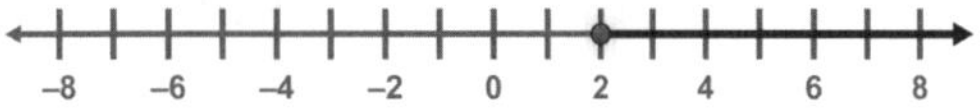

7. Classify each equation as having no solution, exactly one solution, or infinitely many solutions.

A. $7x + 4 = 3(4x - 1)$

B. $2x + 4 = 2(x - 1)$

C. $-x + 4 = 10 - 7x - 6(1 - x)$

D. $7x - 9x + 8 = -x + 8$

E. $-9x + 4 = -9x + 4$

F. $-8x - 2 = -8x$

No solution: __________
Exactly one solution: __________
Infinitely many solutions: __________

8. Complete the parts of the inequality so that its solution is the set of all real numbers.

Determine the single constant or term for each part.

$-8x +$ __________ < 4 (__________ $- 3)$

9. Solve the following inequality.

$\frac{a - 5}{2} \geq 3a$

A. $a \leq 1$

B. $a \leq -1$

C. $a \geq 1$

D. $a \geq -1$

10. Solve the following inequality.

$4x - 3(x + 2) - 5 < 0$

A. $x < -2$

B. $x > -2$

C. $x < 11$

D. $x > 11$

Apply

How Should a Vendor Stock an Ice Cream Cart?

photo: Getty Images

A beachside ice cream vendor has to carefully plan how he stocks his cart to maximize his sales. The vendor knows that cookie ice cream sandwiches and fudge pops are his best sellers, so he decides to stock only those two items. His freezer holds a maximum of 125 ice cream treats. He usually sells at least 45 cookie sandwiches and no fewer than 30 fudge pops each day.

Write a system of inequalities that represents the possible combinations of ice cream treats that the vendor should stock in his cart. Graph the system and test it with at least five points. Make a plan that includes at least five different ways he could stock the cart and not run out of ice cream treats.

Show what you've learned by completing the other performance tasks in the online Apply section.

Apply *(continued)*

Your answer to Apply will be assessed on the following criteria:

1. Writing inequalities that represent the situation
2. Graphing the inequalities
3. Identifying and testing points inside and outside the bounded region to confirm that it is correct
4. Creating a plan with five different ways to stock the cart

Criteria \ Scale	4 Exceeds Criteria	3 Meets Criteria	2 Progressing to Criteria	1 Below Expectations	0 No Expectation
Inequalities	Writes all three inequalities correctly.	Writes two inequalities correctly.	Writes one inequality correctly.	Does not write any inequalities correctly.	Does not attempt task.
Graphs Inequalities	Graphs all three inequalities correctly and shades the correct region (bounded region).	Graphs two inequalities correctly and shades the correct region, or graphs all three inequalities correctly but shades the wrong region.	Graphs one inequality correctly and shades the correct region, or graphs two inequalities correctly but shades the wrong region.	Graphs all of the inequalities incorrectly; doesn't shade or shades the wrong region.	Does not attempt task.
Tests Solution	Tests at least five points to confirm that the shaded region of the graph is correct.	Tests four points to confirm that the shaded region of the graph is correct.	Tests two or three points to confirm that the shaded region of the graph is correct.	Tests one point to confirm that the shaded region of the graph is correct.	Does not attempt task.
Plan	Creates a plan that includes at least five correct options to stock the cart.	Creates a plan that includes three or four correct options to stock the cart.	Creates a plan that includes one or two correct options to stock the cart.	Creates a plan but it does not contain any correct options to stock the cart.	Does not attempt task.

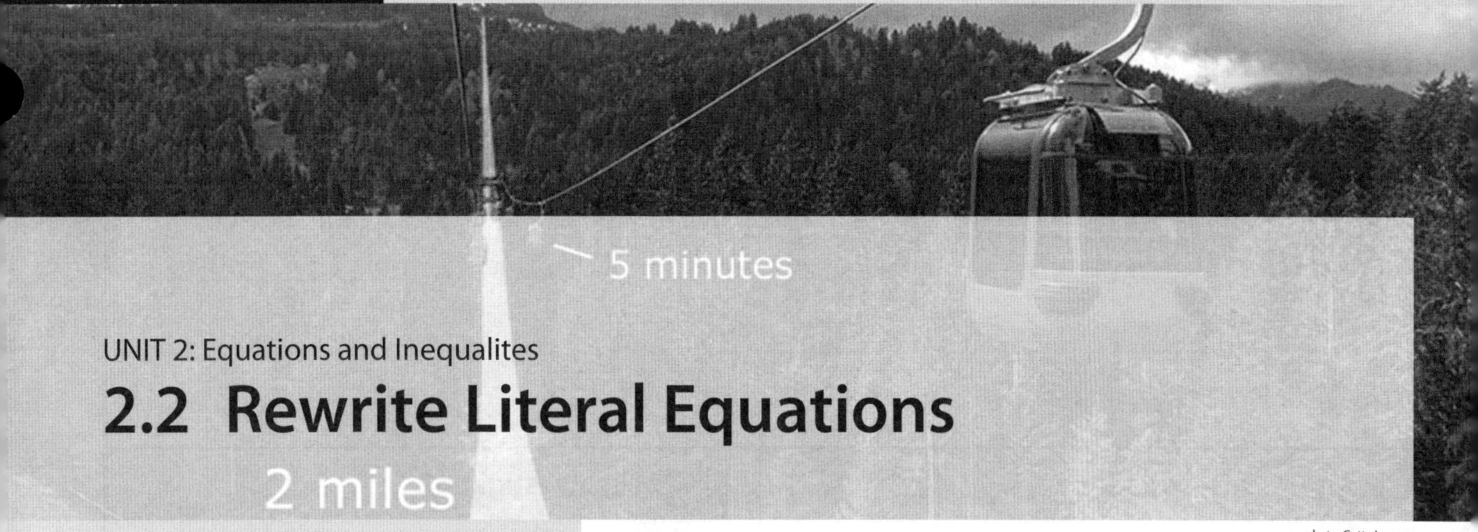

UNIT 2: Equations and Inequalites

2.2 Rewrite Literal Equations

photo: Getty Images

Lesson Objectives

- Solve literal equations for a given variable.
- Transform literal equations to solve real-world problems.

Essential Question

- How are algebraic properties used to transform and solve literal equations?

Investigations

Equations Can Be So Literal

Re-engineer the distance formula to help an amusement park engineer.

Exploring Literal Equations

Refinance the interest formula for specific financial situations.

Installing Aquariums

Manipulate volume formulas to create custom aquariums.

photo: Getty Images

Key Vocabulary

literal equation, solution, variable

Discover

As you complete Engage and the investigations, record the most important ideas you've learned.

Engage	
Investigation 1	**Investigation 2**

Name ______________________________ Date ______________

Check for Understanding

Rewrite Literal Equations: Investigation 1

Bricklayers can estimate the number of bricks they need to build a wall using the formula $n = 7LH$, where L and H are the length and height of the wall in feet, respectively.

1. Which equation shows the bricklayer's formula solved correctly for H?

A. $H = \frac{7n}{L}$ **B.** $H = \frac{L}{7n}$ **C.** $H = n - 7L$ **D.** $H = \frac{n}{7L}$ **E.** $H = \frac{7L}{n}$

2. A wall is 20 feet long and requires 1,470 bricks to build. How high is the wall?

3. Dolbear's Law relates the number of chirps per minute made by Snowy Tree crickets, n, and the air temperature, T, in degrees Fahrenheit: $T = 50 + \frac{n-4}{4}$. To predict the number of chirps at any temperature, rewrite the formula to solve for n.

$n =$ ____________________

4. The formula for the area of a trapezoid is $A = \frac{1}{2}(b_1 + b_2)h$. Which shows how the formula can be rewritten correctly to find another variable? *Select all that apply.*

A. $h = \frac{2A}{b_1 + b_2}$ **B.** $h = \frac{2A - b_1}{b_2}$ **C.** $b_1 = \frac{2A}{h} - 2b_2$

D. $b_2 = \frac{2A}{h} - b_1$ **E.** $b_2 = \frac{2A}{b_1} - h$

5. Is $r = \frac{6.5}{2\pi}$ a literal equation? Explain why or why not.

Name ______________________________ Date ______________

Check for Understanding

Rewrite Literal Equations: Investigation 2

Susan makes and sells cylindrical wax candles. To make a candle, Susan buys a block of wax, melts it, and pours it into molds to make the candles. Susan usually prices a candle based on the volume of wax she uses to make it.

Answer the following questions making sure to include appropriate units on your answers.

1. One of the molds makes a candle with a radius of 4 inches and a height of 1 foot. What are these lengths given to the nearest tenth of a centimeter? (1 inch = 2.54 centimeters)

2. A customer orders a candle with radius of 6 centimeters and height of 20 centimeters. Use $V = \pi r^2 h$ to calculate the amount of wax needed. *Round to the nearest cubic centimeter.*

3. A second customer orders a candle 10 centimeters tall and Susan uses a whole block of wax. The volume of the wax is printed on the box. Which value is unknown?

A. V **B.** π **C.** r **D.** h

4. Rewrite the volume formula to solve for the radius of the candle. Then, use the new formula to find the radius of the candle made for the second customer if 2,450 cubic centimeters of wax are used. *Round your answer to the nearest tenth of a centimeter.*

$r =$ ________________ The measure of the radius is ________________.

5. A third customer asks Susan to make his candle so that the diameter and height are the same length. Rewrite the volume formula to solve for the radius. Explain why the radius depends only on the volume of wax used and no other variable.

Summary

Before you attempt the Practice Exercises, review what you've learned.

Literal Equations

A **literal equation** can be solved for any variable and used to find the value for that variable when other variables' values are known. For example, the formula $A = bh$ can be rearranged so that the equation is solved for B or for h as shown below.

A **literal equation** is an equation that has more than one variable. Formulas are examples of literal equations.

Formula	When to Use
$A = bh$	Use to find the area when you know the length of the base and the height.
$h = \frac{A}{b}$	Use to find the height when you know the area and the length of the base.
$b = \frac{A}{h}$	Use to find the length of the base when you know the area and the height.

EXAMPLE: Solving for a Variable

Solve the formula for perimeter of a rectangle for w.

$P = 2l + 2w$

SOLUTION:

Use the properties of equality to isolate w on one side of the equation.

$P = 2l + 2w$

$P - 2l = 2w$ Subtraction property of equality

$\frac{P - 2l}{2} = w$ Division property of equality

EXAMPLE: Using a Literal Equation

A rectangle has a length of 15.5 inches and a perimeter of 42 inches. What is the width of the rectangle?

SOLUTION:

Use the literal equation $\frac{P - 2l}{2} = w$.

Substitute 15.5 for l and 42 for P.

$$\frac{42 - 2(15.5)}{2} = w$$

$$\frac{42 - 31}{2} = w$$

$$\frac{11}{2} = w$$

$$w = 5.5$$

The width is 5.5 inches.

My Notes

Summary *(continued)*

EXAMPLE: Problem-Solving with Literal Equations

To administer the proper dosage of a drug, doctors sometimes calculate a person's body surface area. One formula that can be used is $B = \sqrt{\frac{wh}{3{,}131}}$, where B is body surface area in square meters, w is the person's weight in pounds, and h is the person's height in inches.

Jessica weighs 130 pounds and has a body surface area of 1.61 square meters. What is Jessica's height in feet and inches?

SOLUTION:

Step 1: Solve the formula $B = \sqrt{\frac{wh}{3{,}131}}$ for h.

$$B = \sqrt{\frac{wh}{3{,}131}}$$

$$B^2 = \frac{wh}{3{,}131}$$

$$3{,}131B^2 = wh$$

$$\frac{3{,}131B^2}{w} = h$$

Step 2: Find the value of h.

$$h = \frac{3{,}131\,B^2}{w}$$

$$= \frac{3{,}131(1.61)^2}{130}$$

$$= 62.42$$

Jessica is about 62 inches tall.

12 inches = 1 foot

$$62 \text{ inches} \cdot \frac{1 \text{ foot}}{12 \text{ inches}} = 5 \text{ feet } 2 \text{ inches}$$

Jessica is 5'2" tall.

My Notes

Practice Exercises

Review what you've learned using these practice problems. For practice problems with feedback, try the Coach and Play items in the Practice section online.

1. To find the area of a triangle, use the formula $A = \frac{1}{2}bh$.

   Which statement correctly explains how you can find the height of a triangle with an area of 40 square units and a base of 12 units?

   A. Use the formula $h = \frac{2A}{b}$ to find that $h = \frac{20}{3}$.

   B. Use the formula $h = \frac{A}{2b}$ to find that $h = \frac{20}{3}$.

   C. Use the formula $h = \frac{A}{2b}$ to find that $h = \frac{5}{3}$.

   D. Use the formula $h = \frac{Ab}{2}$ to find that $h = 240$.

2. A party supply company rents two different sizes of tables:

   - A small square table will seat 4 guests.
   - A large circular table will seat 6 guests.

   The formula $G = 4s + 6c$ calculates the number of guests that can be seated when s square tables and c circular tables are rented.

   Which of these literal equations show the result of rearranging the formula to solve for another variable? *Select all that apply.*

   A. $4s = G - 6c$ **B.** $s = G - 1.5c$

   C. $c = \frac{G - 4s}{6}$ **D.** $c = G - 4s - 6$

   E. $c = \frac{G}{4s + 6}$ **F.** $s = 0.25G - 1.5c$

3. The volume of a cone is represented by the formula $V = \frac{1}{3}\pi r^2 h$. The volume of the cone on the right is 3,014.4 cubic centimeters. Which equation shows the formula solved for the unknown variable?

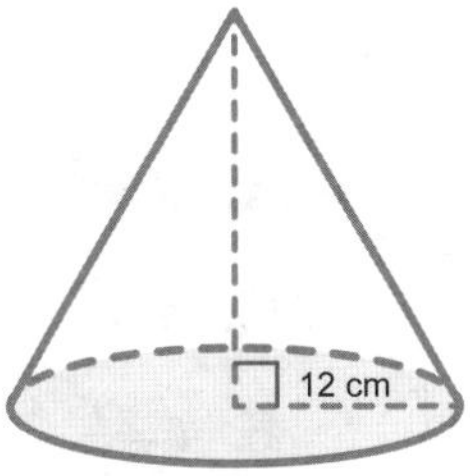

   A. $r = \sqrt{\frac{1}{3}\pi V h}$ **B.** $h = \frac{1}{3}\pi r^2 V$

   C. $h = \frac{3V}{\pi r^2}$ **D.** $r = \sqrt{\frac{3V}{\pi h}}$

4. The area of a trapezoid is represented by the formula $A = \frac{1}{2}h(b_1 + b_2)$.

   A trapezoid has an area of 52 square inches and bases of 8 and 18 inches.

   Rewrite the formula in terms of h to find the height of the trapezoid.

   A. Use the formula $h = \frac{2(b_1 + b_2)}{A}$ to find that $h = 1$.

   B. Use the formula $h = \frac{A}{2(b_1 + b_2)}$ to find that $h = 1$.

   C. Use the formula $h = \frac{A}{b_1 + b_2}$ to find that $h = 2$.

   D. Use the formula $h = \frac{2A}{b_1 + b_2}$ to find that $h = 4$.

5. Raquel lives 150 miles from her aunt. On her way to her aunt's house one day, she stopped to visit a friend. After leaving her friend's house, Raquel noticed that it took her 1.75 hours traveling at an average of 60 miles per hour to complete the trip to her aunt's.

   The formula $D = rt + b$ can be used to represent the total distance, D, from Raquel's house to her aunt's; r is rate of speed in miles per hour; t is time driving; and b is the distance from Raquel's house to her friend's house.

   Which equation shows the formula solved for the unknown variable?

   A. $b = \frac{D}{rt}$ **B.** $b = D - rt$

   C. $D = rt + b$ **D.** $r = \frac{D - b}{t}$

   E. $t = \frac{b - D}{r}$ **F.** $t = \frac{D - b}{r}$

6. The formula for calculating the density of an object is $D = \frac{m}{v}$, where m is mass and v is volume.

   Rewrite the formula in terms of m. Then, rewrite the formula in terms of v.

   $m =$ __________

   $v =$ __________

Practice Exercises *(continued)*

7. A common contest in strongman competitions is the boulder carry, where athletes pick up a large boulder and carry it over a set distance. Samson is calculating the work done by some of the competitors.

   Work performed is calculated by using the formula $W = \frac{(ma) \cdot d}{t}$, where W = work done, m = mass of the object, a = acceleration of the object, d = distance the object is moved, and t = time it takes to move the object.

   The athletes carry 150-kilogram boulders on a course that is 30 meters long.

   Depending on the strength of the athlete, the acceleration may vary, which also affects the total time.

   Considering the formula, which of these variables are constants in this situation? *Select all that apply.*

   A. W

   B. m

   C. a

   D. d

   E. t

8. The equation $m - n = n + r$ can be rewritten as $m = 2n + r$ because of what property? *Select all that apply.*

   A. division property of equality

   B. multiplication property of equality

   C. addition property of equality

   D. distributive property of equality

9. Ohm's law, $I = \frac{V}{R}$, is a formula that is used to calculate the flow of electric current, where I = current, which is the rate of flow; V = voltage, which is the electrical force causing the flow; and R = resistance, which slows the flow.

   Antonio has a lamp with 6 ohms of resistance that requires 2 amperes of current to shine brightly.

   Rearrange the formula to solve for voltage and find the number of volts Antonio needs to power his lamp.

   Formula used: V = __________

   Voltage required: __________ volts

10. The formula for distance is $d = rt$, where d = distance, r = average rate of speed, and t = time.

    Rosa, Marlee, and Anita want to compare their running abilities; however, they have different data.

    Rosa, a sprinter, knows she can run at a speed of 10 meters per second but can only sprint for about 15 seconds. Marlee, a middle-distance runner, can run for 400 meters at 12 meters per second. Anita, a long-distance runner, can run 1,600 meters in 5 minutes and 15 seconds.

    Identify the formula each runner should use for each unknown variable.

    Rosa: __________

    Marlee: __________

    Anita: __________

    A. $d = \frac{t}{r}$ **B.** $d = rt$

    C. $d = \frac{r}{t}$ **D.** $t = \frac{d}{r}$

    E. $t = \frac{r}{d}$ **F.** $t = rd$

    G. $r = \frac{d}{t}$ **H.** $r = dt$

Apply

What Problem Can I Solve with a Formula?

Many real-world situations are modeled using literal equations, or formulas. Find a literal equation that relates at least three variables. Define each variable and state their units. Write a real-world problem that asks about the value of a variable that is not the same as the one isolated in the original formula. Use the techniques of solving literal equations to isolate the new variable. Evaluate your new formula within the context and judge its reasonableness.

By knowing the height of fireworks when they explode, designers create better shows.

Show what you've learned by completing the other performance tasks in the online Apply section.

Apply *(continued)*

Your answer to Apply will be assessed on the following criteria:

1. Defining each variable in your literal equation and stating the units of each variable
2. Writing a real-world problem that involves solving the literal equation for a different variable
3. Solving your literal equation for the desired variable and justifying your steps
4. Using your new equation to answer your real-world problem and demonstrating that your answer is reasonable

Scale / Criteria	4 Exceeds Criteria	3 Meets Criteria	2 Progressing to Criteria	1 Below Expectations	0 No Expectation
Definition of Variables	Correctly defines each variable in the literal equation and states the units of each variable.	Correctly defines all variables, but makes a minor mistake when describing the units of one of the variables.	Correctly defines most variables and states most units, but makes a significant mistake or omission.	Incorrectly defines the variables, or does not include units.	Does not define the variables.
Real-World Problem	Writes a clear and complete description of a real-world problem that involves solving a literal equation for a different variable.	Writes a real-world problem that involves solving a literal equation for a different variable, but the description is unclear.	Writes a real-world problem that involves solving a literal equation for a different variable, but the problem is missing a key piece of information.	Writes a problem that involves a literal equation, but the problem does not reflect the real world or does not involve solving the equation for a different variable.	Does not write a problem.
Desired Variable	Correctly solves the literal equation for the desired variable and justifies each step.	Correctly solves the literal equation for the desired variable, but omits a step or a justification.	Follows a generally correct procedure for solving the literal equation, but makes an error in a step or justification.	Attempts to solve the literal equation, but makes significant algebraic errors.	Does not attempt to solve for the desired variable.
Problem Solution	Correctly interprets the solution in the real-world problem and demonstrates that the answer is reasonable.	Correctly interprets the solution in the real-world problem, but does not correctly demonstrate reasonableness.	Incorrectly interprets the solution in the real-world problem, but does demonstrate reasonableness.	Incorrectly interprets the solution in the real-world problem; does not demonstrate reasonableness.	Does not attempt to interpret the solution.

UNIT 2: Equations and Inequalities

2.3 Solve Absolute Value Equations and Inequalities

photo: Getty Images

Lesson Objectives

- Solve absolute value equations.
- Solve compound inequalities and absolute value inequalities.
- Identify equations and inequalities that have no solution or infinitely many solutions, including absolute value equations and inequalities.

Essential Questions

- How do you solve compound equations and inequalities algebraically and graphically?
- How do you solve absolute value equations and inequalities algebraically and graphically?
- How can you represent and solve a compound inequality?

photo: Getty Images

Investigations

Compound Inequalities

Go fishing to find out how inequalities can set limits. Then use inequalities to limit other situations.

Compound Inequalities

What's the difference between AND and OR? In the realm of inequalities, the difference is huge!

What Values Are Allowed?

Compound problems can have many solutions; not all apply.

Special Cases

Show off solutions and check results with number line graphs.

Absolutely!

An absolute value is a two-case problem. Analyze the cases to find the result.

Are You Absolutely Sure?

Not all solutions are valid. Check the results to find the error disguised as a solution.

Absolute Fun!

Make the connection: absolute value and compound inequalities.

Key Vocabulary

absolute value, compound inequality, inequality, solution, variable

Discover

As you complete Engage and the investigations, record the most important ideas you've learned.

Engage	
Investigation 1	Investigation 2
Investigation 3	Investigation 4
Investigation 5	Investigation 6

Name ____ Date ______________

Check for Understanding

Solve Absolute Value Equations and Inequalities: Investigation 1

1. Hurricane strength is classified using the Saffir-Simpson scale. Category 3 hurricanes have wind speeds greater than 110 miles per hour but at most 130 miles per hour. Which of these show possible wind speeds, w, in this category? *Select all that apply.*

   A. $w > 110$ OR $w \leq 130$ **B.** $w > 110$ AND $w \leq 130$

   C. $110 < w \leq 130$ **D.** $110 > w > 130$

   E. (110, 130)   **F.** (110, 130]

   G. [110, 130)   **H.** [110, 130]

2. Which compound inequality is shown by the number line graph?

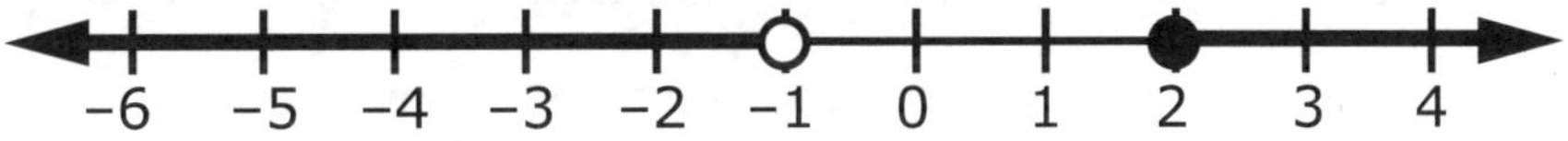

   A. $x < -1$ OR $x > 2$ **B.** $x < -1$ AND $x \geq 2$

   C. $x < -1$ AND $x > 2$ **D.** $x < -1$ OR $x \geq 2$

3. The solutions of two simple inequalities are graphed on the number lines. Which values below are solutions to the compound inequality formed by their union? *Select all that apply.*

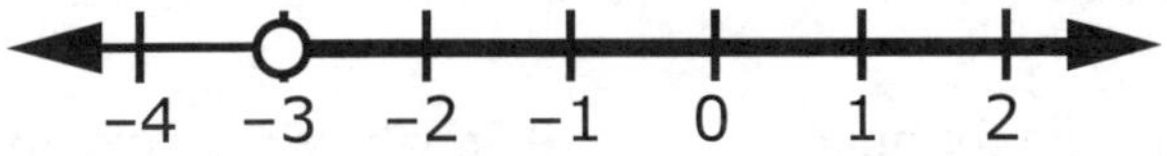

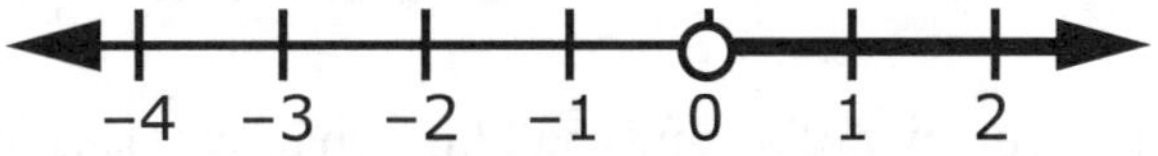

   A. –4   **B.** –3   **C.** –1   **D.** 0   **E.** 1.89   **F.** 2

4. Solve the compound inequality $5x + 6 \leq x + 10$ AND $-2(2x + 3) < 10$. *Show the final solution set algebraically and on a number line.*

5. Erika solved a compound inequality and described her solution set as $-5 > x > -2$. Her friend, Mila, claims Erika's solution set doesn't make sense. Explain what Mila means.

Name ______________________________ Date ______________

Check for Understanding

Solve Absolute Value Equations and Inequalities: Investigation 2

1. To ride the twirl-a-whirl at the carnival, you must weigh at least 40 kilograms. The maximum weight limit of a rider is 110 kilograms. Which compound inequality represents the weights, *w*, of those who aren't eligible to ride?

A. $40 \le w \le 110$ **B.** $w < 40$ OR $w > 110$ **C.** $w < 40$ AND $w > 110$

D. $40 > w > 110$ **E.** $w \le 40$ OR $w \ge 110$ **F.** $w \le 40$ AND $w \ge 110$

2. Match each compound inequality with the best description of its solution set shown on a number line. *Each description may be used once, more than once, or not at all.*

A. $x > -4$ AND $x > 3$ **I.** no solution

B. $x < -4$ AND $x > 3$ **II.** all real numbers

C. $x < -4$ OR $x > 3$ **III.** two intervals of real numbers

D. $x > -4$ OR $x < 3$ **IV.** one interval of real numbers

E. $x < -4$ OR $x < 3$

3. When solving a compound inequality, Micah wrote the solution as $x \ge 50$ OR $x \ge 75$. Which is a simpler way of writing this compound inequality?

A. $50 \le x \le 75$ **B.** $50 \ge x \ge 75$ **C.** $50 \le x \ge 75$ **D.** $x \ge 50$ **E.** $x \ge 75$

4. The velocity of a tennis ball thrown upwards is given by the formula $v = 22.1 - 9.8t$, where v is velocity in meters per second and t is the number of seconds after being thrown. During which times will the velocity be at least 7.4 meters per second but at most 17.2 meters per second? *Complete the inequality.*

______________ $\le t \le$ ______________ seconds

5. Steve and Lynn both solved the same compound inequality. Steve wrote his solution as the interval $(-4, 12)$ while Lynn wrote $[-4, 12]$. How are the values in their solutions similar and how are they different?

Name ______________________________ Date ______________

Check for Understanding

Solve Absolute Value Equations and Inequalities: Investigation 3

Jacob regularly crosses a bridge for which he must pay a toll. The toll to cross the bridge is \$3. A monthly pass costs \$10 and reduces the toll to \$1.50. A yearly pass costs \$360 and permits crossing the bridge for no fee.

1. Write an expression for the monthly cost of each plan in dollars, where n is the number of crossings.

No pass: ____________ Monthly pass: ____________ Yearly pass: ____________

2. If Jacob crosses the bridge 5 times each month, which is the best deal for Jacob and what is his monthly cost?

3. *Complete the statement.* For the monthly pass to be the best deal, Jacob must cross the bridge at least ____________ times but not more than ____________ times each month.

4. Match each compound inequality to the graph of its solution set.

A. $x < -2$ OR $x > 3$

B. $x < -2$ AND $x < 3$

C. $x > -2$ OR $x > 3$

D. $x < -2$ AND $x > 3$

E. $x > -2$ OR $x < 3$

I.
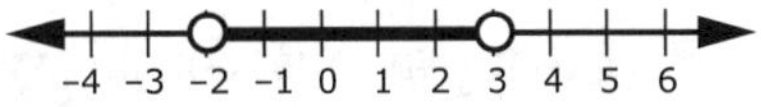

II.
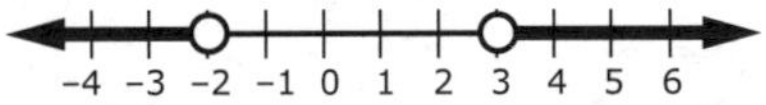

III.
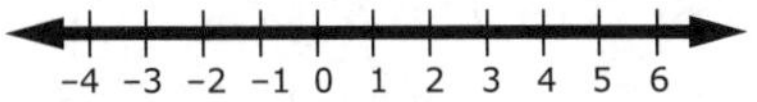

IV.

V.
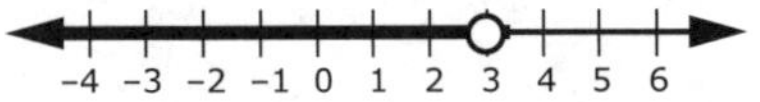

VI.
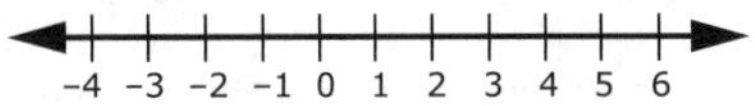

VII.
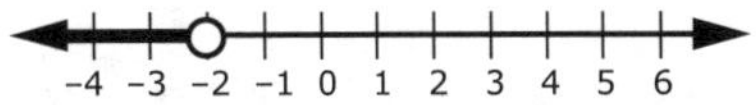

VIII.
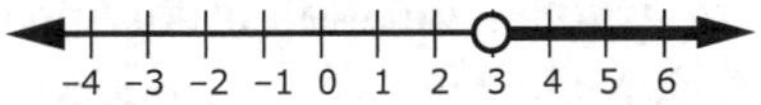

5. When Jodie solved a compound inequality with OR, her first simple inequality simplified to $-1 \geq 5$, while her second inequality was $y \leq 8$. Jodie writes "no solution" as the final answer because $-1 \geq 5$ is a contradiction. Is Jodie correct? If so, explain why. If Jodie is not correct, explain her error and give the correct solution.

Name ______________________ Date ____________

Check for Understanding

Solve Absolute Value Equations and Inequalities: Investigation 4

1. For the equation $|p - 5| = 8$, fill in the blanks to find the value of p using a number line:

There is a distance of ________ units between p and ________ on a number line. To find the value of p, start at ________ and move ________ units in either direction. So, $p =$ __________ or $p =$ __________.

2. Match each equation to its meaning in words. *Not all meanings will be used.*

A. $|x - 4| = 12$

B. $|x - 12| = 4$

C. $|x| - 4 = 12$

D. $|x + 4| = 12$

I. The distance of x from 12 is 4.

II. Twelve less than the distance of x from 0 is 4.

III. The distance of x from −4 is 12.

IV. The distance of x from 4 is 12.

V. Four less than the distance of x from 0 is 12.

3. Which equations show the cases that you must consider when solving the equation $|4x - 11| = 8$? *Select all that apply.*

A. $4x - 11 = 8$ **B.** $4x + 11 = 8$ **C.** $-4x - 11 = 8$ **D.** $-4x + 11 = 8$

4. Solve the absolute value equation: $|3x - 6| - 8 = 40$.

5. The graph shows two solutions to an equation involving absolute value. Explain how you can find an equation with these solutions. Make sure you include the equation in your response.

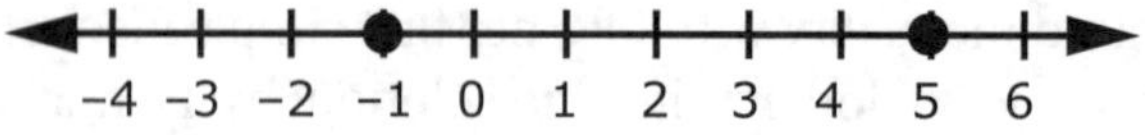

Name ______________________________ Date ____________

Check for Understanding

Solve Absolute Value Equations and Inequalities: Investigation 5

1. When large bags of candies are packaged, the number of candies in each bag must be within 4 pieces of 120. Write an absolute value equation to represent p, the greatest number and least number of candies in each bag. Then, find the values of p.

   Equation: ____________

   Greatest number: ____________ Least number: ____________

2. Classify each absolute value equation based on its number of solutions. *For each equation write* no solution, one solution, *or* two solutions.

   A. $3|x - 8| = 0$ ____________

   B. $10 = 6|x| + 5$ ____________

   C. $|3x + 7| - 4 = -4$ ____________

   D. $|x + 6| + 5 = 3$ ____________

   E. $-3|x - 5| = -6$ ____________

3. Find the two solutions of the absolute value equation $5 - 3|2x + 3| = -10$.

   $x =$ ____________ or $x =$ ____________

4. Which of these are solutions to the equation $|n - 7| = 3n - 1$? *Select all that apply.*

   A. -4 **B.** -3 **C.** $-\frac{3}{2}$ **D.** 2 **E.** 4

5. The number of solutions to $|x - 8| + b = c$ depends on the relationship between b and c. How do b and c compare if there are two solutions? One solution? No solution?

Name ______________________________ Date ____________

Check for Understanding

Solve Absolute Value Equations and Inequalities: Investigation 6

1. At a cheerleading contest, the length of each team's routine, t, must be within 45 seconds of the required time of 300 seconds. Which inequality shows the relationship between t, 45, and 300?

A. $|t - 45| \leq 300$ **B.** $|t - 45| \geq 300$ **C.** $|t - 300| \leq 45$ **D.** $|t - 300| \geq 45$

2. Select two simple inequalities and a connecting word that solves $|3x - 4| \leq 13$.

A. $x \geq \frac{17}{3}$ **B.** $x \geq -\frac{17}{3}$ **C.** $x \leq \frac{17}{3}$ **D.** $x \leq -\frac{17}{3}$

E. $x \geq 3$ **F.** $x \geq -3$ **G.** $x \leq 3$ **H.** $x \leq -3$

I. OR **J.** AND

3. Which graph shows the solution set of $|2a - 5| > 3$?

A. 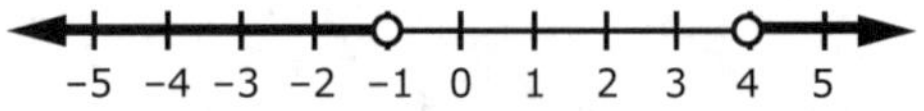

B. 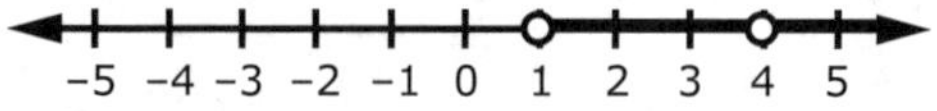

C. 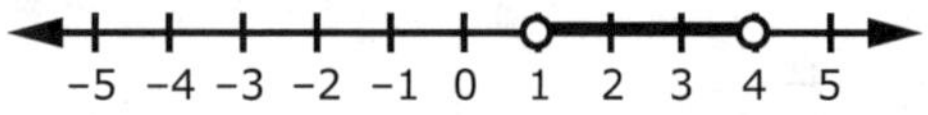

D. 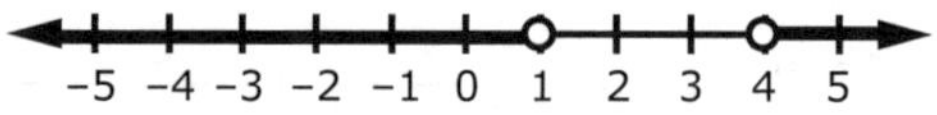

E. 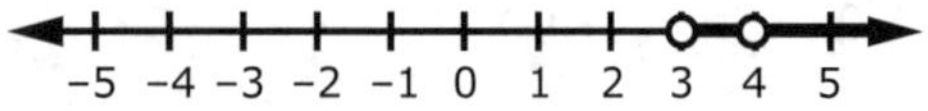

F. 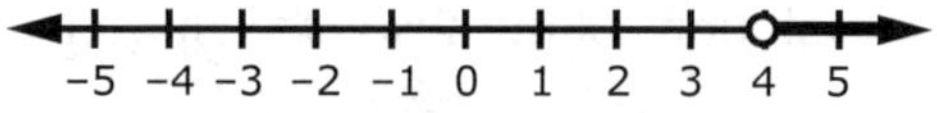

4. Depending on the value of b, the solution set of $|4x - 5| + b > 9$ may be all real numbers. Find an integer value for b so that the inequality's solution for x is all real numbers.

5. When you graph certain absolute value inequalities, the solution set is a single interval. In other absolute value inequalities, the solution set lies to the left and right of a single interval. Explain which type of absolute value inequality leads to each kind of graph.

Summary

Before you attempt the Practice Exercises, review what you've learned.

Solving Compound Inequalities

To solve a **compound inequality**, first solve each of the separate, simple inequalities that make up the compound statement.

> When two inequalities are combined into a single statement, the result is called a **compound inequality**. The two inequalities are connected with either the word AND or the word OR.
>
> The solution of a compound inequality that uses the connecting word AND is all the values that are solutions of both simple inequalities. This compound inequality is called a **conjunction** or an **intersection**.
>
> The solution of a compound inequality that uses the connecting word OR is all the values that are solutions of either simple inequality. This compound inequality is called a **disjunction** or a **union**.

EXAMPLE: Solving a Compound Inequality with OR

Solve and graph the solution of the compound inequality.

SOLUTION:

$3(x+1) \le x-5$	OR	$6x > 4(x-1)$
$3x+3 \le x-5$	OR	$6x > 4x-4$
$3x+3 \le -5$	OR	$2x > -4$
$2x \le -8$	OR	$2x > -4$
$x \le -4$	OR	$x > -2$

Graph both inequalities on the same number line. The solution is all the values that are solutions of either inequality.

Preliminary Graph:

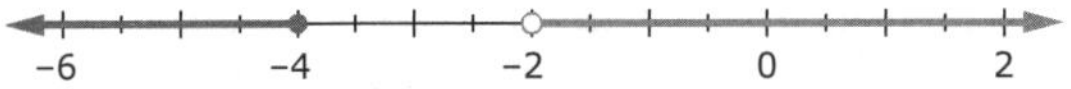

Final Solution:

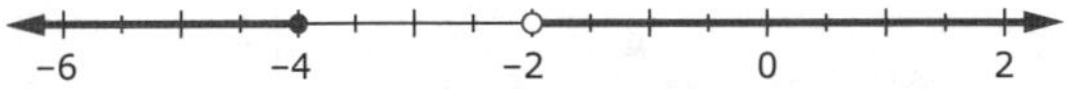

Special Cases

Sometimes one or both of the inequalities in a compound inequality is an identity or a contradiction.

Compound Inequality	Graph
$x < 2$ AND $x > 5$	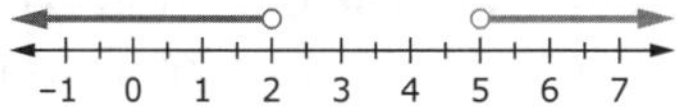 Solution: no solutions
$x > 2$ OR $x < 5$	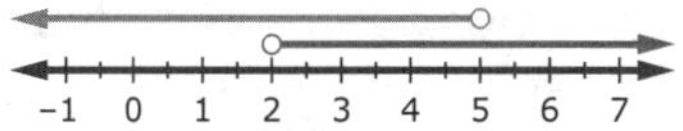 Solution: all real numbers
$x < 2$ AND $x < 5$	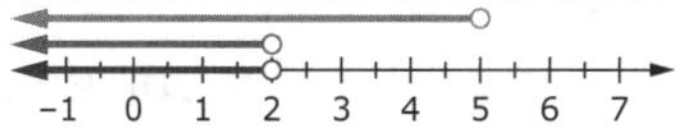 Solution: $x < 2$
$x < 2$ OR $x < 5$	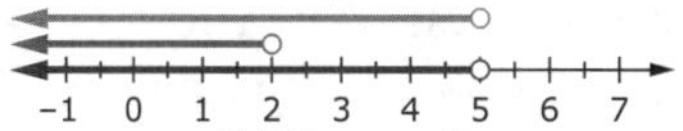 Solution: $x < 5$
$x < 2$ OR $x > 5$	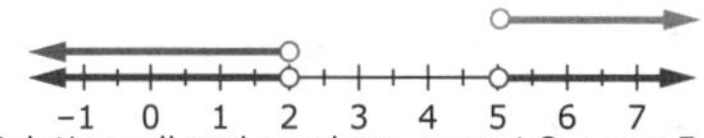 Solution: all real numbers, except $2 \le x \le 5$
$x > 2$ AND $x < 5$	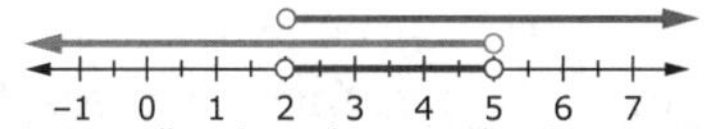 Solution: all real numbers, so that $2 < x < 5$
$x > 2$ AND $x > 5$	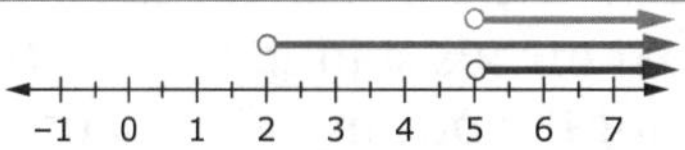 Solution: $x > 5$
$x > 2$ OR $x > 5$	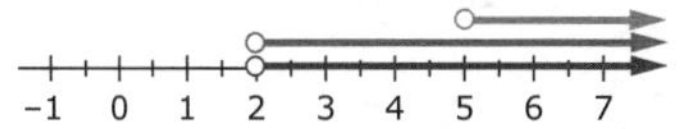 Solution: $x > 2$

My Notes

Summary *(continued)*

Absolute Value Equations

Absolute value equations can be written as compound statements involving OR. Recall the definition of absolute value.

$$|a| = a, \text{ for } a \geq 0, \text{ and } |a| = -a, \text{ for } a < 0$$

So when given the equation $|x| = b$, we must consider the definition of absolute value where $|x| = b$ is rewritten as either $(x) = b$ or $-(x) = b$. And then solve for x in each individual case to find all possible solutions. Ultimately the solution of the absolute value equation $|x| = b$ is $x = b$ OR $x = -b$.

To solve absolute value equations, first isolate the absolute value expression on one side of the equation. Then write the positive and negative cases and solve each equation. Then define each of the positive and negative cases by adjusting signs on the variable side of each case and solve each equation.

It is important to check the solutions in the original equation. Sometimes an absolute value equation appears to have two solutions, but one or more is an extraneous solution. An **extraneous solution** is true for the positive or negative case, but not for the original equation.

EXAMPLE: Solving an Absolute Value Equation

Solve $|2.5x - 1| = 4$.

SOLUTION:

$$|2.5x - 1| = 4$$

$$(2.5x - 1) = 4 \quad \text{OR} \quad -(2.5x - 1) = 4$$

$$(2.5x - 1) = 4 \quad \text{OR} \quad -2.5x + 1 = 4$$

$$2.5x = 5 \quad \text{OR} \quad -2.5x = 3$$

$$x = 2 \quad \text{OR} \quad x = -1.2$$

CHECK:

$$|2.5x - 1| = 4 \qquad |2.5x - 1| = 4$$

$$|2.5(2) - 1| \stackrel{?}{=} 4 \qquad |2.5(-1.2) - 1| \stackrel{?}{=} 4$$

$$|5 - 1| \stackrel{?}{=} 4 \qquad |-3 - 1| \stackrel{?}{=} 4$$

$$|4| \stackrel{?}{=} 4 \qquad |-4| \stackrel{?}{=} 4$$

$$4 = 4 \qquad 4 = 4$$

Both solutions are true.

The solution is $x = 2$ OR $x = -1.2$.

My Notes

Summary *(continued)*

EXAMPLE: Absolute Value Equation with the Variable on Both Sides

Solve $|5x + 2| = 6x$.

SOLUTION:

$$|5x + 2| = 6x$$

$$\begin{array}{rcl} 5x + 2 = 6x & \text{OR} & -(5x + 2) = 6x \\ 5x - 6x = -2 & \text{OR} & -5x - 2 = 6x \\ x = 2 & \text{OR} & -2 = 11x \\ x = 2 & \text{OR} & -\frac{2}{11} = x \end{array}$$

CHECK:

$$\begin{array}{cc} |5x + 2| = 6x & |5x + 2| = 6x \\ |5(2) + 2| \stackrel{?}{=} 6(2) & \left|5\left(-\frac{2}{11}\right) + 2\right| \stackrel{?}{=} 6\left(-\frac{2}{11}\right) \\ |12| \stackrel{?}{=} 12 & \left|\frac{12}{11}\right| \stackrel{?}{=} -\frac{12}{11} \\ 12 = 12 & \frac{12}{11} \neq -\frac{12}{11} \end{array}$$

$x = -\frac{2}{11}$ is an extraneous solution.

So, the solution is $x = 2$.

EXAMPLE: Special Case of Absolute Value Equations

Solve $|4x - 12| = -2$.

SOLUTION:

Because absolute value is never negative, no value of x can make the expression on the left equal to -2.

This equation has no solution.

Absolute Value Inequalities

Absolute value inequalities can be written as compound inequalities with either AND or OR. To solve an absolute value inequality, first isolate the absolute value expression. Then, write it as a compound inequality. Finally, solve it just like any other compound inequality.

When an absolute value inequality is of the form $|ax + b| < c$ with $c > 0$, the compound inequality is $(ax + b) < c$ AND $-(ax + b) < c$. This compound inequality can then be rewritten as: $(ax + b) < c$ AND $(ax + b) > -c$.

When an absolute value inequality is of the form $|ax + b| > c$ with $c > 0$, the compound inequality is $(ax + b) > c$ AND $-(ax + b) > c$. This compound inequality can then be rewritten as: $(ax + b) > c$ OR $ax + b < -c$.

My Notes

Summary *(continued)*

Solving Absolute Value Inequalities

EXAMPLE: Translating to an AND Statement

Solve $|5x + 4| - 5 \leq 4$.

SOLUTION:

First, isolate the absolute value expression.

$$|5x + 4| - 5 \leq 4$$
$$|5x + 4| - 5 + 5 \leq 4 + 5$$
$$|5x + 4| \leq 9$$

Then write it as a compound inequality and solve.

$5x + 4 \geq -9$	AND	$5x + 4 \leq 9$
$5x \geq -13$	AND	$5x \leq 5$
$x \geq -\frac{13}{2}$	AND	$x \leq 1$

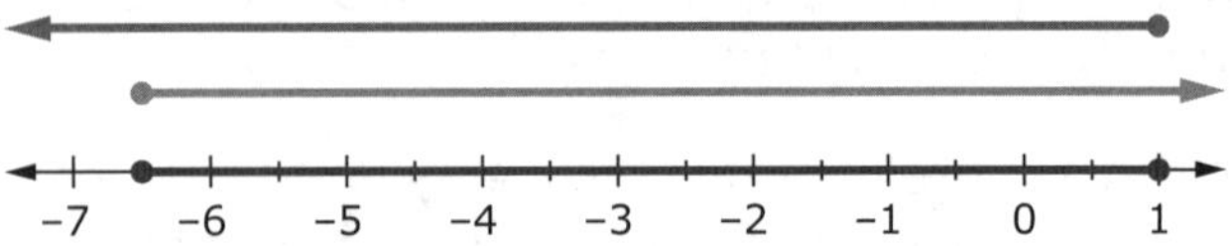

Solving Real-World Problems

EXAMPLE: Solving a Real-World Problem

James is baking a cake. The cake should bake in a 350°F oven, but it will still bake properly if the temperature is within 2.5°F of that temperature. What are the acceptable temperatures for baking the cake?

SOLUTION:

Let t be the temperature of the oven. Write an absolute value inequality to model the situation.

$$|t - 350| = \leq 2.5$$

Then write it as a compound inequality and solve.

$t - 350 \geq -2.5$	AND	$t - 350 \leq 2.5$
$t \geq 347.5$	AND	$t \leq 352.5$

The acceptable temperatures are between or equal to 347.5°F and 352.5°F.

My Notes

Practice Exercises

Review what you've learned using these practice problems. For practice problems with feedback, try the Coach and Play items in the Practice section online.

1. Solve the equation $|2x + 1| = 3$ and then complete the statement below.

The values that make the equation true are

$x =$ __________ and $x =$ __________.

2. The graph shows the solution set for which compound inequality?

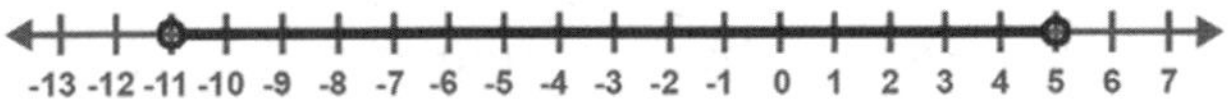

A. $x + 3 < 8$ OR $x + 3 > -8$

B. $x - 3 < -11$ AND $x - 3 > 11$

C. $x + 3 < 8$ AND $x + 3 > -8$

D. $x - 3 < 8$ AND $x - 3 > -8$

3. Solve the inequality $|2x - 4| > 2$ and then complete the statements below.

To solve the inequality, rewrite it into two inequalities:

$2x - 4 <$ __________ OR

$2x - 4 >$ __________.

Write the values of x that make the inequality true:

x is less than __________ or

x is greater than __________.

4. The solution set for the following set of inequalities is represented by a disjunction.

$$2x - 4 > 8$$
$$4 - 5x > 14$$

Solve the inequalities and complete the statement below.

The solution set for the inequalities is any number greater than __________

or any number less than __________.

5. Choose the value(s) for x that solve this equation.

$$\frac{6|x - 2|}{4} = 3x$$

A. $x = 0$ **B.** $x = -2$ or $x = \frac{2}{3}$ **C.** $x = \frac{2}{3}$

D. $x = -2$ **E.** $x = -\frac{2}{3}$ or $x = 2$

6. Solve the inequality $|x + 1| < -2$.

Which graph best represents the solution set of the inequality?

A. 

B. 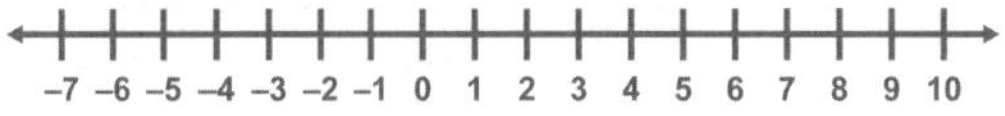

C. 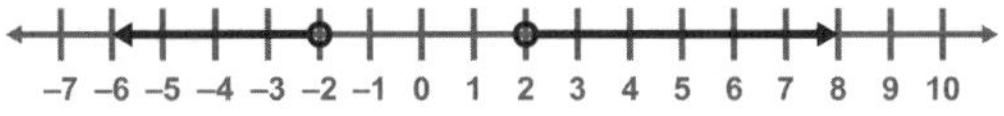

D. 

7. Consider the equation $|2x - 6| + 9 = 5$.

Which statement is true?

A. The equation has exactly one solution.

B. The equation has exactly two solutions.

C. The equation has no solution.

D. The equation has infinitely many solutions.

Practice Exercises *(continued)*

8. Angie and Lee are solving compound inequalities with the same simple inequalities but with different connecting words.

Angie solves $3x - 4 \geq 20$ AND $1 - 4x < 13$.

Lee solves $3x - 4 \geq 20$ OR $1 - 4x < 13$.

Identify the correct solution in interval notation for each compound inequality.

A. $(-\infty, -3)$

B. $(-\infty, 3)$

C. $(-3, +\infty)$

D. $[8, +\infty)$

E. $(-3, 8]$

F. $(3, 8]$

G. $(-\infty, -3) \cup [8, +\infty)$

H. $(-\infty, 3) \cup [8, +\infty)$

I. $(-\infty, +\infty)$

J. no solution

Angie's solution to $3x - 4 \geq 20$ AND $1 - 4x < 13$ is __________.

Lee's solution to $3x - 4 \geq 20$ OR $1 - 4x < 13$ is __________.

9. The values of x shown on the number line are solutions to an absolute value equation.

Which of these verbal descriptions and equations correspond to these solutions? *Select all that apply.*

A. The distance of x from 1 is 4.

B. The distance of x from 4 is 1.

C. The distance of x from –3 is 5.

D. The distance of x from –3 is 8.

E. $|x - 4| = 1$

F. $|x - 1| = 4$

G. $|x + 1| = 4$

H. $|x - 3| = 5$

10. Solve the equation: $3 + 4|x - 2| = 35$.

$x =$ __________ or $x =$ __________

Apply

How Do You Plan for a Healthy Aquarium?

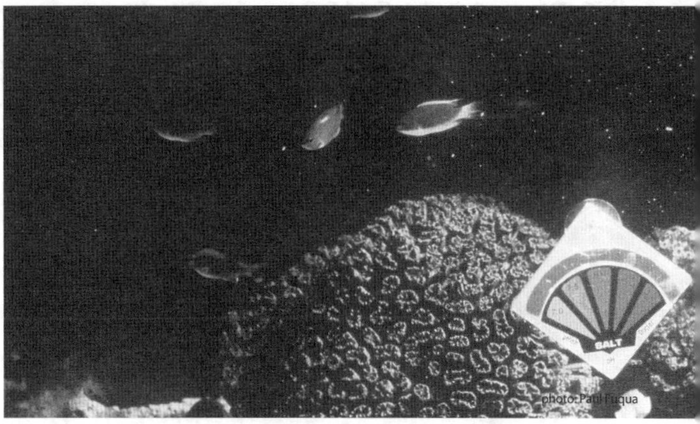
photo: Paul Fuqua

Do you know that fish in an aquarium require lots of care?

Now you get to design your own aquarium! A freshwater "community" tank is often the best choice for beginners. Community fish are fish that do well with many other types of fish and require water conditions that are fairly simple to maintain.

Do research to help you choose the fish, the tank, and the accessories you would like. Explain your choices, using inequalities that represent the facts given below. Include pictures of the fish, a sketch of your design, along with the explanation of your choices. Using compound inequalities, describe the water conditions you must maintain in order to keep your fish healthy.

Basic Aquarium Design Facts

- A freshwater community tank should hold at least 10 gallons (37 liters) of water.
- There should be at least 1 gallon of water per inch of fish. Two 3-inch fish would need 6 gallons of water.
- Each fish needs to be able to swim three times its body length before turning around.
- Fish need a place to hide—rocks, caves, or plants are often used to provide cover.

The fish in this chart are all considered to be community fish.

Common Types of Fish	Approximate Size	pH	Temperature
Albino swordtail	5.5 in. (14 cm)	7.3	77°F (25°C)
Angelfish	6 in. (15 cm)	7	79°F (26°C)
Big spot spiny eel	6 in. (15 cm)	8	82°F (28°C)
Blind cave fish	2.7 in. (7 cm)	7.4	72°F (22°C)
Blue ram	2 in. (5 cm)	6.8	81°F (27°C)
Diamond tetra	2.5 in. (6 cm)	7	79°F (26°C)
Discus	8 in. (20 cm)	6	84°F (29°C)
Elephant nose	11 in. (28 cm)	6.8	79°F (26°C)
Glass fish	2 in. (5 cm)	7.5	82°F (28°C)
Gold nugget pleco	6 in. (15 cm)	7.2	77°F (25°C)
Iridescent shark	10 in. (25 cm)	7	79°F (26°C)
Red swordtail	5 in. (13 cm)	7.3	77°F (25°C)
Zebra pleco	4 in. (10 cm)	7	79°F (26°C)

Show what you've learned by completing the other performance tasks in the online Apply section.

Apply *(continued)*

Your answer to Apply will be assessed on the following criteria:

1. Choosing the fish that you want, explaining why you chose them, and including pictures
2. Finding the required dimensions and volume of the tank you need, supported by inequalities from the Basic Aquarium Design Facts
3. Locating a suitable tank online and sketching it, along with any accessories you chose
4. Describing the water conditions you must maintain in order to keep your fish healthy, using compound inequalities

Scale / Criteria	4	3	2	1	0
	Exceeds Criteria	Meets Criteria	Progressing to Criteria	Below Expectations	No Expectation
Fish Selection	The selection of the fish is accompanied by descriptions, pictures, and reasons for the selection; the water conditions needed by the fish are considered.	The selection of the fish is accompanied by descriptions, pictures, and reasons for the selection.	Either the pictures or the reasons for the selection do not accompany the selection of the fish.	The selection of the fish does not include a reason, and no pictures are included.	Does not attempt task.
Tank Dimensions	Appropriate aquarium for the fish is selected with reason included; choice is supported by inequalities from the Design Facts.	Appropriate aquarium for the fish is selected, but a reason or the supporting inequality is missing.	Appropriate aquarium for the fish is selected, but there is no reason and the supporting inequality is missing.	Makes major mistakes in the selection of the aquarium, which result in the aquarium not fitting the needs of the fish.	Does not attempt task.
Aquarium Description	A suitable tank was found online and sketched, including dimensions; accessories are included and described.	The tank is found and sketched but is missing the dimensions, the accessories, or the description.	The tank is found and sketched but is missing two of the following: dimensions, accessories, or description.	The tank is found and sketched but is missing the dimensions, the accessories, and the description.	Does not attempt task.
Water Conditions	The water temperature and pH are appropriate for the fish; the description for each is accompanied by compound inequalities.	The water temperature and pH are appropriate for the fish; the description for each is not accompanied by compound inequalities.	The water conditions are appropriate for temperature or pH, but not both.	Makes major mistakes in the water conditions, which result in the water conditions not meeting the needs of the fish.	Does not attempt task.

UNIT 3: Functions

3.1 Understand and Interpret Functions

photo: Getty Images

Lesson Objectives

- Explain the difference between functions and relationships that are not functions.
- Use functions to represent real-world situations.
- Identify the domain and range of functions.
- Use function notation to evaluate functions.
- Use function notation to interpret key features such as identifying the value of $f(3)$ given the table of a function and determining x given $f(x) = 5$.

Essential Question

- How can mathematics help us describe and interpret relationships observed in data and real-world phenomena?

Investigations

Telescopes

A telescope's field of view can vary based on several factors. What are the variables?

Man in the Moon?

Look for relationships in space mission data and telescope data. What do they tell you?

Classifying Relationships

Classify mathematical relationships. They aren't all created equal.

The Ins and Outs of Functions

Play a game of Function Quest! Use function characteristics to help you identify a function.

Pseudonyms

Function notation—it may look strange, but it simplifies analysis.

Key Vocabulary

continuous function, dependent variable, domain, function, independent variable, ordered pair, range (of a function), variable

Discover

As you complete Engage and the investigations, record the most important ideas you've learned.

Engage

Investigation 1

Investigation 2

Investigation 3

Investigation 4

Name ______________________________ Date ____________

Check for Understanding

Understand and Interpret Functions: Investigation 1

Each table shows data for different throws of a ball, with input in the left column and output in the right.

Use the tables for questions 1–3.

A.

Height (m)	Horizontal Distance (m)
10	2
20	10
21	20
10	30

B.

Height (m)	Time (s)
10	0.2
18	0.9
14	1.8
10	2
0	2.5

C.

Horizontal Distance (m)	Time (s)
1	6
3	18
4	24
5	30

1. Write the letter of the table for the relationship that represents a function. ____________

2. Choose the equation that models this relationship.

   A. $y = 6x - 4$ B. $y = 6x$ C. $y = 4x + 4$ D. $y = 4x$

3. If the left and right columns were reversed in the tables, which relationships would represent functions?

   ____

4. Choose the graph that represents a function.

I.

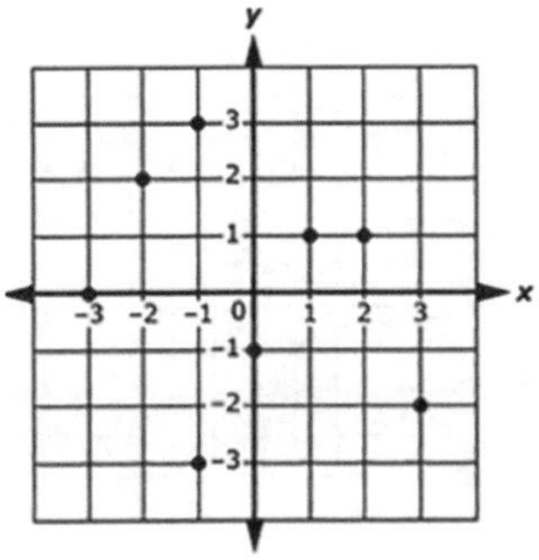

II.

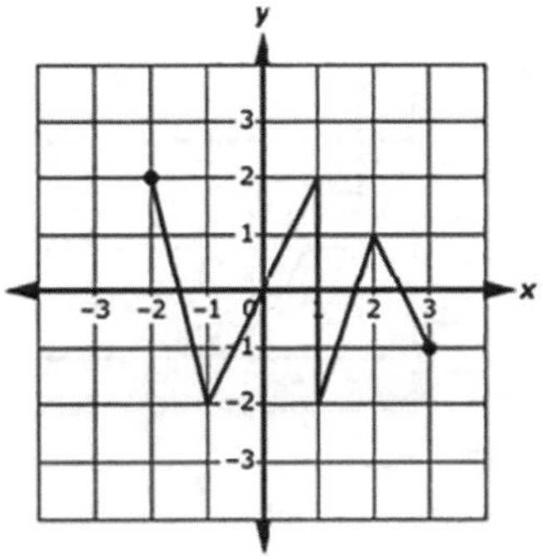

III.

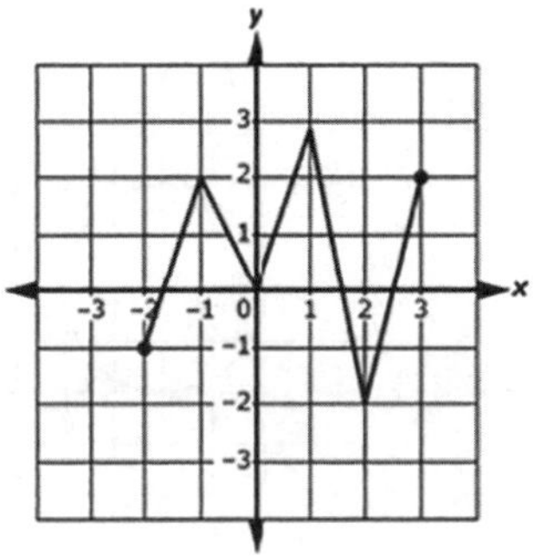

5. Use the graphs above to explain how you know whether a relationship represents a function.

Name ______________________________ Date ______________

Check for Understanding

Understand and Interpret Functions: Investigation 2

1. Complete the table to create a relationship that is

A. a function

x	*y*
−2	2
−1	1
0	3

B. not a function

x	*y*
−3	−1
−2	0
1	1

2. Which of the following relationships describe a function? *Select all that apply.*

A. Input: the amount of fuel a car has used since fill-up
Output: the distance the car has traveled

B. Input: the width of a rectangle
Output: the length of a rectangle

C. Input: the radius of a circle
Output: the circumference of the circle

D. Input: the altitude of an airplane
Output: the distance to the airplane's destination

3. Which of the following graphs represent a function? *Select all that apply.*

A.

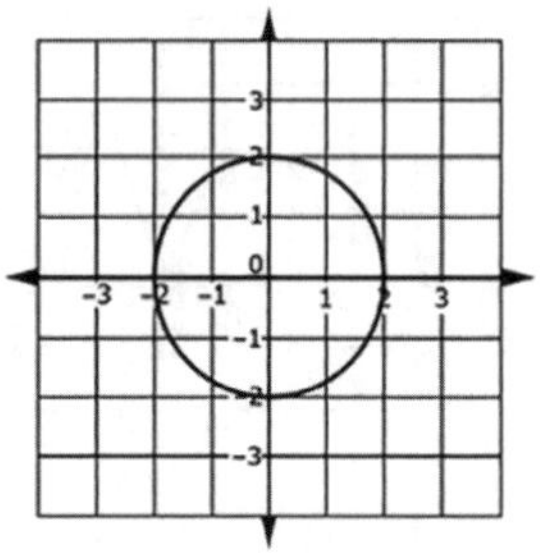

B.

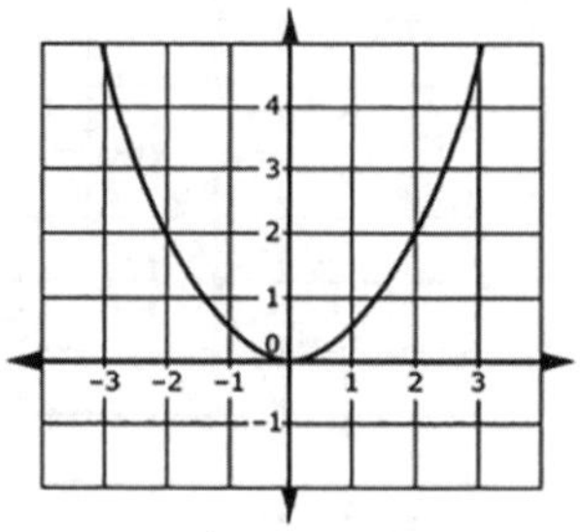

C.

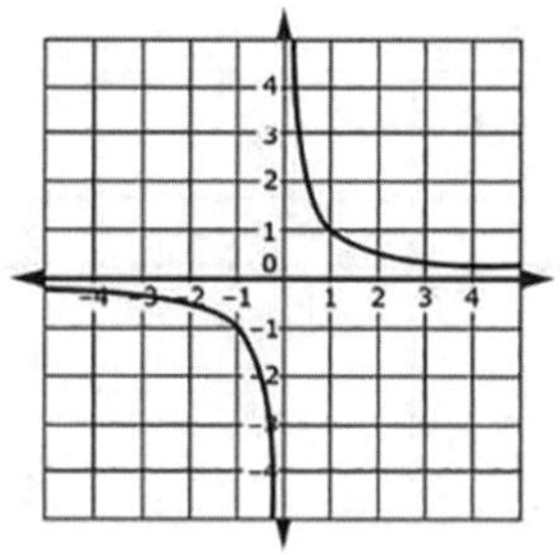

4. Explain why the graph of a horizontal line represents a function and the graph of a vertical line does not represent a function.

Name ______________________________ Date ______________

Check for Understanding

Understand and Interpret Functions: Investigation 3

1. Match the terms on the left with the terms on the right that describe their elements.

A. Domain	**I.** y value
	II. dependent variable
	III. input
B. Range	**IV.** independent variable
	V. x value
	VI. output

Use the following graphs to answer questions 2–3.

I.

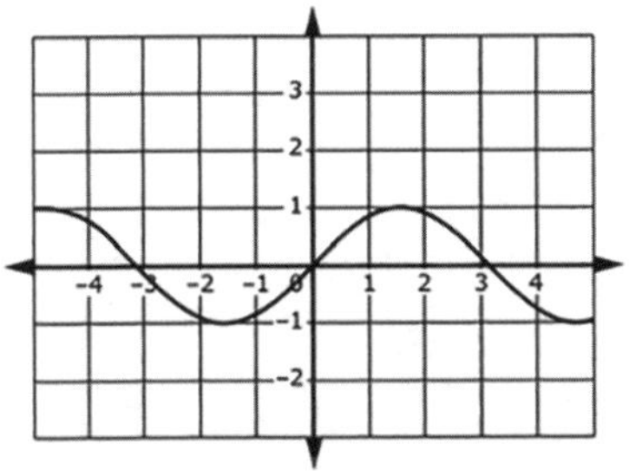

II.

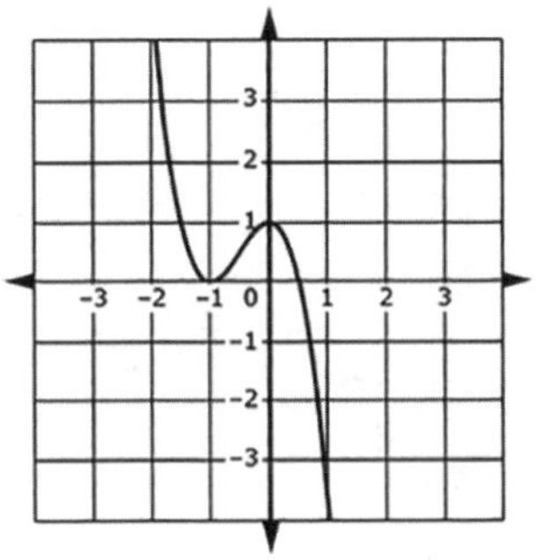

III.

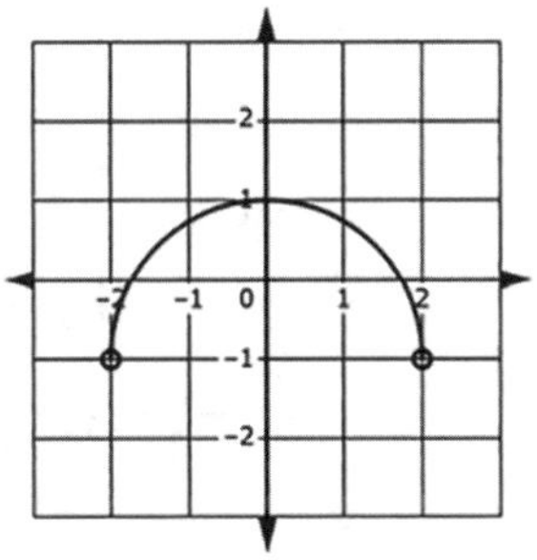

2. Write the numeral(s) for the graphs that have each of the following features. *There may be more than one correct answer.*

   A. The function's maximum value is 1. ______________

   B. The function has a y-intercept at (0, 1). ______________

   C. Every y value corresponds to multiple x values. ______________

   D. The function's output is positive wherever it is increasing. ______________

3. Use set notation to write the domain and range for each of the functions I–III.

   I. ______________ **II.** ______________ **III.** ______________

4. Describe the domain and range for the horizontal line $y = 0$. What is this line called?

Name ______________________________ Date ____________

Check for Understanding

Understand and Interpret Functions: Investigation 4

1. Use the tables below to answer questions A–C.

x	$f(x)$
−3	5
−1	0
0	3
2	−1

x	$g(x)$
−3	−4
0	3
1	−1
2	0

A. $f(2) =$ ________

B. When $f(x) = 0$, $x =$ ________.

C. $f(x) = g(x)$ when $x =$ ________.

2. Use the graph to complete statements A–D with <, =, or >, and then answer E.

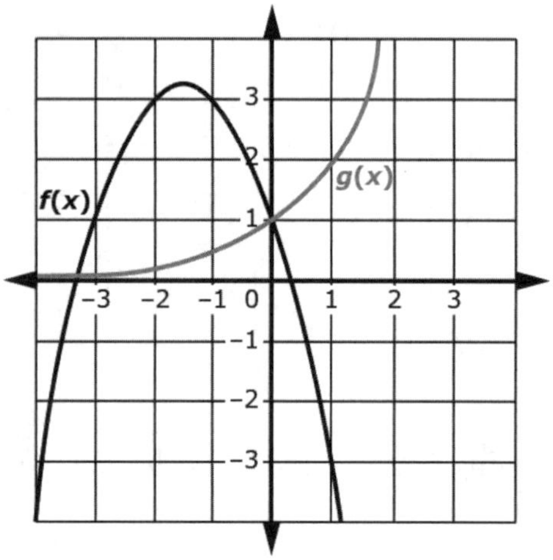

A. $f(-3)$ ________ $g(-3)$

B. $f(1)$ ________ $g(1)$

C. $f(0)$ ________ $g(0)$

D. $f(-3)$ ________ $g(1)$

E. For what value(s) of x is $f(x) = 1$? ________

3. A bee flies 15 miles per hour. Use the formula for distance, rate, and time to answer questions A–C.

A. Write the bee's distance as a function of time. ____________

B. Write and solve an equation to show how far the bee has flown after 20 minutes.

Equation: ____________________ Solution: ____________________

C. Use the equation to show how long it takes the bee to fly 12 miles.

Function: ____________________ Solution: ____________________

Summary

Before you attempt the Practice Exercises, review what you've learned.

> A **function** is a relationship between two sets, the domain and range, so that each element of the domain corresponds to only one element of the range. The **domain** of a function is the set of input values, the **range** of the function is the set of output values.

Functions can be represented as tables, graphs, equations, or ordered pairs. Sometimes there is no algebraic equation that can represent the function.

EXAMPLE: Determining Whether a Relationship Is a Function

Tell whether each relationship is a function. Justify your answers.

A.

Age	Number of Pets
14	2
14	1
15	3
16	2

B. (0, 1)
(2, 2)
(3, 6)
(4, 3)
(5, 3)

C.

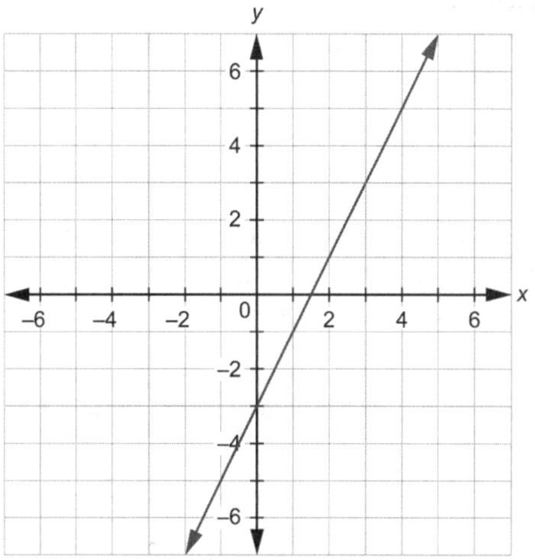

SOLUTION:

A. No: There are two people who are 14 years old who have different numbers of pets. One value in the domain corresponds to more than one value in the range.

B. Yes: Each *x* value in the ordered pairs corresponds to only one *y* value, so the list of ordered pairs represents a function.

C. Yes: Each *x* value corresponds to only one *y* value, so the graph represents a function.

My Notes

Summary *(continued)*

Function Notation

The notation $f(x) = 2x + 5$ is read as "f of x is equal to 2 times x plus 5." This equation means that there is a function called f that has values dependent on x. The inputs are the x values, and the outputs are labeled as $f(x)$.

To evaluate a function at a particular input value, substitute that value for x every time you see an x in the function.

x	$f(x)$
0	6
5	7
7	−3
8	10

EXAMPLE: Reading Function Values from a Table

Read the table to find each value.

1. $f(7)$
2. x when $f(x) = 10$

SOLUTION:

1. Look for the input value 7 in the column of x values. Then read the corresponding output value. $f(7) = -3$
2. Look for the output value 10 in the column for $f(x)$. Then read the corresponding value. $f(8) = 10$

EXAMPLE: Evaluating a Function

The function $A(r) = \pi r^2$ gives the area of a circle with a radius of r. Find the area of a circle with a radius of 6.5 centimeters.

SOLUTION:

Substitute the given input value for r every time you see an r in the function.

$$A(r) = \pi r^2$$

$$A(6.5) = \pi(6.5)^2$$

$$A(6.5) \approx 3.14(42.25)$$

$$A(6.5) \approx 132.665$$

So the area of a circle with a radius of 6.5 centimeters is approximately 132.665 square centimeters.

My Notes

Practice Exercises

Review what you've learned using these practice problems. For practice problems with feedback, try the Coach and Play items in the Practice section online.

1. Which of the following sets of ordered pairs represent a function? *Select all that apply.*

A. {(–1, 5), (2, 8), (0, 6), (–2, 4)}

B. {(–2, 1), (0, 3), (–1, 1), (1, 2)}

C. {(0, 0), (2, 3), (0, –1), (–1, –2)}

D. {(–2, 2), (–1, 1), (0, 2), (1, –1)}

E. {(2, 4), (1, 8), (–1, –2), (2, –1)}

2. Analyze and classify each of the following equations as either a function or not a function.

A. $y = -\frac{3x}{2}$ **B.** $x^2 + y^2 = 16$

C. $x = 2y - 5$ **D.** $y^2 = x - 3$

E. $\sqrt{x} = y - 2$

Function: ___________

Not a function: ___________

3. It takes Becca 7 minutes to type 1 page.

What values do the domain and range represent in the function that describes how long it takes Becca to type a given number of pages?

A. The domain represents the number of pages typed, and the range represents the time it takes to type that many pages.

B. The domain represents the time it takes to type 1 page, and the range represents the number of pages typed.

C. The domain is 0 to 7 minutes, and the range is the number of pages typed.

D. The domain is the number of pages typed, and the range is 0 to 7 minutes.

4. What is the value of x in the function $f(x) = \frac{3}{2}x - 4$ when $f(x) = 5$?

$x =$ ___________

5. To convert temperatures from degrees Celsius to degrees Fahrenheit, the sum of the degrees in Celsius and 32 is multiplied by $\frac{9}{5}$.

Which of the following choices best describes the function notation for converting a temperature given in degrees Celsius, C, to the temperature in degrees Fahrenheit, F?

A. $F(C) = \left(C + 32 \cdot \frac{9}{5}\right)$

B. $F(C) = \frac{9}{5}(C + 32)$

C. $C(F) = \left(F + 32 \cdot \frac{9}{5}\right)$

D. $C(F) = \frac{9}{5}(F + 32)$

6. A school organization is ordering candles for a fundraiser. The cost per box of candles is $75 plus a flat fee of $15 for shipping and handling.

Write a function that can be used to solve for the cost of ordering candles.

Which of the following solutions represents how much it would cost for the organization to order 12 boxes of candles?

A. $f(12) = 75(12) + 15 = 915$

B. $f(75) = 12(75) + 15 = 915$

C. $f(1) = 75(1) + 15 = 90$

D. $f(12) = 75 + 15(12) = 255$

Practice Exercises *(continued)*

7. Study the following graph.

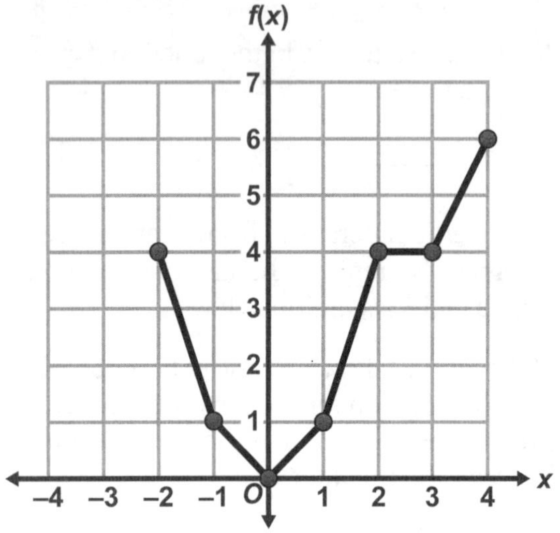

Using the choices below, identify the domain and range for the function.

Domain: ____________

Range: ____________

A. $-4 < x < 4$

B. $-4 \leq x \leq 4$

C. $0 \leq x \leq 6$

D. $-2 \leq y \leq 4$

E. $-2 < x < 4$

F. $0 \leq y \leq 6$

G. $-2 \leq x \leq 4$

H. $0 < y < 6$

I. $-4 \leq x \leq 2$

J. $0 \leq y \leq 4$

8. For the function $f(x) = x^2 + 4$, solve for the whole number value of x when $f(x) = 40$.

$x =$ ____________

9. To celebrate his birth, Damien's parents planted a tree in their yard. The tree's growth each year, in feet, can be modeled by the equation $f(x) = 2.5x + 4$, and the tree is expected to reach full height when Damien is 18 years old. Interpret the values in the function. Determine a reasonable domain and range for the context.

The height of the tree when it was planted was ____________ feet.

The tree grows at a rate of ____________ feet per year.

When Damien is 18 years old, the tree will be ____________ feet tall.

The domain for the function is ____________ $\leq x \leq$ ____________.

The range for the function is ____________ $\leq y \leq$ ____________.

10. The table gives values for the function $f(x) = 3x - 2$.

Use the function to complete the table.

Input	Output
–4	–14
–2	________
0	–2
2	________
4	10

Apply

How Much Does a Million Dollars Weigh?

photo: Getty Images

On a recent game show, a contestant won $1,000,000. That's a ton of money isn't it!

Wait, is it?

Weight of a Million Dollars

Research the weight of a dollar bill.

How much do a million dollars weigh?

What about other denominations of United States currency? Is there a relationship between the denomination of United States currency and the weight of a million dollars? If so, represent any functional relationships algebraically. Describe the values that make sense for the domain.

Show what you've learned by completing the other performance tasks in the online Apply section.

Apply *(continued)*

Your answer to Apply will be assessed on the following criteria:

1. Presenting information using tables and graphs in a clear and organized way
2. Performing all calculations correctly, using technology when appropriate
3. Analyzing the data, representing functional relationships algebraically, and describing values that make sense for the domain
4. Stating reasonable conclusions with appropriate justifications

Scale / Criteria	4 Exceeds Criteria	3 Meets Criteria	2 Progressing to Criteria	1 Below Expectations	0 No Expectation
Presentation	Presents all data information using tables and graphs in a clear and organized way.	Presents most data information using tables and graphs in a clear and organized way.	Presents some data information using tables or graphs in a clear and organized way.	Presentation of data poorly organized; hard to understand information.	Does not collect data.
Calculations	Performs all calculations correctly, using technology and estimates appropriately and understands the implications of round-off error.	Calculations are generally correct but contain minor errors.	Calculations contain major algorithmic or computational errors.	Performs very few calculations correctly.	Does not attempt calculations.
Analysis	Includes strong evidence of analysis; represents functional relationship algebraically and accurately describes values for the domain.	Includes some evidence of analysis; identifies functional relationship and domain.	Includes some evidence of analysis; identifies functional relationship or domain.	Includes some evidence of analysis; does not identify functional relationship or domain.	Does not include analysis.
Conclusions and Justifications	Uses calculations to justify reasonable conclusions.	Uses calculations to justify a mostly reasonable conclusion.	Uses calculations incorrectly to justify a conclusion.	Makes an unreasonable conclusion.	Does not state a conclusion.

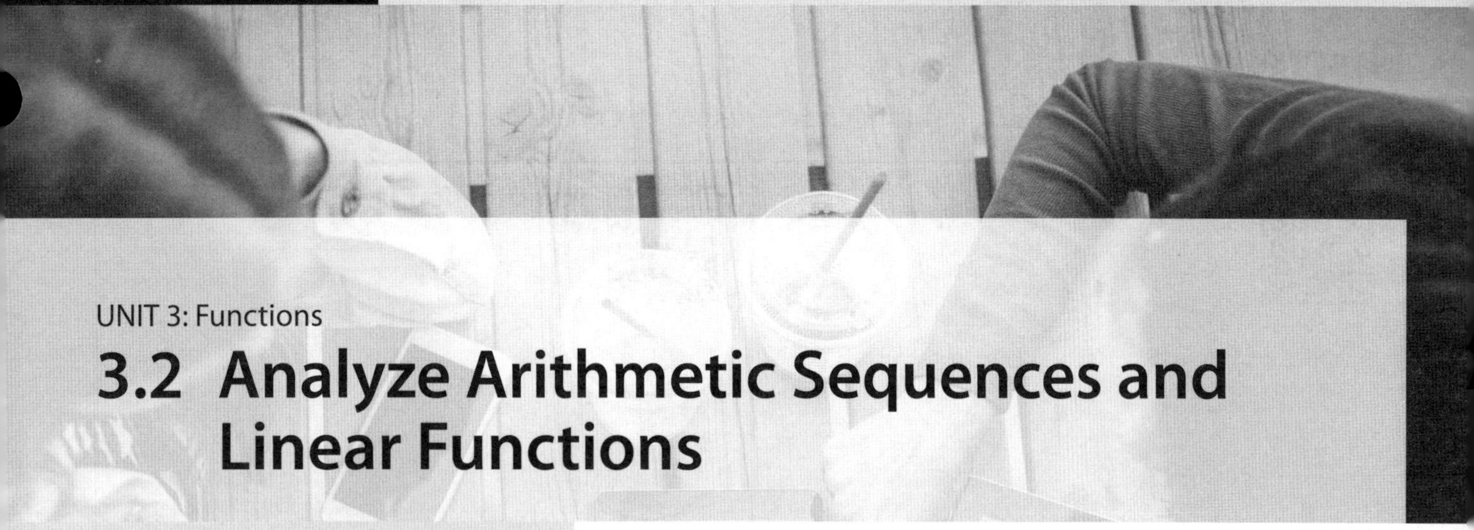

UNIT 3: Functions

3.2 Analyze Arithmetic Sequences and Linear Functions

photo: Getty Images

Lesson Objectives

- Write sequences in next-now and recursive form.
- Relate arithmetic sequences to linear functions.
- Express linear relationships in a variety of forms: next-now, recursive, implicit ($y = mx + b$), and explicit ($f(x) = mx + b$).
- Use functions to represent real-world situations.
- Write equations in slope-intercept form.

Essential Questions

- What are arithmetic sequences and how are they related to linear functions?
- Given limited information, how can you determine the equation of a line?

photo: Getty Images

Investigations

Saving for a Tablet

Patterns make predictions possible.

Thirsty Hikers

Simplify predictions with equations and graphs.

The Pentagon

Sequences are based on previous results. Use a recursive formula to relate *now* to *next*.

Skyscrapers

The buildings get taller as you stroll down the sidewalk. Use functions to describe the city skyline.

Key Vocabulary

arithmetic sequence, domain, function, range (of a function), recursive formula, relation, term (in a sequence)

Discover

As you complete Engage and the investigations, record the most important ideas you've learned.

Engage	Investigation 1
Investigation 2	Investigation 3

Name ______________________________ Date ______________

Check for Understanding

Analyze Arithmetic Sequences and Linear Functions: Investigation 1

1. Fill in the missing terms for the sequence. Then use the sequence for questions A–C.

4, ________, 18, 25, ________, 39, ________, 53, ________

A. What is the starting value? ________

B. What is the common difference? ________

C. Write a Next-Now equation for the sequence. ____________________

Use the sequence of shapes for questions 2–4. A side length for each hexagon is 1 unit.

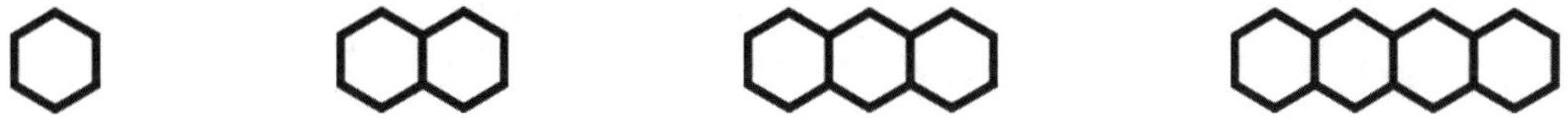

Shape 1 Shape 2 Shape 3 Shape 4

2. Write the sequence for the perimeters of the shapes. ____________________

3. Fill in the table and graph the sequence.

Shape x	Perimeter y
1	
2	
3	
4	

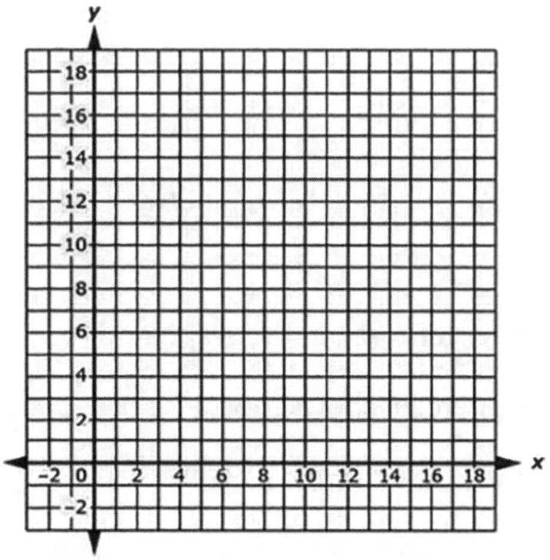

4. Choose the equations that model the sequence.

A. NEXT = NOW + 6 **B.** $y = 4x + 2$ **C.** $f(x) = 4x + 6$

D. $f(x) = 4x + 2$ **E.** NEXT = NOW + 4 **F.** $y = 6x$

5. Danielle claims that both $A = \{-2, -2, -2, -2, \ldots\}$ and $B = \{-2, 2, -2, 2, -2, 2, \ldots\}$ are arithmetic sequences. Explain why you agree or disagree for each set.

Name ______________________________ Date ______________

Check for Understanding

Analyze Arithmetic Sequences and Linear Functions: Investigation 2

Carl has collectible cans that he wants to use to build a tower. He plans to place 12 cans on the bottom row, with two fewer cans for each subsequent row as he builds his tower.

1. Write a Next-Now equation for the number of cans in the next row. ____________

2. Complete the table and graph the number of cans per row. Use the graph for A-B.

Row (n)	Number of Cans $C(n)$

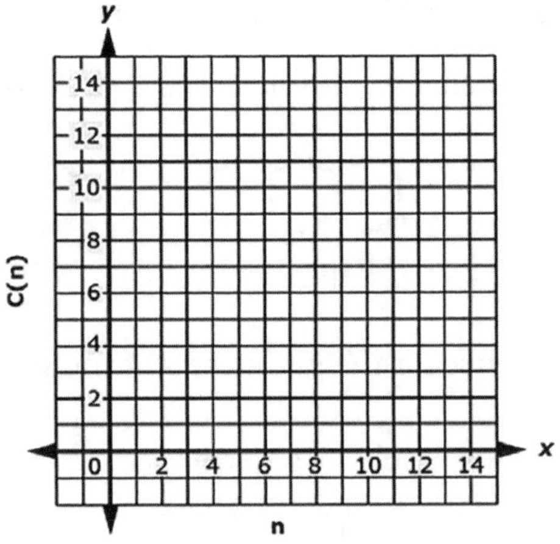

A. Write the recursive formula for the sequence. ____________________

B. Write the function, $C(n)$, that models the number of cans per row in terms of the row number, n.

____

3. Choose the statements about the graph and function that are true. *Select all that apply.*

A. The y-intercept shows the number of cans in the first row.

B. There will be a total of six rows.

C. The slope indicates the change in the number of cans for each subsequent row.

D. The x-intercept will show the number of cans in the last row.

E. If Carl were to use 15 cans in the first row, the x-intercept would show the number of cans in the next-to-last row.

F. If Carl were to use only one less can for each subsequent row, the new slope for the function would be equal to 1.

4. Carl has more cans than will fit in his tower, so he adds another row at the bottom of his original plan. Explain how to adjust the graph and function to model this situation.

Name ______________________________ Date ______________

Check for Understanding

Analyze Arithmetic Sequences and Linear Functions: Investigation 3

1. Write the equation of the line that passes through the points (3,–4) and (–3, 0) in slope-intercept form.

____

A glass elevator allowing views of a city begins a descent from the top floor of its tower to the ground floor. After 8 seconds, it is 744 feet from the ground, and after 16 seconds its height is 648 feet. The elevator's trip can be modeled by an arithmetic sequence.

2. Match each term or expression to its corresponding value. A value may be chosen more than once.

A. common difference	**I.** 8
B. top floor height	**II.** 828
C. $\frac{\Delta y}{\Delta x}$	**III.** –8
D. height after 12 seconds	**IV.** –12
E. slope	**V.** 696
F. y-intercept	**VI.** 840

3. Choose the function(s) modeling the sequence, where n represents the number of seconds. *Select all that apply.*

A. $h(n) = 12n + 828$ **B.** $h(n) = 12n + 840$ **C.** $a_{n+1} = a_n - 12; a_0 = 840$

D. $h(n) = -12n + 828$ **E.** $a_n = a_{n-1} + 12; a_1 = 828$ **F.** $h(n) = -12n + 840$

4. Complete the statements.

A. The elevator's height after 60 seconds is __________ feet.

B. The elevator is at a height of 240 feet after __________ seconds.

C. The elevator reaches the ground floor after __________ seconds.

5. Consider the graph of the function model for the elevator's descent. Describe the domain and range. What would the x- and y-intercepts, and the slope, represent?

Summary

Before you attempt the Practice Exercises, review what you've learned.

What Is an Arithmetic Sequence?

A **sequence** is an ordered list of numbers. Each number in a sequence is a term of that sequence. In the sequence 5, 8, 11, 14, 17, . . . , the number 5 is the first term, the number 8 is the second term, and so on.

In an **arithmetic sequence**, the difference between each term and the previous term is constant. This constant difference is called the common difference of the arithmetic sequence.

EXAMPLE: Identify Arithmetic Sequences

Which sequence is an arithmetic sequence?

A. 98, 86, 74, 62, 50, . . .

B. 3, 6, 12, 24, 48, . . .

My Notes

SOLUTION:

A. Find the difference between each term and the previous term.

$86 - 98 = -12, 74 - 86 = -12,$

$62 - 74 = -12, 50 - 62 = -12$

The sequence has a common difference of –12, so it is an arithmetic sequence.

B. Find the difference between each term and the previous term.

$6 - 3 = 3, 12 - 6 = 6,$

$24 - 12 = 12, 48 - 24 = 24$

The sequence does not have a common difference, so it is not arithmetic.

Modeling Arithmetic Sequences

One way to model an arithmetic sequence is with a next-now equation. In this type of equation, now represents a term of the sequence, and next represents the next term of the sequence.

EXAMPLE: Model an Arithmetic Sequence with a Next-Now Equation

Players can buy game passes at a laser-tag arena. The table shows the cost of passes for up to 5 laser-tag games.

Cost of Game Passes					
Games	1	2	3	4	5
Cost ($)	$8	$15	$22	$29	$36

Write a next-now equation for the arithmetic sequence formed by the costs of the passes.

Summary *(continued)*

SOLUTION:

The arithmetic sequence formed by the costs in dollars is 8, 15, 22, 29, 36. Find the common difference of the sequence.

$15 - 8 = 7$, $22 - 15 = 7$, $29 - 22 = 7$, $36 - 29 = 7$

The common difference is 7, so each term is 7 more that the previous term. To find the next term, add 7 to the now term.

NEXT = NOW + 7

The equation NEXT = NOW + 7 describes the arithmetic sequence, where 8 is the first term of the sequence.

EXAMPLE: Arithmetic Sequences and Linear Equations

A linear function can be used to represent the relationship between the number of the term and the term of an arithmetic sequence. The rate of change of the function will be equal in value to the common difference of the arithmetic sequence.

Write a linear equation that models the functional relationship between the number of the term and the term of the arithmetic sequence formed by the costs of the passes for laser tag games.

SOLUTION:

The arithmetic sequence formed by the costs in dollars is 8, 15, 22, 29, 36, and the common difference is 7. The common difference is equal to the slope.

The y-intercept occurs at the point $(0, b)$. In the sequence, x represents the term number. Find the value of the sequence that would represent term 0.

Term 0 would come just before term 1, so subtract the common difference from the value of term 1: $8 - 7 = 1$. The y-intercept is 1.

So the linear equation that models the arithmetic sequence is $y = 7x + 1$, where x is a whole number greater than 0. In this equation, the input x is a term number, and the output y is a term of the sequence. So an explicit formula for the arithmetic sequence is $f(n) = 7n + 1$, where n is a whole number greater than 0.

My Notes

Summary *(continued)*

Recursive Formulas for Arithmetic Sequences

A A **recursive formula** for a sequence has two parts. The first part gives a starting value for the sequence. The second part tells you how to find the next term of the sequence given the previous term.

EXAMPLE: Model an Arithmetic Sequence with a Recursive Formula

Use function notation to write a recursive formula that models the arithmetic sequence formed by the costs of the passes for laser-tag games.

SOLUTION:

The arithmetic sequence formed by the costs in dollars is 8, 15, 22, 29, 36.

For the first part of the recursive formula, show that the first term is 8. When the input is 1, the output is 8: $f(1) = 8$.

Next, show how to find the next term given a term of the sequence. Start with the next-now equation we wrote earlier: NEXT = NOW + 7.

Let the nth term of the sequence, or $f(n)$, represent now. The next term is then equal to $f(n + 1)$: $f(n + 1) = f(n) + 7$.

So a recursive formula for the arithmetic sequence is $f(1) = 8$; $f(n + 1) = f(n) + 7$, where n is a whole number greater than 0.

My Notes

Practice Exercises

Review what you've learned using these practice problems. For practice problems with feedback, try the Coach and Play items in the Practice section online.

1. –5, –2, 1, 4, 7, 10, . . .

Which of the following NEXT-NOW statements matches the sequence?

A. NEXT = NOW – 3

B. NEXT = NOW – 5

C. NEXT = NOW + 3

D. NEXT = NOW + 5

2. Larry can cut standard boards for framing a house at the rate of 3 boards every minute. The boards he cuts are added to a pile of 17 boards left over from a previous house.

Which recursive rule models Larry's situation?

A. $S(m + 1) = S(m) + 17; S(0) = 3$

B. $S(m + 17) = S(m) + 1; S(0) = 3$

C. $S(m + 3) = S(m) + 17; S(0) = 1$

D. $S(m + 1) = S(m) + 3; S(0) = 17$

3. The first 10 terms of an arithmetic sequence are plotted on the coordinate plane.

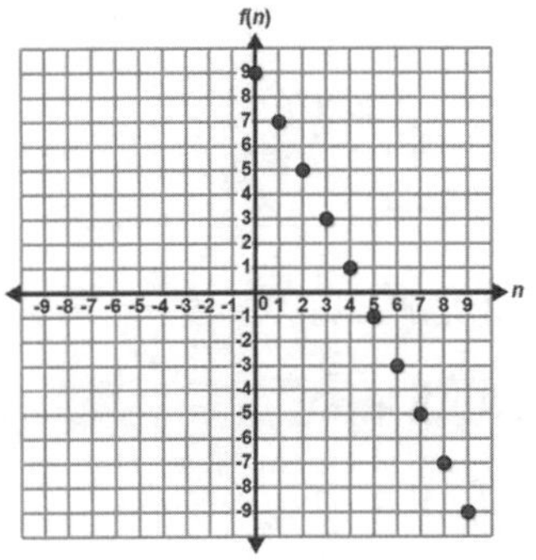

Which recursive formula corresponds to the NEXT-NOW equation?

A. $f(n + 1) = f(n) + 9; f(0) = -2$

B. $f(n + 1) = f(n) - 9; f(0) = 2$

C. $f(n + 1) = f(n) + 2; f(0) = -9$

D. $f(n + 1) = f(n) - 2; f(0) = 9$

4. The function $f(x)$ represents the total cost for x tickets for a carnival, including the entry fee.

$f(x) = 0.50x + 4$

What is the value of $f(20)$, and what does it mean in the context of the situation?

A. $f(20) = 32$; this means that it costs \$20 to purchase 32 tickets and pay the entry fee.

B. $f(20) = 14$; this means that it costs \$14 to purchase 20 tickets and pay the entry fee.

C. $f(20) = 14$; this means that it costs \$20 to purchase 14 tickets and pay the entry fee.

D. $f(20) = 32$; this means that it costs \$32 to purchase 20 tickets and pay the entry fee.

5. The recursive formula for an arithmetic sequence is $f(n + 1) = f(n) + 7; f(0) = 3$.

Which of the following equations are linear models for this sequence? *Select all that apply.*

A. $y = 3x + 7$

B. $f(n) = 3n + 7$

C. $f(n) = 7n + 3$

D. $y = 7x - 3$

E. $f(n + 1) = 7 \cdot f(n) + 3$

F. $y = 7x + 3$

G. $f(n + 1) = 3 \cdot f(n) + 7$

Practice Exercises *(continued)*

6. Two terms of an arithmetic sequence are given in the table.

Term Number	Term Value
4	8
5	2

Which linear function produces these terms?

A. $f(n) = -6n + 32$

B. $f(n) = 8n - 6$

C. $f(n) = -6n + 2$

D. $f(n) = 2n + 6$

7. A full peanut dispenser holds 1,500 peanuts. It dispenses a dozen peanuts each time a lever is pushed.

Which function models the number of peanuts in the dispenser after n presses of the lever?

A. $f(n) = 12n + 1{,}500$

B. $f(n) = -12n + 1{,}500$

C. $f(n) = 1{,}500n - 12$

D. $f(n) = -1{,}500n + 12$

8. A bank is maintaining an inactive account containing $492. Due to a change in bank policy, all accounts are to be charged a $3 fee each month.

Which function gives the amount in the account after n months following the policy change?

A. $a(n) = 3n - 492$

B. $a(n) = -492n + 3$

C. $a(n) = -3n + 492$

D. $a(n) = 492n - 3$

9. Andrew, Beatrice, and Charles are saving up for a trip.

Andrew is saving money in a manner that can be modeled by the recursion rule: $A(n + 1) = A(n) + 30$; $A(0) = 60$.

Beatrice is saving money according to the function $B(n) = 40n + 20$.

Charles is keeping track of his savings in a ledger; part of a page is shown in the table.

Third Month	$145
Fourth Month	$180
Fifth Month	$215
Sixth Month	$250

Who started with the most money? __________

Who is saving at the fastest rate? __________

10. An anchor chain has 32 links lying on a ship's deck, while the rest of the chain is below deck height. Each time the raise-anchor control is activated, 9 more links come over the edge to lie on the deck.

Which of the following representations model the number of chain links on the deck after n activations of the raise-anchor control? *Select all that apply.*

A. $c(n + 1) = c(n) + 9$; $c(0) = 32$

B. $c(n) = 9n + 32$

C. $c(n + 1) = c(n) + 32$; $c(0) = 9$

D. NEXT = NOW + 9

E. NEXT = NOW + 32

F. $c = 32n + 9$

Apply

photo: Getty Images

How Many Seats Will Be in the Next Row?

In many stadiums and arenas, seating is arranged in an arithmetic sequence.

Stadium Seating

Use the Internet to research seating plans of stadiums and concert halls.

- Find diagrams for two seating arrangements that each form an arithmetic sequence.
- Use the diagrams to write the arithmetic sequence for the arrangements, in both implicit and explicit forms.
- Decide which venue would make the most profit based on the number of seats and the price of the tickets.
- Compare the seating of the two venues and determine how many rows or seats one venue needs to add to match or exceed the profit of another venue.
- Think about how many rows there are at each price and where it would be most profitable to add a row based on how many seats there would be.

Show what you've learned by completing the other performance tasks in the online Apply section.

Apply *(continued)*

Your answer to Apply will be assessed on the following criteria:

1. Finding two seating arrangements that each form arithmetic sequences
2. Writing implicit and explicit forms of each sequence
3. Comparing and contrasting the seating arrangements
4. Determining which venue to add seats to and how many to match or exceed profits of the other; showing where it would be most profitable to add a row

Criteria \ Scale	4	3	2	1	0
	Exceeds Criteria	**Meets Criteria**	**Progressing to Criteria**	**Below Expectations**	**No Expectation**
Research	Performs correct and useful Internet research to gather valid and usable information.	Performs Internet research to gather mostly valid and usable information.	Performs Internet research to gather sometimes valid and usable information.	Performs Internet research to gather mostly invalid and unusable information.	Does not perform research.
Writing Functions (sequences)	Formulates correct functions to represent each seating arrangement.	Formulates mostly correct functions to represent each seating arrangement.	Formulates some correct functions to represent each seating arrangement.	Formulates functions for each seating arrangement that are mostly incorrect.	Does not write functions.
Comparing and Contrasting	Uses the information to correctly and reasonably compare and contrast the seating arrangements and profits.	Uses the information to mostly correctly and reasonably compare and contrast the seating arrangements and profits.	Uses the information to somewhat correctly and reasonably compare and contrast the seating arrangements and profits.	Uses the information to incorrectly compare and contrast the seating arrangements and profits.	Provides no comparison.
Computation	Correctly computes the number of seats to add and where it will be most profitable.	Correctly computes the number of seats to add and where it will be most profitable but makes a small computation mistake.	Correctly computes the number of seats to add or where it will be most profitable.	Incorrectly computes the number of seats to add and where it will be most profitable.	Does not perform the computations.

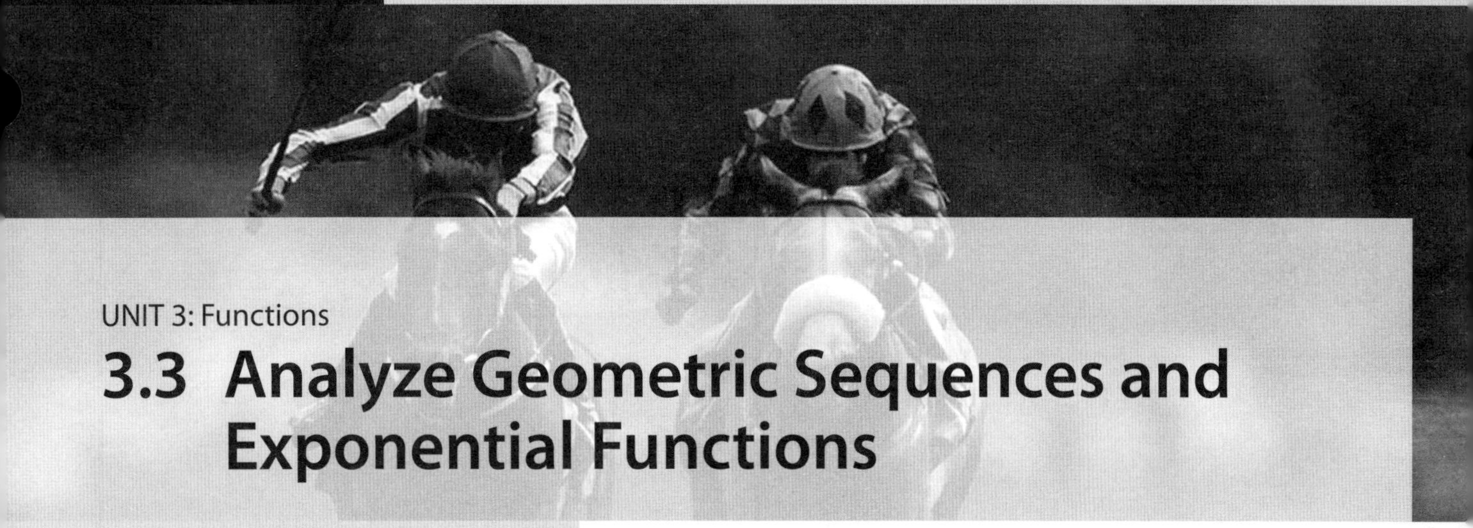

UNIT 3: Functions

3.3 Analyze Geometric Sequences and Exponential Functions

photo: Getty Images

Lesson Objectives

- Relate geometric sequences to exponential functions.
- Express exponential relationships in a variety of forms: next-now, recursive, implicit ($y = a\,b^x$), and explicit ($f(x) = a\,b^x$).
- Describe functions using multiple representations: verbally, numerically in tables, algebraically.
- Compare linear and exponential functions.

Essential Question

- What are geometric sequences, and how are they related to exponential functions?

photo: Getty Images

Investigations

Off to the Races!

Analyze the stages of a horse race. Is the sequence arithmetic or geometric?

Elvis's Race

Elvis's race data has a pattern. Can it be described in algebraic terms?

Modeling Half-Life

What makes decay geometric? Look to the sequence for clues.

Growth Rates

Recursive, linear, exponential? They can all be described with algebraic equations.

Key Vocabulary

common ratio, exponential function, geometric sequence, growth factor, recursive formula, term (of a sequence)

Discover

As you complete Engage and the investigations, record the most important ideas you've learned.

Engage	Investigation 1
Investigation 2	Investigation 3

Name ______________________________ Date ______________

Check for Understanding

Analyze Geometric Sequences and Exponential Functions: Investigation 1

The following pattern was created by repeating divisions of a triangle by the same factor at each stage. The first few stages are shown below.

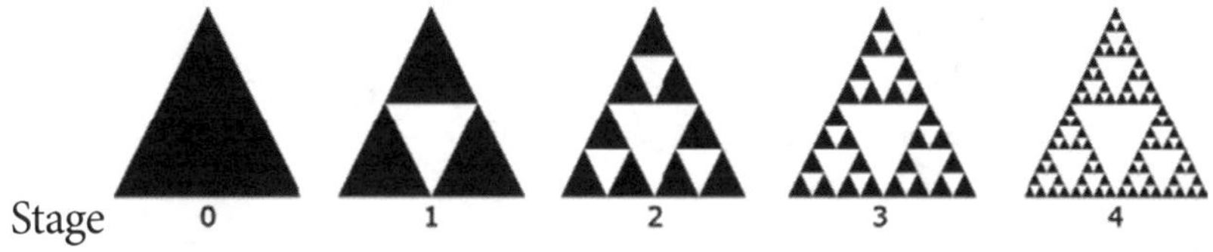

1. Enter the number of shaded triangles at Stages 0-3 into the table and graph the values.

Stage (n)	Number of Shaded Triangles $T(n)$
0	1
1	
2	
3	

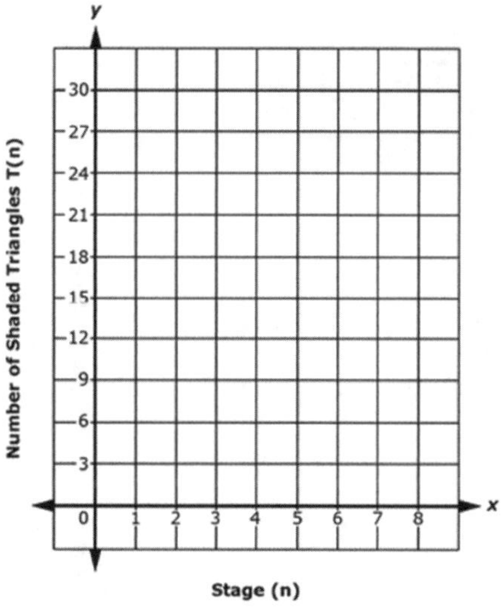

2. *Fill in the blanks.* The common ratio of the sequence is _____. The value for Stage 4 will be _____.

3. Choose the equations that model the sequence. *Select all that apply.*

   A. $T(n+1) = T(n)^3$ B. $T(n) = 3^n$ C. NEXT = NOW · 3

   D. $T(n) = 3n$ E. NEXT = NOW · n^3 F. $T(n+1) = T(n) \cdot 3$

4. *Circle the correct words in the sentence.* The sequence is [arithmetic / geometric], modeled by a(n) [linear / quadratic / exponential] function.

5. Jerome claims that the set $A = \{-2, -2, -2, -2, \ldots\}$ is an arithmetic sequence, and Sonia claims that it is a geometric one. Is either or both of them correct, and why or why not?

Name ______________________________ Date ____________

Check for Understanding

Analyze Geometric Sequences and Exponential Functions: Investigation 2

The following pattern was created by repeating divisions of a triangle by the same factor at each stage. The first few stages are shown below.

Stage

1. The total triangle area at Stage 0 is 8. Fill in the table and graph the data for the area of the shaded triangles at each stage. *Round decimals to two places, if necessary.*

Stage (n)	Approximate Total Area $A(n)$
0	8
1	
2	
3	
4	
5	

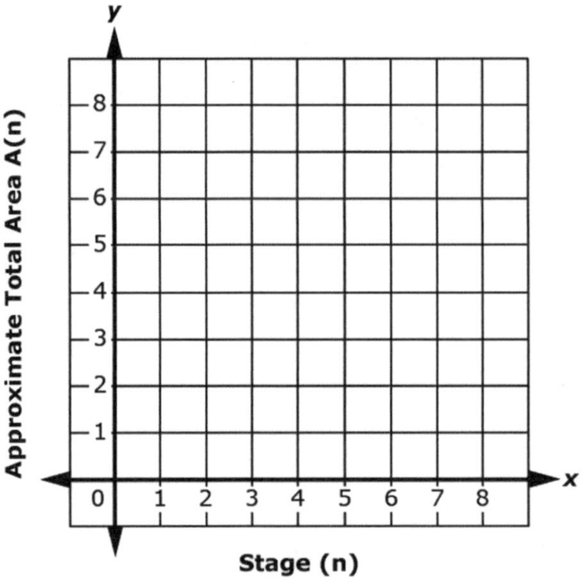

2. What is the common ratio for the sequence? ____________

3. **A.** Write the Next-Now equation for the sequence. ______________________

   B. Recursive formula: ______________________

4. At what stage will the total shaded area be less than 1? ____________

5. Explain how the start value and common ratio can be used to determine the sequence value, $S(n)$, for any term number, n, of the sequence.

Name ______________________________ Date ____________

Check for Understanding

Analyze Geometric Sequences and Exponential Functions: Investigation 3

Carmen posted a video online. The next day, she began recording the number of new "likes" and "shares" in the tables below, though she forgot to write in the values for Day 2.

Use the tables below to answer questions 1–4.

Day (n)	New Likes $L(n)$
1	6
2	
3	54
4	162

Day	New Shares $S(n)$
1	12
2	
3	28
4	36

1. Assuming a constant difference or ratio, write the Next-Now equations for each of the relationships.

   A. New Likes: ____________________ **B.** New Shares: ____________________

2. How many new likes and shares did Carmen's video get on Day 2?

   A. Number of New Likes: ____________ **B.** Number of New Shares: ____________

3. How many new likes and shares did Carmen's video get on the day she posted it?

   A. Number of New Likes: ____________ **B.** Number of New Shares: ____________

   C. How would the likes and shares on the day she posted be shown on the graph of each relationship?

   ____

4. Write the recursive formula and function that model each relationship.

   A. New Likes recursive formula: ____________________

   B. New Likes function: ____________________

   C. New Shares recursive formula: ____________________

   D. New Shares function: ____________________

5. A ball is dropped from a height of 64 feet. The maximum height between the first 4 bounces forms a geometric sequence, recorded as follows: 48 feet, 36 feet, 28 feet, and 20.25 feet. Identify which height was recorded in error and explain how you know.

Summary

Before you attempt the Practice Exercises, review what you've learned.

When a sequence of numbers follows a pattern of multiplying a fixed number from one term to the next, the sequence is called a **geometric sequence**.

Geometric Sequences

In a geometric sequence, the ratio of each term to the previous term is constant. This constant ratio is called the **common ratio** of the geometric sequence.

You can model a geometric sequence with a next-now equation. In this type of equation, *now* represents a term of the sequence, and *next* represents the next term of the sequence.

My Notes

EXAMPLE: Model a Geometric Sequence with a Next-Now Equation

A biologist is creating a model of how an infectious disease affects a population of cattle. The table shows the predicted number of new infections in each of the first 5 weeks.

Cattle Disease					
Week	1	2	3	4	5
New Infections	5	15	45	135	405

Write a next-now equation for the geometric sequence formed by the numbers of new infections.

SOLUTION:

The geometric sequence formed by the number of new infections is 5, 15, 45, 135, 405. Find the common ratio of the sequence.

$$\frac{15}{5} = 3, \frac{45}{15} = 3, \frac{135}{45} = 3, \frac{405}{135} = 3$$

The common ratio is 3, so each term is 3 times the previous term. To find the *next* term, multiply the *now* term by 3.

NEXT = 3 · NOW

The equation NEXT = 3 · NOW describes the geometric sequence, where 5 is the first term of the sequence.

The ordered pairs from a geometric sequence lie on an exponential curve. So you can use an exponential equation whose domain is a subset of the integers to represent a geometric sequence.

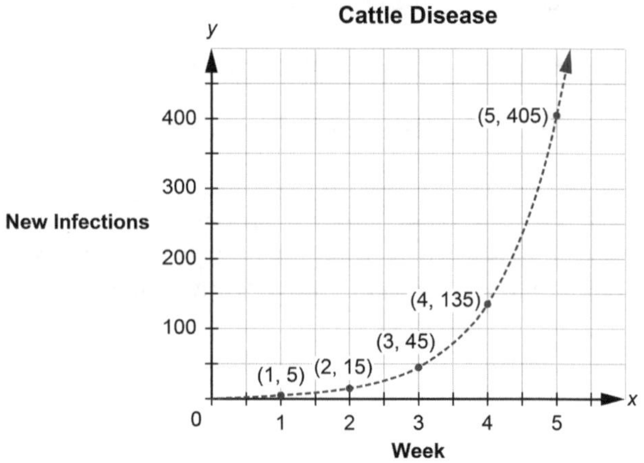

Summary *(continued)*

Exponential Functions

An **exponential function** can be written in the form

$y = ab^x$, where $a \neq 0$, $b > 0$, and $b \neq 1$.

While the value of a linear function changes by the same amount for equal increases in x, the value of an exponential function changes by the same factor for equal increases in x.

The graph of an exponential function is nonlinear.

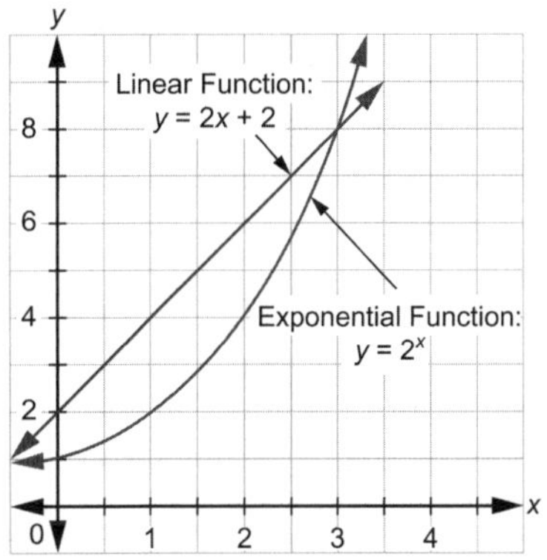

EXAMPLE: Writing an Exponential Equation

Write an exponential equation that models the relationship between the term number and the term of the geometric sequence formed by the number of new cattle infections.

SOLUTION:

The graph shows the ordered pairs for the relationship. The common ratio of the sequence is 3, which means that each week, the number of new infections increases by a factor of 3. So the base b in the exponential equation that models the sequence is 3.

Now find the value of the constant a.

$y = ab^x$	Write an exponential equation.
$5 = a \cdot 3^1$	Substitute 3 for the base b. Substitute the ordered pair (1, 5) for x and y.
$5 = a \cdot 3$	Simplify the right side.
$\frac{5}{3} = a$	Divide both sides by 3.

The value of a is $\frac{5}{3}$.

So the exponential equation that models the relationship is $y = \frac{5}{3}(3^x)$, where x is a whole number greater than 0. In this equation, the input x is a term number, and the output y is a term of the sequence.

EXAMPLE: Model a Geometric Sequence with a Recursive Formula

Use function notation to write a recursive formula that models the geometric sequence formed by the number of new infections.

My Notes

Summary *(continued)*

SOLUTION:

The geometric sequence formed by the number of new infections is 5, 15, 45, 135, 405.

For the first part of the recursive formula, show that the first term is 5.

$f(1) = 5$	When the input is 1, the output is 5.
Next, show how to find the next term given a term of the sequence.	
NEXT = 3 · NOW	Start with the next-now equation we wrote earlier.
$f(n+1) = 3 \cdot f(n)$	Let the *n*th term of the sequence, or $f(n)$, represent now. The next term is then equal to $f(n+1)$.

So a recursive formula for the arithmetic sequence is

$f(1) = 5; f(n+1) = 3 \cdot f(n)$, where n is a whole number greater than 0.

My Notes

Modeling Linear and Exponential Functions

EXAMPLE: Write Equations to Model Functions

Write an equation to model the function represented by each table.

A.	x	2	3	4	5	6
	y	18	15	12	9	6

B.	x	0	1	2	3	4
	y	6	12	24	48	96

SOLUTION:

A.	For each unit increase in x, the y values decrease fairly slowly. $15 - 18 = -3$, $12 - 15 = -3$, $9 - 12 = -3$, $6 - 9 = -3$ The sequence has a common difference of −3, so it is an arithmetic sequence. The linear function has a slope of −3. Now find the y-intercept.	
	$y = mx + b$	Write the slope-intercept form of a linear equation.
	$18 = -3(2) + b$	Substitute −3 for the slope. Substitute the ordered pair (2, 18) for x and y.
	$18 = -6 + b$	Simplify the right side.
	$24 = b$	Add 6 to both sides.
	The y-intercept is 24. So the equation that models the function in the table is $y = -3x + 24$.	

B.	For each unit increase in x, the y values increase rapidly. $\frac{12}{6} = 2$, $\frac{24}{12} = 2$, $\frac{48}{24} = 2$, $\frac{96}{48} = 2$ The sequence has a common ratio of 2, so it is a geometric sequence. The exponential function has a base of 2. Now find the value of the constant a.	
	$y = ab^x$	Write an exponential equation.
	$6 = a \cdot 2^0$	Substitute 2 for the base b. Substitute the ordered pair (0, 6) for x and y.
	$6 = a \cdot 1$	Simplify the right side.
	The value of a is 6. So the equation that models the function in the table is $y = 6 \cdot 2^x$.	

Practice Exercises

Review what you've learned using these practice problems. For practice problems with feedback, try the Coach and Play items in the Practice section online.

1. The table shows selected values from an arithmetic sequence.

Index	Term
2	31
5	43
6	47
10	63

What is the recursive rule for the arithmetic sequence in the table?

A. $f(n+1) = f(n) + 4;\ f(0) = 23$

B. $f(n+1) = f(n) + 8;\ f(0) = 39$

C. $f(n+1) = f(n) + 12;\ f(0) = 19$

D. $f(n+1) = f(n) + 16;\ f(0) = 15$

2. The function $f(x)$ models a geometric sequence where x is the term number and $f(x)$ is the xth term of the geometric sequence. A graph of $f(x)$ and some of its points are shown at right.

Refer to the graph to find the common ratio of the geometric sequence and the y-intercept of the graph of $f(x)$.

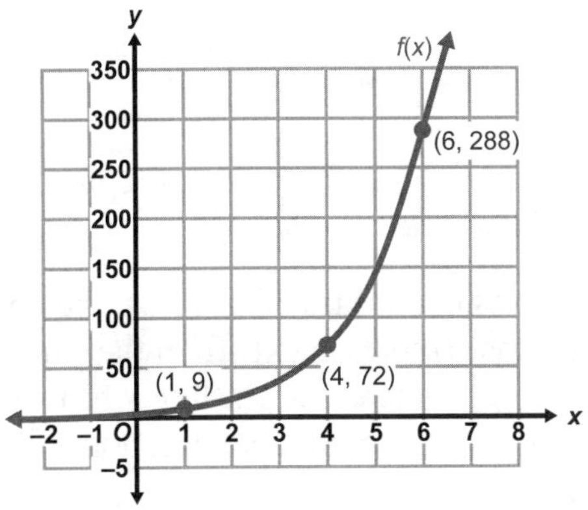

The common ratio of the geometric sequence is ___________.

The y-intercept of the graph of $f(x)$ is (__________, __________).

3. The graph shows three ordered pairs of $f(x)$. This function models a sequence where x is the term number and $f(x)$ is its corresponding term.

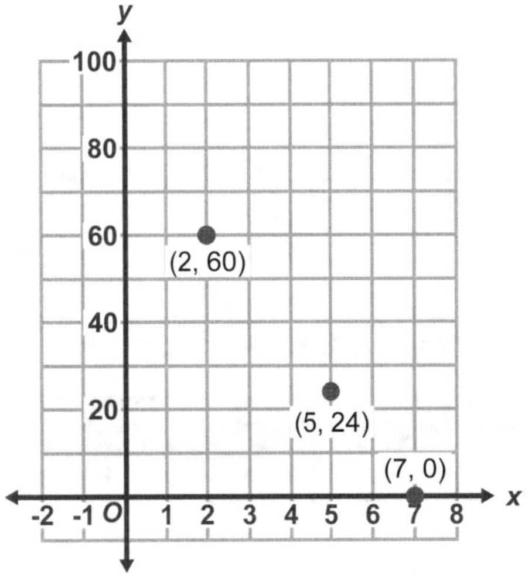

What is the y-intercept of the graph of $f(x)$?

(__________, __________)

4. Look at the table at following table.

Which of the following statements are true of the sequence that is modeled by the function represented in the table? *Select all that apply.*

x	y
1	2
2	4
3	8
4	16

A. The equation is $y = 2^x$, because the given values of x are positive integers and the values of y are powers of 2.

B. The equation is $y = x^2$, because the given values of x are positive integers and the values of y are perfect squares.

C. In the sequence, the value of the term $n = 6$ will be 64.

D. In the sequence, the value of the term $n = 7$ will be 256.

Practice Exercises *(continued)*

5. Given the values for $f(3)$ and $f(4)$ in an exponential sequence, match each statement to an equivalent expression.

The common ratio: __________

$f(6)$, given the common ratio, b: __________

$f(2)$, given the common ratio, b: __________

$f(0)$, given the common ratio, b: __________

A. $\frac{1}{b} \cdot f(3)$ **B.** $b \cdot f(3)$ **C.** $b^2 \cdot f(4)$

D. $\frac{f(3)}{f(4)}$ **E.** $\frac{f(4)}{f(3)}$ **F.** $\frac{1}{b^3} \cdot f(3)$

6. Compare the functions shown in the tables.

x	$f(x)$
1	3
2	9
3	15
4	21

x	$g(x)$
1	4
2	8
3	16
4	32

Which statement is true?

A. The function $g(x)$ is an exponential function that can be expressed as $g(x) = 2 \cdot 2^x$.

B. The function $f(x)$ is an exponential function whose equation is $f(x) = 3^x$.

C. The functions $f(x)$ and $g(x)$ are both exponential.

D. The function $g(x)$ is an exponential function whose equation $g(x) = 4^x$.

7. Consider the following geometric sequence.

4, 8, 16, 32, 64

Which of the functions below model the sequence? *Select all that apply.*

A. $f(n) = 2^n \cdot 2$ **B.** $f(n) = 2n$

C. $f(n + 1) = 2^n$ **D.** $f(n + 1) = 2 \cdot f(n)$

8. Look at the following table.

x	1	2	3	4
y	5	25	125	625

Which of the following statements are true of the sequence that is modeled by the function represented in the table? *Select all that apply.*

A. The equation is $y = 5x$, because the given values of x are positive integers and the values of y are multiples of 5.

B. The equation is $y = 5^x$, because the given values of x are positive integers and the values of y are powers of 5.

C. In the sequence, the value of the term $n = 6$ will be the value at $n = 4$ multiplied by 5^2.

D. In the function, the average rate of change between $x = 2$ and $x = 6$ is 3,900.

9. Compare the functions shown on the graph.

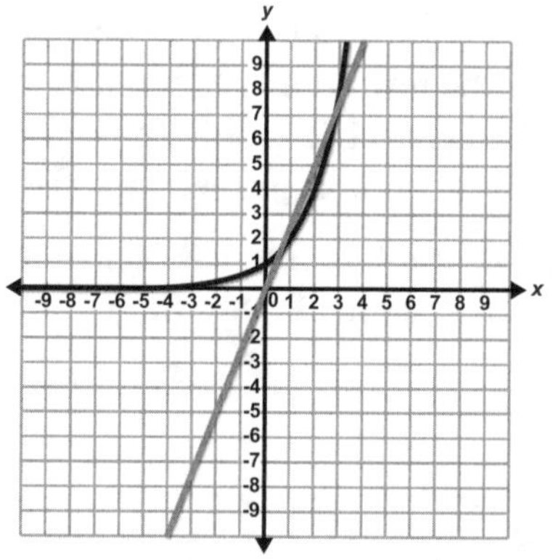

Which of the following descriptions represent these functions? *Select all that apply.*

A. For positive values of x, at first, the line increases more quickly than the curve. Then, as x increases, the linear function exceeds the exponential function.

B. When $x = 1$ and $x = 2$, the linear function has a greater value than the exponential function.

C. For positive values of x, at first, the line increases more quickly than the curve. Then, as x increases, the exponential function exceeds the linear function.

D. As x increases, the functions both increase at varying rates.

Apply

How Does Your City Grow?

It is estimated that by 2050, the world population will reach 9.6 billion people. Predictions for population growth can vary from city to city.

Plotting the Curve of Best Fit: View this video segment to see students, with the help of their graphing calculators, use world population data from the last 50 years to predict the population in the year 2050.

Population Growth

- Choose a city, and research the population for that city in the year you were born, the year you entered Kindergarten, and the year you entered Grade 6.
- Use your birth year and Kindergarten data to create an arithmetic sequence as well as a geometric sequence that could represent the yearly growth of your city.
- Create mathematical models in the form of an equation, table, and graph for each sequence.
- Use the model that yields results closest to your actual Grade 6 data to predict the population of your city in the year you will graduate from high school. Justify your reasoning.

Show what you've learned by completing the other performance tasks in the online Apply section.

Apply *(continued)*

Your answer to Apply will be assessed on the following criteria:

1. Conducting research on the population growth of a city of your choice
2. Creating arithmetic and geometric sequences, and writing mathematical models that represent each sequence
3. Presenting clear, organized, and detailed information using algebraic, graphical, and numerical support
4. Stating reasonable conclusions with appropriate justification based on mathematical analysis

Scale / Criteria	4	3	2	1	0
	Exceeds Criteria	**Meets Criteria**	**Progressing to Criteria**	**Below Expectations**	**No Expectation**
Conducting Research	Presents clear and convincing evidence of research about the population of a city over time; collects all relevant data.	Presents clear and convincing evidence of research about the population of a city over time; collects data with minor errors or gaps.	Presents adequate evidence of research about the population of a city over time; collects data with minor errors or gaps.	Presents limited evidence of research about the population of a city over time.	Does not attempt.
Creating Mathematical Models	Creates arithmetic and geometric sequences based on accurate calculations, using technology and estimates appropriately and understands the implications of round off error; writes mathematical models that accurately represent each sequence.	Creates arithmetic and geometric sequences and writes mathematical models; errors are evident in either the sequences or the mathematical models; calculations are generally correct but contain minor errors.	Creates arithmetic and geometric sequences and writes mathematical models; errors are evident in both the sequences and the mathematical models; calculations contain major algorithmic or computational errors.	Creates sequences or mathematical models; performs very few calculations correctly.	Does not attempt.
Presenting Results	Uses appropriate algebraic, graphical, and numerical support to present information in a clear, organized, and detailed way.	Uses appropriate algebraic, graphical, and numerical support to present information in a clear, organized, and detailed way with minor errors.	Uses some appropriate algebraic, graphical, and numerical support to present information in a clear and organized way with major errors.	Uses incorrect algebraic, graphical, and numerical support to present information with major errors.	Does not attempt.
Stating Conclusion and Justification	Uses data analysis to provide a convincing justification for choice of mathematical model; uses accurate calculations to make reasonable conclusions about population.	Uses data analysis to provide a justification for choice of mathematical model; uses calculations to make mostly reasonable conclusions about population.	Justification is poorly articulated; uses calculations incorrectly to justify a conclusion.	Makes an unreasonable conclusion.	Does not attempt.

UNIT 4: Graphs of Functions

4.1 Analyze Graphs of Functions

photo: Getty Images

Lesson Objectives

- Identify the domain and range of linear and exponential functions.
- Graph functions by hand and using technology, and identify key features of the graphs.
- Explore the slopes of parallel and perpendicular lines.
- Determine the rate of change of a linear function.
- Explore the effect of replacing $f(x)$ by $f(x) + k$ or $f(x + k)$ on the graph of linear and exponential functions.
- Use function notation to interpret key features.

Essential Questions

- How can you determine whether a real-world scenario represents a linear or exponential function?
- How can you describe changes in real-world scenarios as transformations of linear or exponential functions?
- How can you discern between the domain and range of a real-world scenario versus the domain and range of the algebraic function that models the behavior?

Investigations

Roller Coaster

Get specific about the design of a roller coaster. How high can it go?

Density

How dense can it be? It's all a matter of mass and volume.

Underwater Cameras

Compare rental plans. Adjust functions for changing prices and watch the graphs change.

Transforming Graphs

You've transformed linear functions. Now it's time to take on exponential functions. Make the shift!

Protect the Pigs

Build parallel and perpendicular fences to protect the pigs from the wolves.

Key Vocabulary

function, input, output, rate of change, slope, slope-intercept form, standard form, volume, x-intercept, y-intercept

Discover

As you complete Engage and the investigations, record the most important ideas you've learned.

Engage	
Investigation 1	**Investigation 2**
Investigation 3	**Investigation 4**

Name ______________________________ Date ____________

Check for Understanding

Analyze Graphs of Linear and Exponential Functions: Investigation 1

As their names suggest, Black Ironwood and Leadwood are heavy.

Use the following table and graph showing the volume and mass for each type of wood to answer questions 1–3.

Black Ironwood	
Volume (cm^3)	Mass (g)
6	8.16
8.5	11.56
11	14.96
13.5	18.36

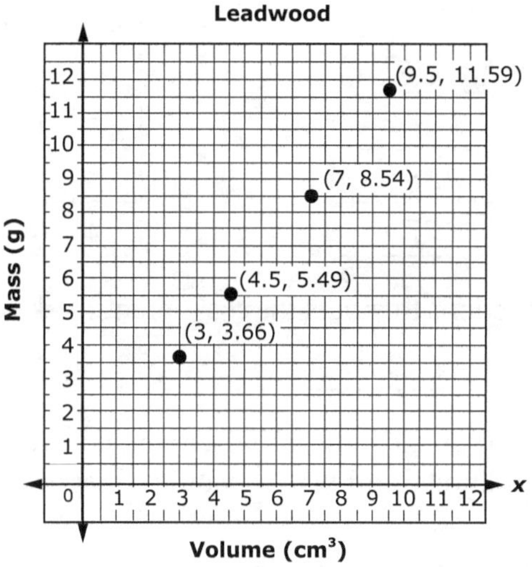

1. Determine the density, D, and write a function, $m(V)$, for each type of wood that represents the relationship between its mass and volume.

   A. Black Ironwood: D = ______________ Function: ______________

   B. Leadwood: D = ______________ Function: ______________

2. Choose the statements that are true. *Select all that apply.*

   A. The density of Black Ironwood is greater than the density of Leadwood.

   B. Leadwood will likely float in water, but Black Ironwood will sink.

   C. Both functions are linear.

   D. The slope of each function is the ratio of mass to volume for each type of wood.

   E. The domain for each function is all real numbers.

   F. The graphs of each function are steeper than the graph for the density of water.

3. The density in customary units of Black Ironwood is approximately 84.5 lb/ft^3. Explain why this value is different from the one you calculated, and why they are equivalent.

Name ________________________________ Date ______________

Check for Understanding

Analyze Graphs of Linear and Exponential Functions: Investigation 2

The costs for ski and snowboard equipment rentals at Powder Paradise are shown in the tables below.

Use the tables to answer questions 1–2.

Ski Rentals	
Days	Cost ($)
3	$93
5	$135
7	$177

Snowboard Rentals	
Days	Cost ($)
1	$42
4	$114
6	$162

1. Write two functions to model the cost of renting skis $S(x)$ and renting a snowboard $B(x)$.

A. Skis: ______________ **B.** Snowboard: ______________

C. In each function, the initial fee would be represented on the graph by the ______________ and the additional cost per day would be represented on the graph by the ______________.

2. Use function notation to show the cost for two days for each type of equipment.

A. Skis: ______________ **B.** Snowboard: ______________

3. For how many days can each type of equipment be rented if the budget is $200?

A. Skis: ______________ **B.** Snowboard: ______________

4. Write a function to show the total cost for 2 ski rentals and 1 snowboard rental, and use it to solve for the total cost of their rental for 5 days.

A. Function: ______________ **B.** Total for 5 days: ______________

5. Why is the total cost of renting a snowboard initially less than the total cost of renting skis, but then is greater after a number of days? How would this be shown in the graphs of the functions?

Name ______________________________ Date ______________

Check for Understanding

Analyze Graphs of Linear and Exponential Functions: Investigation 3

1. Mobile device subscriptions in Waterview have grown according to the table below. Graph the values.

Year	Total Number of Mobile Users (×1,000)
0	14
1	28
2	56
3	112

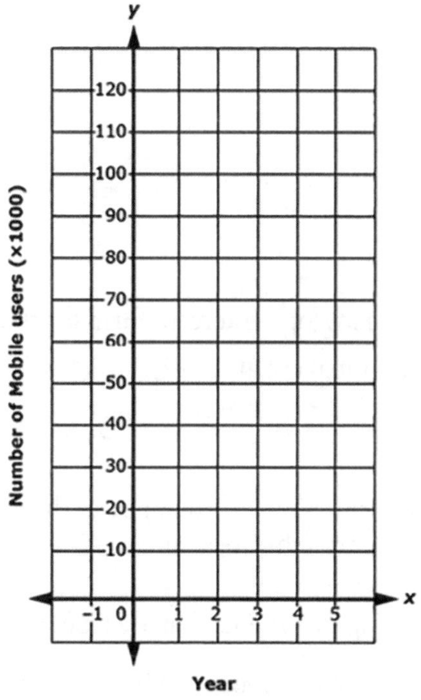

Use the table and graph to answer questions 2–4.

2. Identify the values for *A* and *B* from the table to write a function, $M(x)$, that models the growth of mobile device users in Waterview.

   A. Base: ______________ **B.** Initial Number of Users (thousands): ______________

   C. Function: ______________

3. Predict the number of new subscriptions for Year 4 (thousands). ______________

4. Choose the statements that are true from the following. *Select all that apply.*

   A. If the number of users for Year 0 changed to 5, the graph of the function would shift right by 5.

   B. There were 14,000 new users in Year 1.

   C. The average rate of change between Years 4 and 5 is greater than the average rate of change between Years 2 and 3.

   D. The number of users in the year prior to Year 0 would have been 2,000.

   E. The domain for the function is non-negative integers.

5. How would starting with the date one year earlier affect the function and graph? Explain.

Name ______________________________ Date ____________

Check for Understanding

Analyze Graphs of Linear and Exponential Functions: Investigation 4

1. Match the equation of each line to the equation for the one that is perpendicular to it.

A. $y = 5x - 3$	**I.** $-5x + y = 3$
B. $y = \frac{5}{6}x - 2$	**II.** $6x + 5y = 25$
C. $y = -\frac{1}{5}x + 7$	**III.** $x + 5y = 1$
D. $y = -\frac{5}{6}x + 5$	**IV.** $6x - 5y = -10$

2. Write the equations in slope-intercept form for the lines that are parallel and perpendicular to $y = \frac{2}{3}x + 5$ with a y-intercept at $(0, -2)$.

   A. Parallel: ______________ **B.** Perpendicular: ______________

3. *Circle the correct answer.* The line that passes through $P(2, -2)$ and $Q(1, -1)$ is [parallel / perpendicular] to the line that passes through $R(6, 3)$ and $S(-3, -6)$.

4. Choose the following statements that are true. *Select all that apply.*

   A. The product of the slopes of a line and another line perpendicular to it is -1.

   B. If the graph of a line is increasing, one perpendicular to it will necessarily be decreasing.

   C. Changing the equation for a line from $y = 7x - 4$ to $y = 7x + 4$ will shift its graph upward and to the right.

   D. Two lines in the coordinate plane that are perpendicular to the same line are parallel to each other.

   E. If two lines are perpendicular, their x-intercepts will have x-coordinates that are additive inverses, or opposites.

5. Given line s on a graph containing point $P(a, b)$, explain how you can begin from P and locate a point on a line, t, that is perpendicular to s through P.

Summary

Before you attempt the Practice Exercises, review what you've learned.

Graphs and Equations

Linear and exponential functions can model the curves found in many real objects.

Writing an Equation from a Graph or Table

EXAMPLE: Write Equations from Graphs or Tables

Write an equation for the function that is modeled. Then give the domain and range for the function.

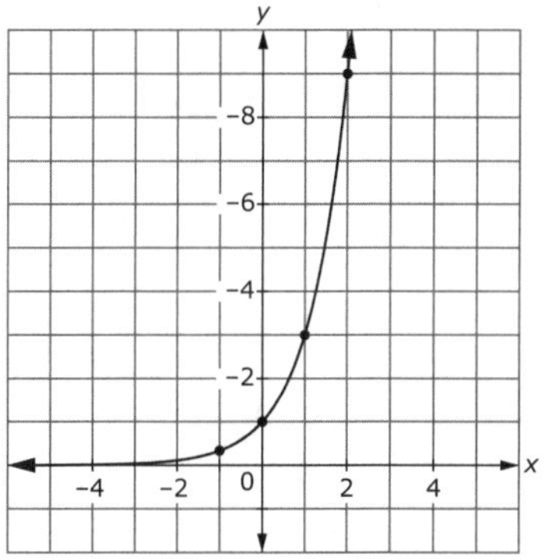

SOLUTION:

The graph appears to be an exponential function.

Make a table of values and look for a common ratio.

x	y
-1	$\frac{1}{3}$
0	1
1	3
2	9

$\frac{1}{\frac{1}{3}} = 3$ $\frac{3}{1} = 3$ $\frac{9}{3} = 3$

The ratio between each pair of successive y values is 3, so the function is exponential.

Use the common ratio to write an equation: $y = 3^x$.

The domain is $-\infty < x < \infty$.

The range is $0 < y < \infty$.

Comparing Linear Functions

The slope-intercept form of a linear equation is written as $y = mx + b$.

- (x,y) is a point the line goes through.
- m is the slope of the line.
- b is the y-intercept of the line.

EXAMPLE: Compare Linear Functions

In some cities, companies run car-sharing businesses.

At Share-a-Lift, the cost is relative to the number of miles driven, $R(x)$. At Car Share Plus, the cost is dependent on the number of hours the car is rented, $P(h)$. Both companies charge an annual membership fee, and at both companies, the relationship between the domain and the range is linear.

My Notes

Summary *(continued)*

Share-a-Lift	
Miles, x	Cost, $R(x)$
20	$46.00
45	$59.75

Car Share Plus	
Hours, h	Cost, $P(h)$
1	$31.25
8	$75.00

Assume that your commute to work is 30 miles one way. You work a 5-hour shift, so you will need the car for at least 6 hours. Which company should you rent from?

SOLUTION:

First write an equation for each relationship.

Share-a-Lift

Use the slope formula to find the slope.

$$\frac{59.75 - 46.00}{45 - 20} = \frac{13.75}{25} = 0.55$$

Then use slope-intercept form to find the y-intercept.

$R(x) = mx + b$

$46 = 0.55(20) + b$

$46 = 11 + b$

$35 = b$

The cost at Share-a-Lift can be modeled by $R(x) = 0.55x + 35$, where x is the number of miles and $R(x)$ is the total cost. The slope, 0.55, represents the charge per mile, and the y-intercept, 35, represents the annual membership fee.

Car Share Plus

Use the slope formula to find the slope.

$$\frac{75.00 - 31.25}{8 - 1} = \frac{43.75}{7} = 6.25$$

Then use slope-intercept form to find the y-intercept.

$P(h) = mx + b$

$75 = 6.25(8) + b$

$75 = 50 + b$

$25 = b$

The cost at Car Share Plus can be modeled by $P(h) = 6.25h + 25$, where h is the number of hours and $P(h)$ is the total cost. The slope, 6.25, represents the charge per hour, and the y-intercept, 25, represents the annual membership fee.

Next, use the equations to compare the cost at Share-a-Lift for a 60-mile round-trip commute and the cost at Car Share Plus for renting a car for 6 hours.

Share-a-Lift	Car Share Plus
$R(x) = 0.55x + 35$	$P(h) = 6.25h + 25$
$R(x) = 0.55(60) + 35$	$P(h) = 6.25(6) + 25$
$R(x) = 33 + 35$	$P(h) = 37.50 + 25$
$R(x) = 68.00$	$P(h) = 62.50$

The cost of renting the car at Share-a-Lift to drive 60 miles is more expensive than the cost to rent the car at Car Share Plus for 6 hours. I would rent from Car Share Plus.

My Notes

Summary *(continued)*

Comparing Exponential Functions

EXAMPLE: Analyzing Exponential Functions

Researchers apply an antibiotic to a colony of bacteria and a different antibiotic to another colony. The populations of the two colonies over time are shown in the graph. The populations decay at exponential rates.

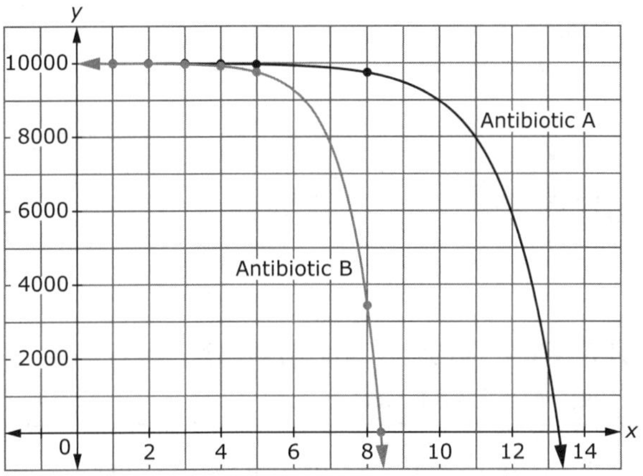

Which antibiotic appears to eliminate the bacteria faster?

SOLUTION:

The curve for Antibiotic B is decaying faster than the curve for Antibiotic A. The graph shows that the colony treated with Antibiotic B is eliminated at a time of just more than 8 periods while Antibiotic A eliminates the other colony at a time of just more than 13 periods.

Antibiotic B appears to be more effective.

Transforming Linear Functions

EXAMPLE: Transformations of Linear Functions

Start with the function $f(x) = -4x + 5$. Describe how the following transformations affect the graph.

A. $g(x) = (-4x + 5) + 6$

B. $g(x) = -4(x - 2) + 5$

C. $g(x) = -6(-4x + 5)$

SOLUTION:

A. Adding 6 to the original function rule affects the output values. The graph will be shifted up 6 units.

B. Subtracting 2 from the x value (the input) shifts the graph 2 units to the right.

C. Multiplying the original function by −6 affects the slope of the graph. The slope will change from negative to positive, and the slope will be steeper.

My Notes

Summary *(continued)*

EXAMPLE: Transformations of Exponential Functions

Start with the function $f(x) = 3^x$. Describe how the following transformations affect the graph.

A. $g(x) = 3^x - 2$

B. $g(x) = -4 \cdot 3^x$

SOLUTION:

A. Subtracting 2 from the original function affects the output values. The graph will be shifted down 2 units. The domain changes from all positive real numbers to all real numbers greater than −2.

B. Multiplying the original function by −4 affects the shape of the graph. The graph changes from increasing to decreasing, and the function decreases at a faster rate.

Parallel and Perpendicular Lines

Two parallel lines have the same slope. Two perpendicular lines have slopes that are opposite reciprocals.

My Notes

EXAMPLE: Write an Equation for a Parallel Line

Write an equation for a line that is parallel to $y = \frac{1}{2}x - 4$ and goes through the point (−2, 3).

SOLUTION:

The slope of $y = \frac{1}{2}x - 4$ is $\frac{1}{2}$, and you know one point that the parallel line contains. Use point-slope form to write an equation for the parallel line.

$$y = \frac{1}{2}x + 4$$
$$y_2 - y_1 = m(x_2 - x_1)$$
$$y - 3 = \frac{1}{2}(x - (-2))$$
$$y - 3 = \frac{1}{2}x + 1$$
$$y = \frac{1}{2}x + 4$$

EXAMPLE: Write an Equation for a Perpendicular Line

Write an equation for a line that is perpendicular to $y = -3x - 5$ and goes through the point with coordinates (3, −1).

SOLUTION:

The slope of $y = -3x - 5$ is −3, so the slope of a perpendicular line will be $-\left(\frac{1}{-3}\right) = \frac{1}{3}$.

$$y - y_1 = m(x - x_1)$$
$$y + 1 = \frac{1}{3}(x - 3)$$
$$y + 1 = \frac{1}{3}x - 1$$
$$y = \frac{1}{3}x - 2$$

Slope-Intercept Form and Standard Form

A common form for a linear equation is called **standard form**.

The standard form of a linear equation is written as $Ax + By = C$.

- A, B, and C are integers.
- A and B are not both equal to 0.
- A is nonnegative.

Practice Exercises

Review what you've learned using these practice problems. For practice problems with feedback, try the Coach and Play items in the Practice section online.

1. Which of the following lines is perpendicular to the line defined by the equation $y = \frac{4}{3}x + 3$ and has the same y-intercept?

A. $y = -\frac{3}{4}x - 3$ **B.** $y = \frac{3}{4}x - 3$

C. $y = \frac{4}{3}x - 3$ **D.** $y = -\frac{4}{3}x + 3$

E. $y = -\frac{3}{4}x + 3$ **F.** $y = \frac{3}{4}x + 3$

2. Consider the function $y = 3^x + 5$.

Complete the statements regarding the function and its graph.

A. domain	**B.** range	**C.** x
D. y	**E.** horizontal	**F.** vertical
G. 0	**H.** 3	**I.** 5

The __________ of the function is all real numbers greater than 5.

The __________ of the function is all real numbers.

The __________ -intercept of the graph is (0, 6).

The graph has a __________ asymptote at $y =$ __________.

3. Consider these two linear equations.

$y = ax - 5$

$-2x + y = 3$

Which value of a will make the two lines parallel?

A. -2 **B.** 2 **C.** $\frac{1}{2}$ **D.** $-\frac{1}{2}$

4. A line is parallel to the line $y = 3x - 8$ and passes through the point $(-1, 6)$.

For the graphs of the two lines to be parallel, they must have the same __________.

The equation of the parallel line is
$y =$ __________ $x +$ __________.

5. Consider the functions shown. The graphs of the functions are perpendicular lines.

$f(x) = ax - 2$

$g(x) = -\frac{3}{2}x + \frac{1}{2}$

What is the value of a?

$a =$ ______________

6. The table represents a function.

x	y
1	5
3	9
5	13
7	17

Which of the following statements are true?

Select all that apply.

A. The average rate of change, or slope, for the interval shown in the table is 4.

B. The y-intercept of the function model is (0, 3).

C. The average rate of change, or slope, for the interval shown in the table is 2.

D. The value of y decreases as the value of x increases.

E. On a graph of the function, when $y = 8$, the value of x will be 2.

F. The equation for the function is $y = 2x + 3$.

Practice Exercises *(continued)*

7. Shown is the graph of a function, $f(x)$.

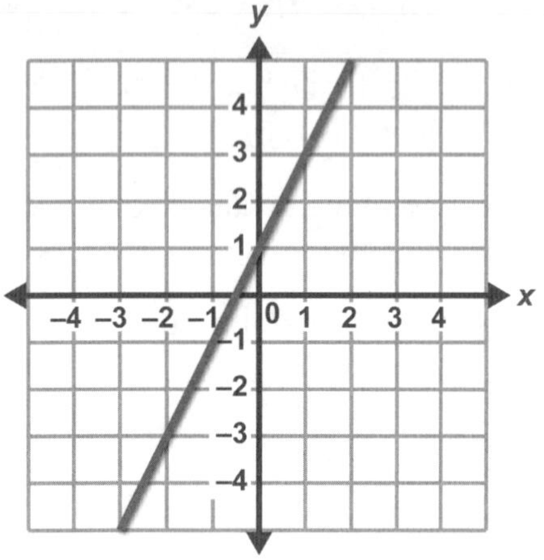

How will the graph of $f(x)$ be affected if it is transformed to $g(x) = -2x + 1$?

A. $g(x)$ will be perpendicular to $f(x)$.

B. The average rate of change will double.

C. The new function will be decreasing rather than increasing.

D. The graph will be translated down by 2 units.

8. Maria pours 5 gallons of water into a container, but the container immediately starts to leak.

After 3 hours, there are 3 gallons of water left in the container.

After 6 hours, there is 1 gallon of water left in the container.

Which statement is true about the relationship, if t represents the number of hours the container is leaking?

A. The average rate of change is $\frac{2}{3}$, and a reasonable domain is $0 \le t \le 5$.

B. The average rate of change is $-\frac{2}{3}$, and a reasonable domain is $0 \le t \le 7.5$.

C. The average rate of change is $\frac{2}{3}$, and a reasonable domain is $0 \le t \le 6$.

D. The average rate of change is $-\frac{2}{3}$, and a reasonable domain is $0 \le t \le 6$.

9. Danica's homework contains a table with values from a linear function, but she spilled her drink on the corner, and the output value for $x = 2$ was faded out.

x	0	1	2
y	5	1	

What must the missing value in the table be?

The missing value is ________________.

10. What are the domain and range of the function below?

$f(x) = 2^x + 8$

A. domain: all real numbers
range: all real numbers

B. domain: all real numbers
range: $f(x) < 8$

C. domain: all real numbers
range: $f(x) > 8$

D. domain: $x \ge 0$
range: $f(x) \ge 8$

11. Describe the relationship between the graphs of the following two equations.

$3x + 2y = 10$
$-2x - 3y = 13$

A. Parallel lines

B. Perpendicular lines

C. Parallel and perpendicular lines

D. Neither parallel nor perpendicular lines

Apply

How Can You Use Functions to Make an Image?

Architects, fashion designers, and engineers are among the many professionals who rely upon CAD, or computer-aided design, to help them envision new products. You can create simple images in a similar way by graphing pieces of functions.

photo: Discovery Education

Using graphs of linear and/or exponential functions, and vertical line segments if necessary, draw a recognizable figure. You may use graph paper or a graphing software program.

- State the equation and the domain of each function as it is shown in your image.
- Include at least one pair of non-vertical parallel or perpendicular lines in your design and explain the relationship between the slopes of those lines.
- Describe how some pieces of your image are related as transformations of each other.

Show what you've learned by completing the other performance tasks in the online Apply section.

Apply *(continued)*

Your answer to Apply will be assessed on the following criteria:

1. Creating a recognizable figure with functions and vertical segments
2. Writing the correct equation and domain for each piece of the image
3. Including parallel or perpendicular segments and discussing their slopes
4. Explaining how transformations to the functions relate the pieces of the graph

Scale / Criteria	4	3	2	1	0
	Exceeds Criteria	**Meets Criteria**	**Progressing to Criteria**	**Below Expectations**	**No Expectation**
Recognizable Figure	Figure is recognizable.	Figure is mainly recognizable.	Figure is somewhat recognizable.	Figure is not recognizable.	Does not draw a figure.
Correct Equations	All of the equations and domains are correct.	There are one or two minor errors in the equations or their domains.	There are many minor errors or a major error in the equations or their domains.	Equations are attempted but are mostly incorrect.	Does not write equations.
Parallel and Perpendicular Lines	Includes at least one pair of non vertical parallel or perpendicular lines; relationship between the slopes is correctly explained.	Includes at least one pair of non vertical parallel or perpendicular lines; has minor errors in the explanation of slopes.	Includes at least one pair of non vertical parallel or perpendicular lines; has major errors in the explanation of slopes.	Includes at least one pair of non vertical parallel or perpendicular lines; relationship between the slopes is not explained.	Does not include parallel or perpendicular lines.
Transformations	Explanation of how the graph pieces and their functions are related by transformation is accurate and understandable.	Explanation of how the graph pieces and their functions are related by transformation is mostly accurate and understandable.	Explanation of how the graph pieces and their functions are related by transformation is not very accurate or understandable.	Explanation of how the graph pieces and their functions are related by transformation is incorrect.	Does not explain work.

UNIT 4: Graphs of Functions

4.2 Compare Graphs of Linear and Exponential Functions

photo: Getty Images

Lesson Objectives

- Use functions to represent real-world situations.
- Graph functions by hand and using technology, and identify key features of the graphs.
- Describe functions using multiple representations.
- Explain "changing rate of change" and identify graphs that exhibit this feature.

Essential Questions

- How can you analyze the rate of change over multiple intervals of data to determine whether a function is linear or exponential?
- How do the data in an arithmetic sequence support the graphic and algebraic structure of the model?
- How do the data in a geometric sequence support the graphic and algebraic structure of the model?
- What are differences between the domain and range of the algebraic function and the domain and range of the context that it models?

Investigations

Battery Decay

Batteries run out of juice over time. Can the decay be modeled as a linear or exponential function?

Basketball

Tournaments have a way of eliminating losing teams. Is the model linear or exponential?

Quilts

Geometric designs are functional too. Use functions and graphs to plan a quilt design.

Geometric Patterns

What does it look like? Geometric patterns make linear and exponential functions visible.

Key Vocabulary

common difference, common ratio, function, input, output, rate of change, slope, slope-intercept form, standard form, x-intercept, y-intercept

Discover

As you complete Engage and the investigations, record the most important ideas you've learned.

Engage	Investigation 1
Investigation 2	Investigation 3

Name ________________________________ Date ______________

Check for Understanding

Compare Graphs of Linear and Exponential Functions: Investigation 1

1. In a doubles tennis tournament for 128 players, teams of 2 play sets of games, called matches, to determine a winning team that will move on to the next round. Table 1 shows the number of matches left after each round. Table 2 shows the players left in the tournament after each match in Round 1.

Complete the tables to answer questions 2–4.

Table 1	
Round	Number of Matches Left after Round
Start of Tournament	
1	
2	
3	

Table 2	
Match	Number of Players Left after Each Match
Start of Tournament	
1	
2	
3	

2. Circle the correct terms and fill in the blanks to complete the following statements.

   A. The sequence in Table 1 is [arithmetic / geometric] and its common [difference / ratio] is ____________________.

   B. The sequence in Table 2 is [arithmetic / geometric] and its common [difference / ratio] is ____________________.

3. Write a function, $M(x)$, for the number of matches played at each round, x, and another for $P(x)$, for the number of players, x, at the end of each match.

   A. Number of matches played: ____________________

   B. Number of players at end of each match: ____________________

4. Determine a reasonable domain and range for functions modeling each situation.

   A. $M(x)$ Domain: ____________________ Range: ____________________

   B. $P(x)$ Domain: ____________________ Range: ____________________

5. Explain why the common ratio for a decreasing exponential function and the common difference for a decreasing linear function have opposite signs.

Name ______________________________ Date ______________

Check for Understanding

Compare Graphs of Linear and Exponential Functions: Investigation 2

Elena and Erik began with a basic quilt pattern and added pieces at each stage as shown in the tables.

Use the data to answer questions 1–3.

Elena's Quilt	
Stage Number	Number of Triangles
1	3
2	6
3	9
4	12

Erik's Quilt	
Stage Number	Number of Squares
1	3
2	9
3	27
4	81

1. Which type of algebraic model, linear or exponential, describes each pattern and why?

A. Elena's pattern is ______________ because it has a common ______________.

B. Erik's pattern is ______________ because it has a common ______________.

2. Write a function for Elena's triangles pattern, $T(n)$, and a function for Erik's squares pattern, $S(n)$, where n is the stage number.

A. ______________ **B.** ______________

3. *Fill in each blank.*

A. At Stage 6, Elena's quilt will have $T(6) =$ ______________ triangles.

B. At Stage 5, Erik's quilt will have $S(5) =$ ______________ squares.

C. The y-intercept of the function model for Elena's quilt is (_____,_____) , and the y-intercept of the function model for Erik's quilt is (_____,_____).

D. Erik's quilt will have the same number of pieces as Elena's does at Stage 3, or when S(_______) = _______.

E. Elena's quilt will have 42 pieces at Stage ______________.

4. If Erik's quilt will be complete with 243 squares, and Elena's quilt will be complete with 42 triangles, what are a reasonable domain and range for each of the function models? Explain what the domain and range mean in the context.

Name ______________________________ Date ______________

Check for Understanding

Compare Graphs of Linear and Exponential Functions: Investigation 3

The tables below show points for two functions.

Analyze the data to answer questions 1–3.

x	$f(x)$
1	6
2	12
3	24
4	48

x	$g(x)$
1	11
2	19
3	27
4	35

1. Choose the interval(s) over which the average rate of change for $f(x)$ is greater than the average rate of change for $g(x)$. *Select all that apply.*

A. (1, 2) **B.** (1, 3) **C.** (2, 4) **D.** (3, 4)

2. Indicate whether each function is linear or exponential and write the common difference or ratio. *Circle the correct terms and fill in the blanks.*

A. $f(x)$ is [linear / exponential] and has a common [difference / ratio] of ______________.

B. $g(x)$ is [linear / exponential] and has a common [difference / ratio] of ______________.

3. Enter the coordinates for the y-intercept for each function.

A. $f(x)$: (_____,_____)

B. $g(x)$: (_____,_____)

4. Choose the correct statement(s) from among the following. *Select all that apply.*

A. An increasing linear function will always have a greater rate of change over the first few intervals than an increasing exponential function.

B. An arithmetic sequence's average rate of change is constant over all intervals.

C. The common ratio for a geometric sequence is its average rate of change over all intervals.

D. The common ratio for a geometric sequence is the base for the exponential function that models it.

5. Is the starting value of an arithmetic or geometric sequence the same as the initial value, or y-intercept of its function model? Why or why not?

Summary

Before you attempt the Practice Exercises, review what you've learned.

Linear vs. Exponential Models

The growth of a quantity may follow a **linear** or **exponential pattern**.

When a quantity grows at a constant rate, the pattern is **linear**. Linear decay also has a constant rate of change.

When a quantity grows by rates of change that continually increase by the same factor, the pattern is **exponential**. In exponential decay, the rates of change continually decrease by the same factor.

EXAMPLE: Explore Relationships in a Graph

Talia deposited \$50 in a savings account that earns interest every month. The parts of the graphs shown model how her money grows over time. Observe the rate of change over each interval of the function's domain. For each interval, explain whether the relationship is linear or exponential and support your answer by describing the rates of change.

SOLUTION:

In order to justify a curve's tendency to be linear or exponential, we must calculate and compare intervals of domain to determine the rate of change.

If the rate of change over different intervals remains the same, the graphic behavior can be justified to have a linear tendency.

If intervals of the domain reflect changing rate of change, the equation is not linear. Through further analysis we can determine whether the rate of change reflects increasing or decreasing exponential behavior.

Graph A

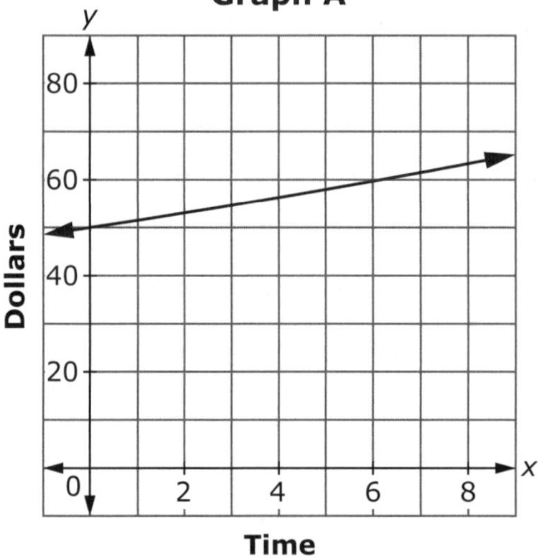

Graph A looks like it could be linear. It looks like the rate of change over [0, 6] is approximately $\frac{10}{6} \approx 1.67$.

To prove it is truly linear, we can further analyze other rates of change for other intervals:
The interval of [3, 6] can be represented by points (3, 55) and (6, 60).

The rate of change for this interval is approximately: $\frac{5}{3} \approx 1.67$.

Since both intervals gave us the same rate of change, we have evidence that this graph demonstrates linear behavior. More intervals could be checked to strengthen our evidence.

My Notes

Summary *(continued)*

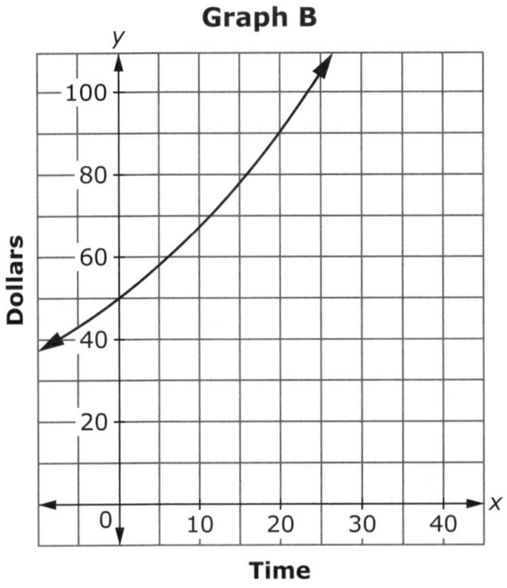

Graph B appears as though it could be linear or exponential. It looks like the rate of change over [0, 20] is approximately $\frac{40}{20} = 2$. Using the approximate value of (10, 67), we see that the rate of change over [0, 10] is approximately $\frac{17}{10} = 1.7$. Using the approximate values of (10, 67) and (20, 90), we see that the rate of change over [10, 20] is approximately $\frac{23}{10} = 2.3$. After analyzing these rates of change, we can see that this graph is not linear. If we must choose between linear and exponential, we choose exponential. Since the graph is increasing, the graph demonstrates increasing exponential behavior.

EXAMPLE: Determine a Model for Decay

Iodine-125 is a radioactive isotope that is used in medical tests and radiation therapy. Radioactive materials decay over time. A certain lab experiment begins with 0.3 grams of I-125. The table shows the amount of I-125 remaining over time.

Time Periods	Grams of I-125
0	0.3
1	0.15
2	0.075
3	0.0375
4	0.01875

Determine what type of mathematical model fits the data and write an equation for the model. Explain the reasonable domain for the context.

SOLUTION:

First, observe the relationship between each row of data and the one before it.

The number of days increases at a constant rate. Each number of grams is half the amount of the row above it, representing a common ratio of $\frac{1}{2}$. This is a geometric sequence, which can be modeled by an exponential equation.

Next, write an equation.

The starting amount is 0.3 grams, and the amount decays by half during each time period.

$y = 0.3\left(\frac{1}{2}\right)^x$

Then, determine the contextual domain. Time cannot be negative, so the domain is non-negative real numbers. The amount starts at 0.3 grams and will continue to decrease as it approaches 0 grams, so the range is $0 < y \leq 0.3$.

My Notes

Summary *(continued)*

EXAMPLE: Determine a Model for Growth

Observe how the perimeter of the figure changes in each step of this sequence.

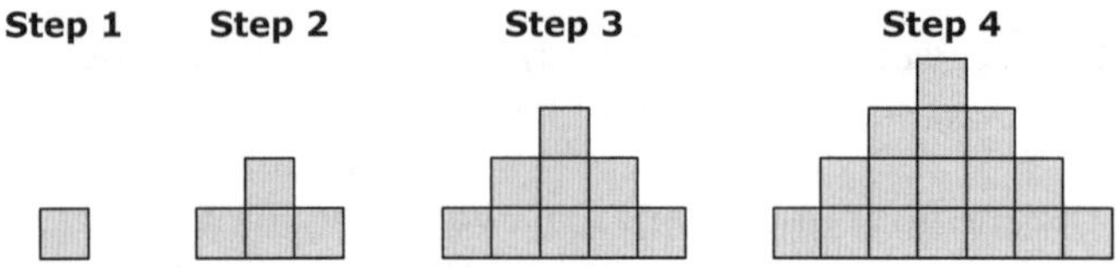

Determine what type of mathematical model fits the perimeter sequence and write an equation for the model. Explain the reasonable domain for the context.

SOLUTION:

First, make a table to record the perimeter at each step in the sequence. Then note any patterns you see.

Step	Perimeter
1	4
2	10
3	16
4	22

The perimeter of each figure is 6 units more than the figure in the step before. This means it is an arithmetic sequence with a common difference of 6. An arithmetic sequence can be modeled by a linear equation.

Next, write an equation.

The slope of the equation is equal to the common difference, 6. Use one point and the slope to find the y-intercept.

$y = 6x + b$

$10 = 6(2) + b$

$10 = 12 + b$

$-2 = b$

$y = 6x - 2$

Now, determine the contextual domain.

The step number and the perimeter must be counting numbers, so the domain is defined by the natural numbers. The range is a subset of the natural numbers. More specifically, the range is the set of numbers that consists of 4 plus non-negative multiples of 6. In other words, the range contains any natural number that can be obtained by adding a non-negative multiple of 6 to 4.

My Notes

Practice Exercises

Review what you've learned using these practice problems. For practice problems with feedback, try the Coach and Play items in the Practice section online.

1. Timmy's Taxi Service charges a flat fee of $1.50 per ride, plus an additional fee for each mile of travel. A 5-mile trip costs $11.50, and a 9-mile trip costs $19.50.

Which statement is true about the relationship between miles and cost?

A. The relationship is linear because it increases at a constant rate of change of $2 per mile.

B. The relationship is linear because it increases at a constant rate of change of $8 per mile.

C. The relationship is exponential because in increases at a changing rate of change of $2 per mile.

D. The relationship is exponential because it increases at a changing rate of change of $8 per mile.

2. Anthony buys a brand-new truck for $20,000. He looks up the prices of used models and finds that after 1 year, his truck will be worth $16,000, and after 2 years, it will be worth only $12,800.

What is the average rate of change per year for the value of the truck? ____________________

Is the rate of change a common difference or a common ratio? ____________________

3. Celia is looking at an image of a function graph. She thinks the function is either linear or exponential, but the image is zoomed in so far it's hard to tell from the shape of the graph.

She sees that the points $\left(\frac{1}{2}, 1\right)$ and (1, 2) are on the graph of the function.

Which additional point would confirm the function could be exponential?

A. $\left(\frac{3}{2}, 3\right)$

B. (2, 4)

C. $\left(\frac{3}{2}, 4\right)$

D. (0, 3)

E. (2, 3)

4. Examine the table of values.

x	y
2	8,000
3	2,000
4	500

Which statement is true about the relationship shown by the data in the table?

A. The relationship is linear because the rate of change over the domain is constant.

B. The relationship is linear because the y value is divided by 4 each time the x value increases by 1.

C. The relationship is exponential because the y value is multiplied by 4 each time the x value increases by 1.

D. The relationship is exponential because the rate of change over the domain changes by a common factor.

Practice Exercises *(continued)*

5. A function has the following ordered pairs: $\left(-1, \frac{3}{2}\right)$, (2, 12), and (4, 48).

Which of the following statements describes the graph of the function?

A. The graph is linear with a constant difference of 4 and a y-intercept at (0, 6).

B. The graph is exponential with a constant difference of 2 over multiple intervals and a y-intercept at (0, 3).

C. The graph is linear with a constant ratio of 2 and a y-intercept at (0, 0).

D. The graph is exponential with a constant ratio of 2 over multiple intervals and a y-intercept at (0, 3).

E. The graph is exponential with a constant ratio of 2 over multiple intervals and a y-intercept at (0, 0).

6. Shelly has a grand opening for a furniture store, and she is pleased with the number of customers who visit on the first day. One week later, she has 100 customers, and after week 2, she has 400 customers.

Shelly can use a function in the form of $f(x) = ab^x + c$ to model the number of customers she has compared with the number of weeks, x.

Which statements are correct? *Select all that apply.*

A. The value of c is 25.

B. The value of a is 4.

C. The value of b is 25.

D. The value of b is 4.

E. The value of c is 4.

7. The graph $y = f(x)$ contains the points (1, 2), (2, –1), (3, –4), and (4, –7). Which of the following could be the equation of $f(x)$?

$f(x) =$ ____________

A. $3x + 5$ **B.** $3x - 5$ **C.** $-3x + 5$ **D.** $-3x - 5$

8. Luis and his family are visiting the state fair. He started his day with \$150.00 in cash. After spending \$78.00 on entry fees and food, he and his family are buying tickets for rides at \$1.50 each.

Write a function that describes the average rate of change in the amount of money, y, Luis has after he buys x tickets.

A. $y =$ ____________ $x +$ ____________

Give the coordinates of the point that describes the maximum number of tickets Luis can buy.

B. (____________, ____________)

9. Jamaal is examining points on the graphs of two unknown functions. Function A has these points: (–2, –4), (1, 2), (4, 8). Function B has these points: $\left(-1, \frac{1}{2}\right)$, (1, 2), (3, 8).

Complete the following statements.
When $x = 5$, the value of function A is ____________ and the value of function B is ____________. Function ____________ has a greater value at $x = 5$.

10. Isabel knows that a certain function is either linear or exponential. She knows the graph contains the points (0, 4) and (1, 1).

Write the missing coordinate that the function includes if it is linear. Then write the missing coordinate that the function includes if it is exponential.

If the function is linear, it includes the point (2, ____________).

If the function is exponential, it includes the point (2, ____________).

Apply

How Does a Fractal Grow?

photo: Discovery Education

A fractal is a pattern that repeats itself on a smaller and smaller scale. You can see examples of fractals in nature, including plants such as Queen Anne's lace and ferns. There is also a type of fractal called a mathematical fractal, with patterns that repeat on to infinity, unlike natural fractals whose patterns only repeat a limited number of times. How would you model the growth of a fractal pattern with an exponential function?

Research mathematical fractals and choose one to explore. Determine a measurement that you can calculate for each stage, such as the change in length, the perimeter, or the number of branches of the fractal.

- Draw the first five stages of your fractal (Stage 0 through Stage 4).
- Calculate the change in measurement for each stage. Make a table to track the changes.
- Graph the change in measurement as a function of the stage number.
- Write a function to model the change in measurement as a function of the stage number.
- Discuss what type of function makes the best model and make a prediction for the changes in Stage 10 of your fractal.

Show what you've learned by completing the other performance tasks in the online Apply section.

Apply *(continued)*

Your answer to Apply will be assessed on the following criteria:

1. Choosing a fractal and drawing the first five stages of the fractal
2. Finding and recording the change in measurement for the first five stages
3. Graphing the changes in measurement as a function of the stages
4. Modeling the data for the first five stages with a function and using it to predict a future value

Criteria \ Scale	4 Exceeds Criteria	3 Meets Criteria	2 Progressing to Criteria	1 Below Expectations	0 No Expectation
Fractal and Measurement	Chooses a fractal and sketches the first five stages of the fractal correctly.	Chooses a fractal and sketches the first five stages of the fractal, with minor errors.	Chooses a fractal and sketches the first five stages of the fractal, with significant errors.	Chooses a pattern that is not a fractal and sketches fewer than five stages.	Does not attempt the task.
Table	Correctly lists the measurement for each stage in a table.	Lists the measurement for each stage in a table, with minor errors.	Lists the measurement for each stage in a table, with significant errors.	Lists the measurement for each stage in a table, with major errors.	Does not attempt the task.
Graph	Correctly graphs the change in measurement as a function of the stages.	Graphs the change in measurement as a function of the stages, with minor errors.	Graphs the change in measurement as a function of the stages, with significant errors.	Graphs the change in measurement as a function of the stages, with major errors.	Does not attempt the task.
Function and Prediction	Correctly writes a function to model the change in measurement as a function of the stages; uses it to make a correct prediction.	Writes a function with minor errors; uses it to make a correct prediction.	Writes a function with major errors; uses it properly, but the prediction is incorrect.	Writes a function with major errors; does not use it properly to make a prediction.	Does not attempt the task.

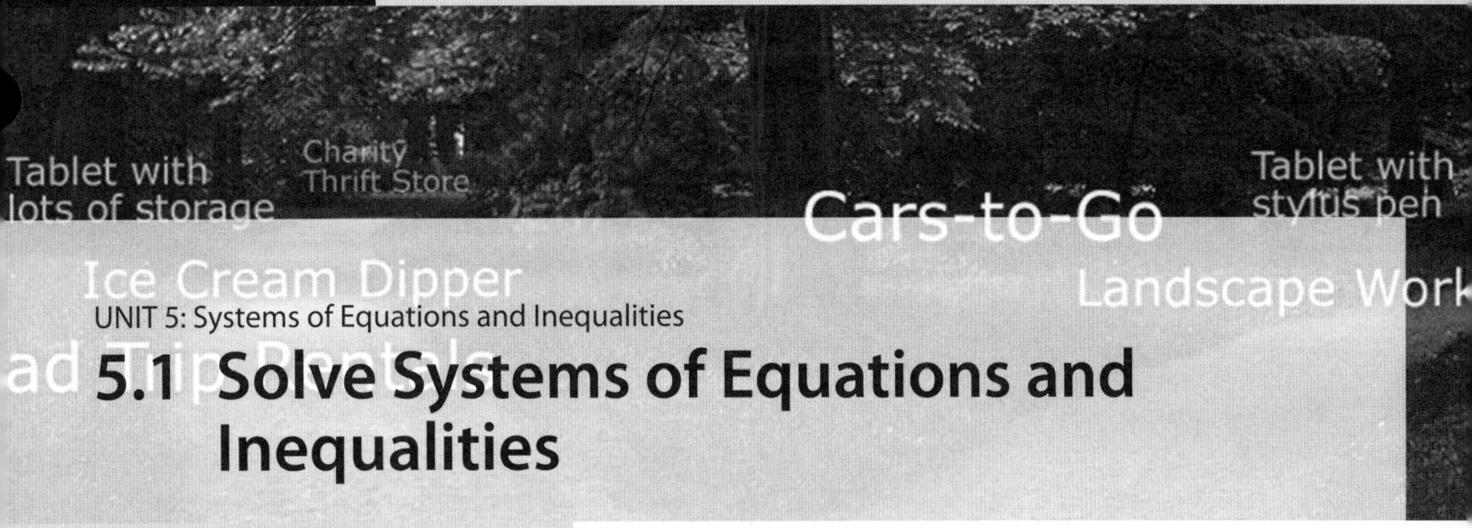

UNIT 5: Systems of Equations and Inequalities

5.1 Solve Systems of Equations and Inequalities

photo: Getty Images

Lesson Objectives

- Solve systems of linear equations graphically and algebraically.
- Solve systems of inequalities graphically.
- Strategically convert between various forms for a linear equation, depending on the situation.
- Use graphical representations of inequalities to interpret constraints.

Essential Question

- How can systems of linear equations and inequalities be used in decision making?

photo: Getty Images

Investigations

Decision, Decisions

Road Trip! Use equations and graphs to choose the best rental car option.

Road Trip! – the Sequel

So many solution methods! Compare them side by side. Will they always work?

The Tortoise and the Hare

How many ways could the story end? Use systems of equations to analyze race results.

Which Tablet Should Kara Buy?

So many things to consider! Help Kara choose a tablet that meets her needs.

Charity Fundraiser

Linear inequalities are better models for some situations. What does the graph tell you?

Systems of Linear Inequalities

What's the point? Can systems of inequalities help you find a solution?

Key Vocabulary

boundary, elimination method, intersecting lines, simultaneous equations, solution, substitution method, system of equations

Discover

As you complete Engage and the investigations, record the most important ideas you've learned.

Engage	**Investigation 1**
Investigation 2	**Investigation 3**
Investigation 4	**Investigation 5**

Name ______________________________ Date ____________

Check for Understanding

Solve Systems of Equations and Inequalities: Investigation 1

1. Five Star Carz rents cars for \$28 plus \$0.05 per mile, and GoCars rents cars for \$25 plus \$0.07 per mile. Write a system of equations that represents the scenario. Then use substitution to create one equation, and use it to determine the number of miles that will make the cost equal at both car rentals.

A. System: $\begin{cases} \_\_\_\_\_\_\_\_\_\_ \\ \_\_\_\_\_\_\_\_\_\_ \end{cases}$

B. Substitute: ____________

C. At what number of miles will the cost the same? ____________

2. Use substitution to create a single equation for each system, and then use the equation to solve the system.

A. $\begin{cases} y = 5x - 3 \\ y = 3x - 5 \end{cases}$

Equation: ____________

Solution: (_____,_____)

B. $\begin{cases} x - y = 6 \\ 4x - y = 3 \end{cases}$

Equation: ____________

Solution: (_____,_____)

3. Graph the system and estimate the solution. Then solve for the exact solution.

$\begin{cases} 5x - y = -2 \\ x - y = -5 \end{cases}$

A. Estimated Solution:

(_____,_____)

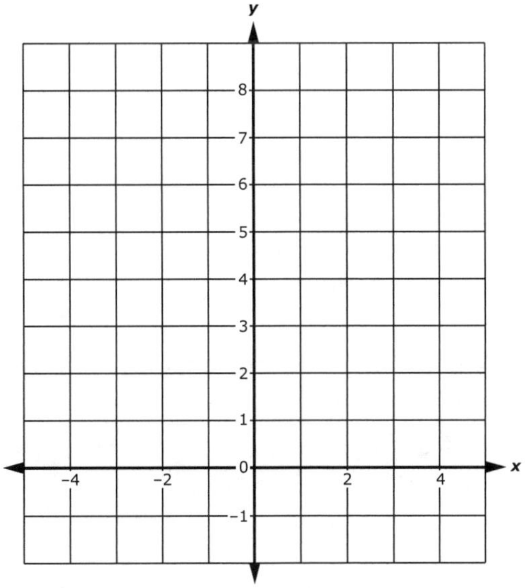

B. Exact Solution:

(_____,_____)

4. Explain what the solution to the equation $f(x) = g(x)$ means, both graphically and algebraically.

Name ______________________ Date ______________

Check for Understanding

Solve Systems of Equations and Inequalities: Investigation 2

1. The speeds of the tortoise and hare for three different races are shown. Determine when, if ever, the hare will catch up to the tortoise. Then answer question 2.

Race	Tortoise's Speed	Tortoise's Distance When Hare Begins	Hare's Speed	Hours Until Hare Catches Tortoise
A	3.2 mph	4 miles	6.4 mph	
B	2.5 mph	3 miles	4 mph	
C	1.6 mph	2 miles	1.6 mph	

2. Enter the letter(s) for the races to answer the following.

A. If the distance for each race is 5 miles, which race(s) will the tortoise win? ______________

B. At what distance will race A and race B end in a tie? ______________

C. If the tortoise has walked at 2 mph for 3 miles when the hare begins, how fast must the hare run to tie a 6-mile race? ______________

3. Match each system to the number of solutions it has.

A. $\begin{cases} y = 6x - 2 \\ y = 4x - 2 \end{cases}$

B. $\begin{cases} 3x + 6y = 9 \\ 2x + 4y = 6 \end{cases}$

C. $\begin{cases} y = -4x - 9 \\ 8x + 2y = 7 \end{cases}$

I. No Solution

II. One Solution

III. Infinitely Many Solutions

4. Given an equation, $y = mx + b$, explain how to create an equation for a second line so that the system will have no solution or infinitely many solutions.

Name ______________________________ Date ______________

Check for Understanding

Solve Systems of Equations and Inequalities: Investigation 3

1. On his old phone, Carson had stored 324 videos and 670 photos, using 22,266 MB of storage. On his current phone, he has 486 videos and 1,255 photos, using 33,849 MB.

   A. Write two equations that can be used to find the average video and photo sizes.

   Equation 1: ______________

   Equation 2: ______________

   B. To use linear combination with the simplest value, multiply Equation 1 by ______________.

   C. The combined equation is then ______________.

   D. The average video file size is _____ MB and the average photo file size is _____ MB.

   E. New phone prices increase significantly with each increase in storage capacity. To increase his usage by a similar amount, should Carson choose a storage option of 32 GB, 64 GB, or 128 GB? (1 GB = 1,000 MB) ______________.

2. Given two equations, $3.6x + 5.2y = -6$ and $-1.4x - 2.6y = -0.6$, match each step in the solution for x to the property that justifies it.

A.	$\begin{cases} 3.6x + 5.2y = -6 \\ -2.8x - 5.2 = 1.2 \end{cases}$	I.	addition property of equality
B.	$0.8x = -4.8$	II.	division property of equality
C.	$x = -6$	III.	multiplication property of equality

3. For the system below, choose the smallest values by which each equation could be multiplied to create an equivalent system that can be combined to a single equation.

System	Multiply Equation by	Equivalent System
$\begin{cases} 3x + 4y = 8 \\ 2x + 3y = 4 \end{cases}$	• ________ • ________	______________ ______________

4. When might you choose to solve by substitution, linear combination using one multiplication, or linear combination using two multiplications?

Name ______________________________ Date ____________

Check for Understanding

Solve Systems of Equations and Inequalities: Investigation 4

1. For their charity fundraiser, the photography club is doing a photo shoot at the fair. They will sell individual prints, *x*, for \$6 each and group prints, *y*, for \$9.50 each.

A. Write an inequality that represents the number of each type of print the club must sell to collect more than \$300. ____________________

B. Based on their resources, the club can create no more than 80 of both types of prints. Write an inequality based on this constraint. ________________________

2. Convert each of the following inequalities to slope-intercept form.

A. $3x + 6y > 12$ __________________ **B.** $4x - 2y \leq 10$ __________________

3. Write the inequality shown in the graph below in part A, and graph the inequality given in part B.

A. Write the inequality. __________

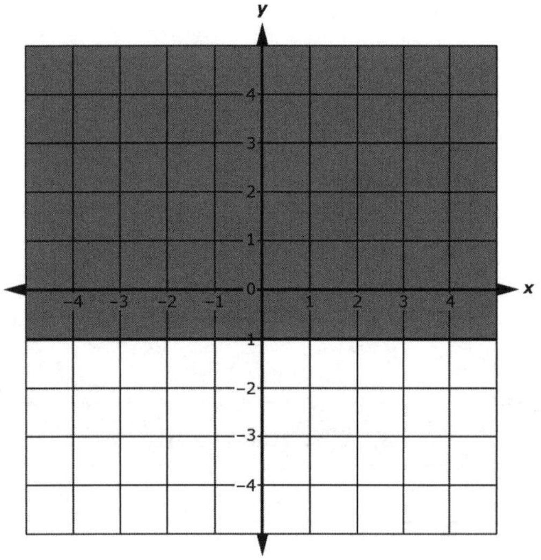

B. Graph $2x + y < 6$ on the coordinate plane.

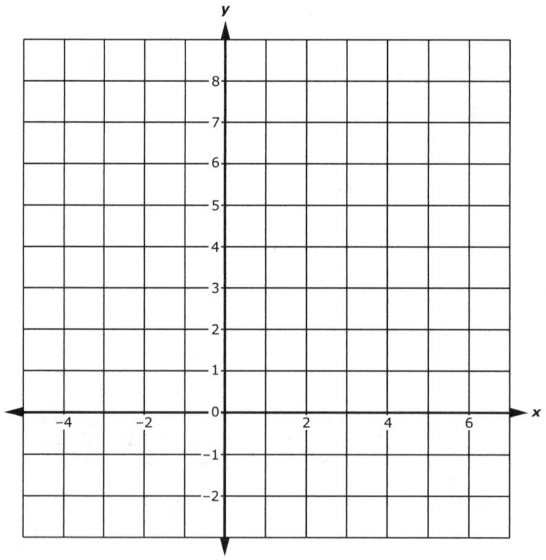

4. Given an inequality in slope-intercept form, explain how you know whether to use a dashed or solid line, and whether to shade above or below the line.

Name ______________________________ Date ____________

Check for Understanding

Solve Systems of Equations and Inequalities: Investigation 5

1. Russell will make wooden jewelry boxes for the craft fair. Small ones, x, will sell for \$10, and large, y, for \$15. He needs to sell at least \$180 worth, but he has time to make no more than 15 boxes total.

   A. Write two inequalities for a system that represents the constraints.

   ______________________ ______________________

   B. Choose the possible solutions that represent the numbers of boxes Russell can make. *Circle all that apply.*

   $(1, 10)$ $(2, 12)$ $(5, 9)$ $(6, 10)$ $(3, 9)$ $(7, 8)$ $(15, 0)$ $(0, 12)$

2. Write the equations for the system that matches the given solution set, and then graph the solution set that matches the given system.

   A. Write the inequalities for the system that is graphed below.

   ____

   ____

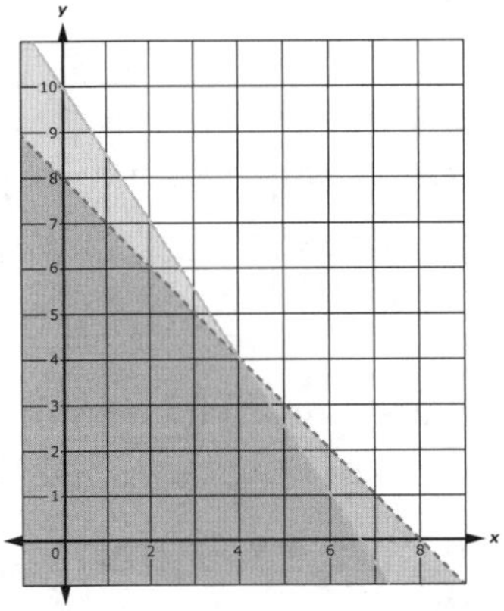

   B. Graph the solution set for the given system:

$$\begin{cases} y > 2x - 10 \\ x + 2y < 12 \end{cases}$$

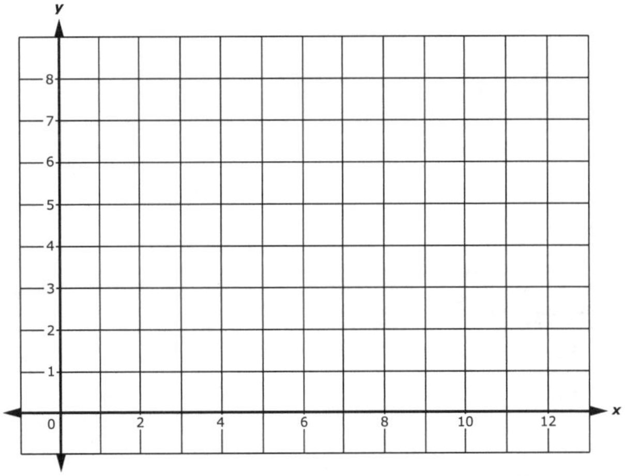

3. Explain what the x- and y-intercepts mean when they are within a solution set for an inequality. What are the minimum possible values for the domain and range of the inequalities in a system that make sense within a context such as the craft fair?

Summary

Before you attempt the Practice Exercises, review what you've learned.

Systems of Linear Equations

A system of linear equations in two variables is a set of two or more linear equations. A solution to the system is an ordered pair that satisfies each equation in the system.

Linear systems with two equations and two variables can have no, one, or infinitely many solutions.

Solving a System of Equations Graphically

EXAMPLE: Solve a System of Equations Graphically

Solve the system graphically. Then verify the solution algebraically.

$$5x + 0.5y = 19$$

$$-2x - 1.5y = -18$$

SOLUTION:

Graph each line in the system.

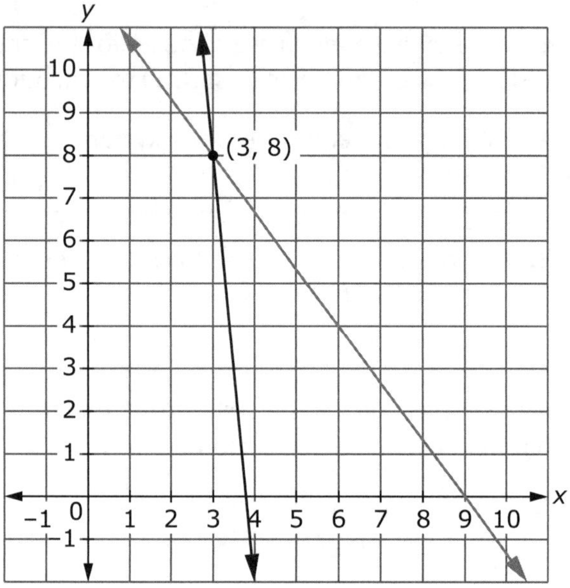

The lines appear to intersect at the point (3, 8).

To verify the solution, substitute 3 for x and 8 for y in each equation.

$5x + 0.5y = 19$	$-2x - 1.5y = -18$
$5(3) + 0.5(8) \stackrel{?}{=} 19$	$-2(3) - 1.5(8) \stackrel{?}{=} -18$
$15 + 4 \stackrel{?}{=} 19$	$-6 - 12 \stackrel{?}{=} -18$
$19 = 19$	$-18 = -18$

Both statements are true, so (3, 8) is the solution to the system.

Solving a System of Equations Algebraically

Systems can be solved algebraically using substitution or elimination.

EXAMPLE: Solve a System Using Substitution

The perimeter of a rectangular garden is 16 feet. The length of the garden is $\frac{1}{2}$ foot more than twice the width. What are the dimensions of the garden? Solve the system using substitution.

My Notes

Summary *(continued)*

SOLUTION:

Step 1: Write a system.

$2l + 2w = 16$

$l = 2w + \frac{1}{2}$

Step 2: Solve the system.

Substitute $2w + \frac{1}{2}$ into the first equation for l.

$$\begin{aligned} 2\left(2w + \tfrac{1}{2}\right) + 2w &= 16 \\ 4w + 1 + 2w &= 16 \\ 6w + 1 &= 16 \\ 6w &= 15 \\ w &= \tfrac{5}{2} \end{aligned}$$

Substitute $\frac{5}{2}$ into either equation for w.

$$\begin{aligned} l &= 2w + \tfrac{1}{2} \\ &= 2\left(\tfrac{5}{2}\right) + \tfrac{1}{2} \\ &= 5 + \tfrac{1}{2} \\ &= \tfrac{11}{2} \end{aligned}$$

The length of the garden is $\frac{11}{2}$ or $5\frac{1}{2}$ feet, and the width is $\frac{5}{2}$ feet or $2\frac{1}{2}$ feet.

EXAMPLE: Solve a System Using Linear Combination with Multiplication

Solve the system using elimination.

$-0.2x + 1.5y = -1$

$x + 4.5y = 1$

SOLUTION:

$-0.2x + 1.5y = -1$

$x + 4.5y = 1$

Multiply one of the equations by a value so that one of the variables will be eliminated.	$0.6x - 4.5y = 3$ $x + 4.5y = 1$
Add the equations. (Add the x-terms, add the y-terms, and add the constants.)	$1.6x + 0y = 4$
If not already done, write the equation with only one variable term.	$1.6x = 4$
Solve for the variable.	$x = \frac{5}{2}$
Substitute the value from the previous step into one of the original equations and solve for the other variable.	$(2.5) + 4.5y = 1$ $4.5y = -1.5$ $y = -\frac{1}{3}$
Write the solution as an ordered pair.	$\left(\frac{5}{2}, -\frac{1}{3}\right)$

My Notes

Summary *(continued)*

When one of the equations in a system is already solved for one of the variables or when one of the variables has a coefficient of 1, it may be more efficient to use substitution.

In this system, the first equation is solved for x.	In this system, the coefficient of y in the second equation is 1.
$x = 2y - 17$ $2x + 3y = 10$	$0.5x - 3.5y = 2$ $2.5x + y = 15$

Systems of Linear Inequalities

A system of linear inequalities in two variables is a set of two or more linear inequalities. A solution to the system is an ordered pair that satisfies each inequality in the system.

To solve a system of linear inequalities, graph each inequality in the system. The intersection of the half-planes is the solution to the system.

EXAMPLE: Solve a System of Inequalities

Tomika has two jobs. Her job at the mall pays \$8.50 per hour, and her job babysitting pays \$10 per hour. She wants to work no more than 25 hours this week. How many hours at each job could she work if she wants to earn at least \$125 this week?

SOLUTION:

Step 1: Write a system. Let m be the number of hours she works at her mall job and b be the number of hours she works at her babysitting job.

$$8.5m + 10b \geq 125$$
$$m + b \leq 25$$

Step 2: Solve the system by graphing. The solutions lie in the overlap of the shaded regions.

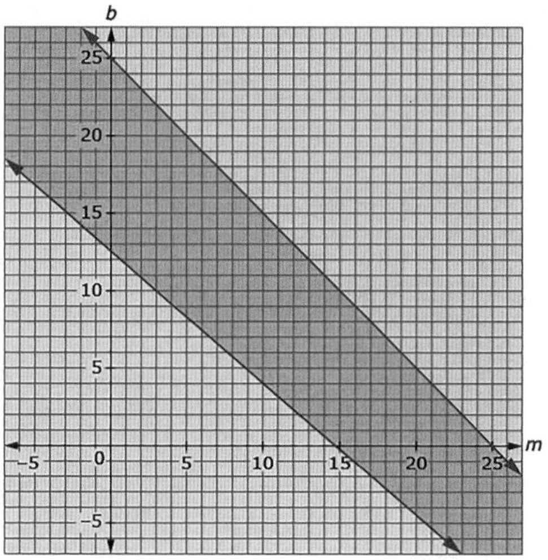

Tomika can work any number of hours in the shaded region. For example, the point (16, 2) is in the shaded region, so she could work 16 hours at the mall and 2 hours babysitting and meet her earnings goal.

EXAMPLE: A System of Linear Inequalities with No Solution

Solve the system graphically.

$$-0.5x + 1.3y \geq 5.5$$
$$x - 2.6y \geq 2$$

SOLUTION:

Graph both inequalities. The graphs of the solutions do not appear to intersect. Compare the slopes of both boundary lines. Since the slope of each boundary lines is $\frac{5}{13}$, the lines are parallel. This system has no solution.

My Notes

Practice Exercises

Review what you've learned using these practice problems. For practice problems with feedback, try the Coach and Play items in the Practice section online.

1. Find the exact solution for the following system of equations.

   $2x + 3y = 4$

   $8x + 6y = 11$

   Solution: (__________, __________)

2. Use an algebraic method to find the solution to the system of equations.

   $y = \frac{1}{3}x - 6$

   $y = -3x + 4$

   $(x, y) =$ (__________, __________)

3. Mary Beth decided to use elimination to solve a system of equations. Complete Mary Beth's work to solve the system.

   $-3(-x + 2y = -6)$

   $3x + 6y = 6$

   A. $3x - 6y =$ __________

   $3x + 6y = 6$

   B. $6x =$ __________

   C. $(x, y) =$ (__________, __________)

4. Consider the following system of equations:

   $y = -2x + 7$

   $2x + y = -3$

   Complete the solution to the system by choosing from the options provided.

   $2x +$ (__________) $= -3$

   __________ $= -3$

   The system of equations has __________.

   A. -3 **B.** 7 **C.** $-2x + 7$ **D.** $2x + y$

   E. infinitely many solutions

   F. no solution

   G. one solution

5. A hockey team will play no more than 10 games during their regular season. For each game, they are awarded 2 points for a win, and 1 point for a tie. To make the playoffs, a team will need to score at least 12 points during their regular season.

   Let x represent the number of games the team will win and y the number of ties.

   On the coordinate plane, graph the solution set for the number of wins and ties that will qualify a team for the playoffs.

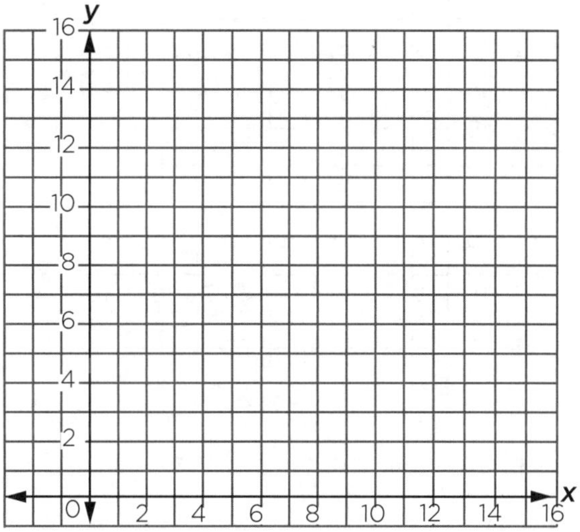

6. Use a graphing calculator or pencil and paper to graph the following equations.

   $y = x + 2$

   $y = 2x - 4$

   Complete the statement below.

   The point of intersection on the graph of the given equations is the ordered pair (______, ______).

Practice Exercises *(continued)*

7. A soccer team is selling shirts and ball caps to raise money to buy new uniforms. Its profit is \$8 for each shirt and \$5 for each ball cap. The team needs to raise at least \$400.

   Let x represent the number of shirts sold and let y represent the number of caps sold.

   On the coordinate plane, graph an inequality to represent the constraint. Then shade the region that represents the solutions to the inequality.

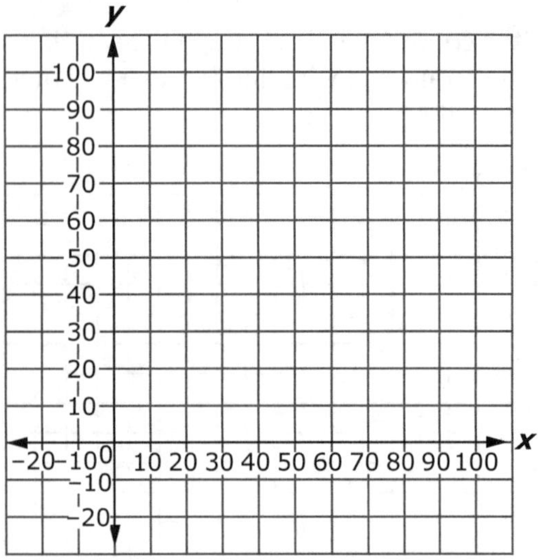

8. Graph the system of inequalities.

   $2y > -x + 1$
   $y + 0.5x < -4$

   Which is a solution of the system? *Select all that apply.*

   A. (0, 0)

   B. (2, 3)

   C. (10, −10)

   D. (−6, 2)

   E. There is no solution.

9. Graph the system of inequalities.

   $y < 3x + 6$
   $x - 2y > -7$

   Which is a solution of the system? *Select all that apply.*

   A. (0, 0)

   B. (2, 5)

   C. (5, −6)

   D. (−7, −4)

   E. There is no solution.

10. Let x represent the number of hours and y represent the number of minutes Dave spent on his history project. When solving the following system of equations, what is the number of hours Dave spent working on his project?

    $x - 2y = 2$
    $3x + y = 6$

    A. 0 hours

    B. 1 hours

    C. 2 hours

    D. 3 hours

11. Let x represent the number of batteries Mrs. King has for her calculators, y. How many batteries does Mrs. King have?
    $x = 4y$
    $3y = -2x + 44$

    A. 4

    B. 7

    C. 13

    D. 16

Apply

Can Selling the Pizza Bring in the Dough?

photo: Getty Images

Different clubs at school can sign up to sell food at after school events. The clubs hope to make some money for their own activities, but most are happy just to break even.

The Mathletes want to sell slices of pizza during a pep rally. They will need to buy the pizzas, plates, and napkins. They hope to do better than break even. Their goal is to earn an extra $50 for a Senior Prize.

Research local costs of pizza, plates, and napkins from at least two sources. Create two pricing options and model the income and expenses of both with equations and graphs. Compare and analyze your models so that you can advise the Mathletes on their plan. How much pizza, and at what price, do they need to sell in each plan to (1) break even and (2) reach their goal?

Show what you've learned by completing the other performance tasks in the online Apply section.

Apply *(continued)*

Your answer to Apply will be assessed on the following criteria:

1. Estimating the number of plates and napkins needed and researching and recording the costs of the pizza, plates, and napkins, with at least two sources cited
2. Writing and graphing systems of equations to represent income and expenses and interpreting the graphs in the context of the problem
3. Solving the systems and interpreting the solutions in the context of the problem
4. Finding the number of pizzas that need to be sold and the necessary price per slice for each plan and writing a proposal to the club

Scale / Criteria	4	3	2	1	0
	Exceeds Criteria	Meets Criteria	Progressing to Criteria	Below Expectations	No Expectation
Cost	Presents costs from two cited sources, estimates a reasonable number of paper goods, and includes an explanation.	Presents costs from two cited sources and estimates a fairly reasonable number of paper goods.	Presents the cost from one cited source and estimates a fairly reasonable number of paper goods.	Presents costs but does not support them with citations; estimates an unreasonable number of paper goods.	Does not attempt task.
Equations	Writes and graphs a system of equations to represent income and expenses; interprets the graphs in the context of the problem.	Makes a mistake in one equation; graphs match the equations.	Determines equations, but has inaccuracies while determining income and graphing the equations.	Determines a plan but has major inaccuracies in the equation and graph for the plan.	Does not attempt task.
Success Point for Plans	Uses the information from the equations and graphs to determine what needs to happen not to lose money and to make $50.	Makes minor errors when using the information from the equations and graphs to determine what needs to happen not to lose money and to make $50.	Makes errors determining what needs to happen not to lose money and to make $50; errors result in incorrect information about the results of a particular plan.	Makes major errors using the information from the equations and graphs, making the results of the inquiries totally inaccurate.	Does not attempt task.
Solution and Justification	Proposes a reasonable solution and justifies its reasonableness mathematically.	Proposes a reasonable solution but may make minor errors in justifying its reasonableness.	Proposes a solution which may contain some inaccuracies or has an incomplete or inaccurate justification.	Proposes a plan that has major inaccuracies and its justification also contains inaccuracies.	Does not attempt task.

UNIT 5: Systems of Equations and Inequalities

5.2 Use Systems in Decision Making: Linear Programming

photo: Getty Images

Lesson Objectives

- Use systems of inequalities to represent real-world situations involving constraints.
- Interpret and solve systems of inequalities involving constraints.

Essential Question

- How can you use linear programming to budget resources?

photo: Getty Images

Investigations

The Hardware Store

Maximize profit! Considering the variables, what's possible?

The Hardware Store Revisited

What are the profit limits? Can you constrain the variables?

Hay Is for Horses

Good nutrition has many variables, even for horses. Help a farmer minimize costs.

Cake Shop

How many cake decorators does it take to complete orders without making a mess of the budget?

Gigi's Bakery

Considering costs of ingredients, how can Gigi get the most bread from her bread?

Key Vocabulary

boundary, feasible region, intersection, solution

Discover

As you complete Engage and the investigations, record the most important ideas you've learned.

Engage	
Investigation 1	**Investigation 2**
Investigation 3	**Investigation 4**

Name ______________________________ Date ____________

Check for Understanding

Use Systems in Decision Making: Linear Programming: Investigation 1

A camp manager needs to hire counselors, *c*, and counselors-in-training, or trainees, *t*, to staff summer camps. She will pay each counselor $16 per hour and each trainee $12 per hour.

Answer questions 1–3 to help her decide how many of each she should hire to minimize her costs for staff.

1. Write an inequality for each constraint, and write the objective function.

A. Each counselor should have at least 1 trainee. ____________

B. The manager will hire no more than 14 trainees. ____________

C. The manager will need at least 18 counselors and trainees. ____________

D. The camp has room for at most 22 staff. ____________

E. Write the objective function, *C*. ____________

2. Graph the system of inequalities and shade the area containing all possible solutions. Then list the corner points (vertices) of this area (the feasible region).

A. Graph of the system:

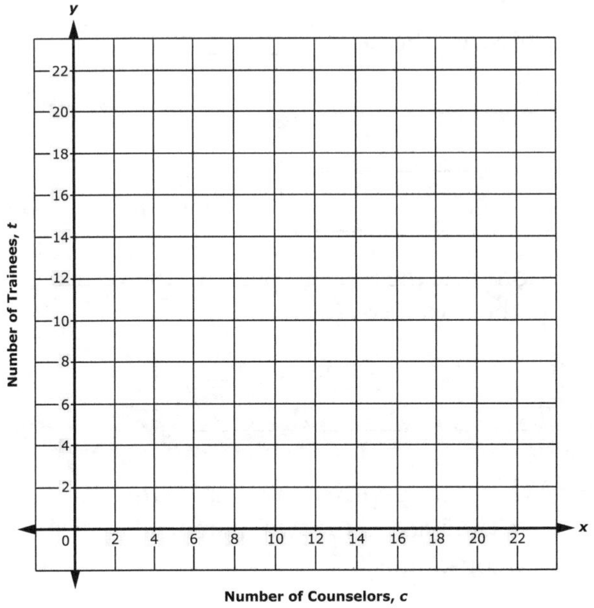

B. Vertices of the feasible region:

I. ____________ **II.** ____________

III. ____________ **IV.** ____________

3. How many of each type of staff should the manager hire to minimize cost?

A. Number of counselors: ____________ **B.** Number of trainees: ____________

4. If the vertices of a feasible region had non-integer coordinates for parameters that must have integer values, what values should be used to evaluate the objective function?

Name ______________________________ Date ______________

Check for Understanding

Use Systems in Decision Making: Linear Programming: Investigation 2

A deli wants to minimize the cost of turkey and Swiss cheese for their most popular sandwich, while advertising that it is high in protein and low in calories.

- One ounce of their turkey has 4 grams of protein and 30 calories.
- One ounce of Swiss cheese has 6 grams of protein and 120 calories.
- A sandwich must have at least 1 ounce but no more than 2 ounces of turkey.
- Each sandwich will have at least 1 ounce of Swiss cheese.
- A sandwich must have no more than 200 calories and at least 12 grams of protein.
- The deli's cost for turkey is 37.5¢ per ounce and for Swiss cheese, 26.8¢ per ounce.

1. Define the quantities and write a system of inequalities to represent the constraints. *Circle the correct units of measure and fill in the blanks.*

   A. Let x = the [cost / calories / ounces / grams of protein] of turkey.

   Let y = the [cost / calories / ounces / grams of protein] of Swiss cheese.

   B. Write the constraints as inequalities. Use a compound inequality if appropriate.

   Ounces of turkey: ______________ Ounces of Swiss cheese: ______________

   Total number of calories: ______________________________

   Total grams of protein: ______________________________

   C. Write the objective function. C = ________________________

2. Shade the feasible region on the graph.

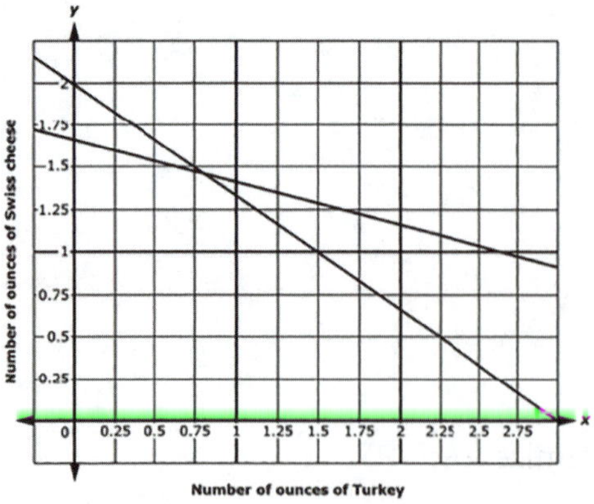

3. Complete the ordered pairs for the vertices of the feasible region.

   (_____, 1.33), (_____, 1.42),

   (_____, 1), (_____, 1), (_____, 1.17)

4. What is the approximate minimum cost per sandwich of the two ingredients, given the constraints?

   ____

5. Explain why the vertices of a feasible region are the only points that can minimize or maximize the objective function for a set of constraints.

Name ______________________________ Date ____________

Check for Understanding

Use Systems in Decision Making: Linear Programming: Investigation 3

A facility trains service dogs for a period of one year prior to selling the dogs. Consider the following constraints on the current group.

- The staff can train up to 16 adult dogs and 12 puppies at a time.
- Adult dogs require 600 hours of training per year and puppies require 800.
- An adult dog eats 1.2 pounds of food per day, and a puppy eats 0.8 pounds.
- The facility can provide up to 12,000 hours of training per year and 16.8 pounds of food per day.
- Their profit is $3,500 per adult dog and $2,500 per puppy.

1. Define the appropriate quantities for the variables.

Let x = the ______________________________.

Let y = the ______________________________.

2. Write the inequalities that represent the constraints.

______________________ ______________________

______________________ ______________________

______________________ ______________________

3. Graph the solution set for the system. Then fill in the blanks to complete parts B-D.

A. Graph and lightly shade the feasible region.

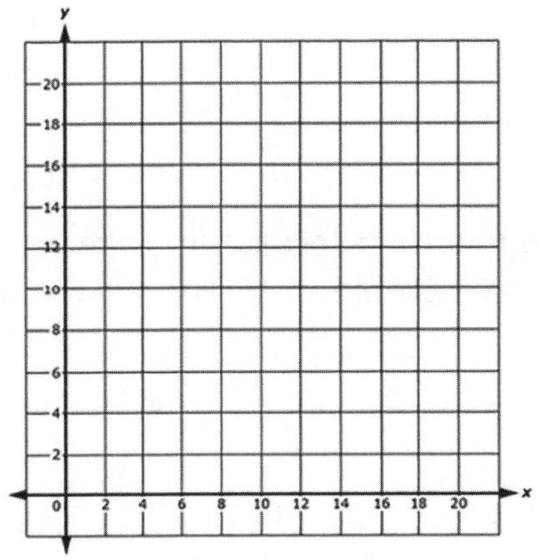

B. Identify the vertices of the feasible region.

(_____, _____), (_____, _____),
(_____, _____), (_____, _____),
(_____, _____)

C. Write the objective function.

$P =$ ______________________

D. The facility should train _____ adult dogs and _____ puppies for a maximum profit of $__________.

4. Is it possible that the coordinates that optimize an objective function for a set of constraints will be on one of the axes? Why or why not?

Name ______________________________ Date ______________

Check for Understanding

Use Systems in Decision Making: Linear Programming: Investigation 4

A bakery will bake and decorate ginger cookies and sugar cookies to sell at the winter festival. They have created a system of inequalities for the constraints and the objective function:

- g represents the dozens of ginger cookies the bakery can make.
- s represents the dozens of sugar cookies the bakery can make.
- P represents the profit they can make from selling the cookies.
- $g \geq 0$ and $g \leq 35$
- $s \geq 0$ and $s \leq 50$
- Constraint for mixing and baking time: $10g + 12s \geq 360$
- Constraint for decorating time: $5g + 8s \leq 480$
- Objective Function: $P = 48g + 36s$

1. Match each description to its value in the constraints.

A. mixing and baking time for a dozen sugar cookies

B. time required to decorate a dozen gingerbread cookies

C. maximum number of sugar cookies that can be made

D. maximum time allowed for decorating cookies

E. profit from the sale of a dozen gingerbread cookies

I. 5 minutes
II. 6 hours
III. 8 minutes
IV. 8 hours
V. $8
VI. 10 minutes
VII. 12 dozen
VIII. 12 minutes
IX. $36
X. $48
XI. 50 dozen

2. When the bakery graphed the system and found the feasible region, they decided to make 16 dozen gingerbread cookies. If they sold all the cookies they made, and the profit was $1,200, how many dozen sugar cookies did they make? *Fill in the blank.*

They made _________ dozen sugar cookies.

3. Explain why the vertices of the feasible region are the only points that can maximize or minimize an objective function.

Summary

Before you attempt the Practice Exercises, review what you've learned.

Linear Programming

In a linear programming problem, the goal is to maximize or minimize the outcome based on the conditions of the problem.

> An **objective function** is an algebraic statement that expresses a plan to maximize or minimize the problem within the constraints. **Constraints** are a system of linear inequalities that set the limitations on where the optimal solution may be found.

Setting up a linear programming problem requires three steps:

- Define the variables.
- Write the objective function.
- Define the constraints.

EXAMPLE: Setting Up a Linear Programming Problem

A bakery sells regular banana bread and gluten-free banana bread. The cost to make a loaf of banana bread is $1.50, while the cost to make a loaf of gluten-free banana bread is $4.00. The baker can make as many as 24 loaves of each type of bread per day, and he knows that he can sell at least 6 gluten-free loaves. Regular banana bread sells for $4.50 per loaf, while gluten-free bread sells for $9.50 per loaf. The baker will place a $48.00 order for ingredients to make the two types of bread. Set up the constraints and the objective function that will allow the baker to find his maximum profit.

Set up the constraints and the objective function that will allow the baker to find his maximum profit.

SOLUTION:

First, define the variables.

x = loaves of regular banana bread

y = loaves of gluten-free banana bread

Next, write the objective function. It costs $1.50 to make a loaf of regular banana bread, and it sells for $4.50, so the profit is $4.50 – $1.50 = $3.00. It costs $4.00 to make a loaf of gluten-free banana bread, and it sells for $9.50, so the profit is $9.50 – $4.00 = $5.50. The objective function for the baker's profit is $P = 3x + 5.50y$.

Then, define the constraints.

The baker can make no more than 24 loaves of each type of bread.	$x \leq 24$ $y \leq 24$
He cannot make a negative number of regular loaves of bread.	$x \geq 0$
He will make at least 6 loaves of gluten-free bread.	$y \geq 6$
It costs $1.50 to make a loaf of regular bread and $4.00 to make a loaf of gluten-free bread. The total cost of making the loaves cannot be more than $48.00.	$1.5x + 4y \leq 48$

My Notes

Summary *(continued)*

Once the variables, objective function, and constraints are defined, the linear program is ready to be graphed and then solved.

> The region of all possible points that is defined by the constraints (the inequalities) is called the **feasible region.**

The points along the boundary lines of the feasible region where two or more constraints intersect are called the vertices or corner points.

The solutions to the objective function are found at the corner points of the feasible region.

After setting up the linear programming problem, solving the problem requires four steps:

- Graph the constraints.
- Identify the coordinates of each corner point.
- Evaluate the objective function at each corner point.
- Compare those values to find the minimum or maximum value, depending on the goal of the problem.

My Notes

EXAMPLE: Solving a Linear Programming Problem

How many loaves of regular banana bread and how many loaves of gluten-free banana bread should the baker make to earn the most profit? What is the most profit he can make?

SOLUTION:

First, graph the constraints.

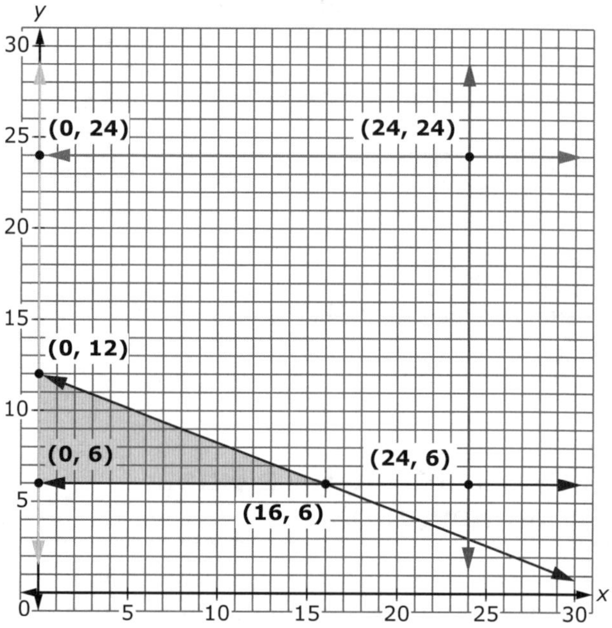

Then, identify the corner points. Finally, evaluate the objective function at each corner point.

Note that (0, 24), (24, 6), and (24, 24) are corner points, but they are outside the feasible region.

x	y	$P = 3x + 5.5y$
0	6	33
0	12	66
16	6	81

If the baker orders the supplies to make 16 loaves of regular banana bread and 6 loaves of gluten-free banana bread and sells all the loaves he makes, he will make the maximum profit of $81.00.

Practice Exercises

Review what you've learned using these practice problems. For practice problems with feedback, try the Coach and Play items in the Practice section online.

1. A local farmer's cooperative has arranged to grow tomatoes on a 2,100-square-foot plot of land to donate to the local community kitchen.

The club plans to grow two types of tomatoes—bush tomatoes, which take up 25 square feet of space per plant, and climbing tomatoes, which take up 20 square feet per plant. The goal is to grow at least twice as many bush type as climbing type.

Let x represent the number of bush plants and let y represent the number of climbing plants. Choose the inequalities that represent the constraints for this situation. *Select all that apply.*

A. $x \geq 0, y \geq 0$ **B.** $x \geq 2y$

C. $20x + 25y \leq 2{,}100$ **D.** $25x + 20y \leq 2{,}100$

E. $2x \geq y$

2. A local community group has arranged to grow tomatoes and peppers on a 210-square-foot plot of land. Tomatoes take up 4 square feet of space per plant, and peppers take up 2 square feet. The goal is grow at least twice as many tomatoes as peppers.

Let x represent the number of tomatoes and let y represent the number of peppers. Determine which of the following are possible solutions (x,y). *Select all that apply.*

A. (20, 10) **B.** (20, 20)

C. (30, 60) **D.** (102, 45)

E. (30, 15)

3. Aaron is hosting a birthday party for a friend. He wants to have ice cream and frozen yogurts. One serving of ice cream has 7 grams of fat, 30 mg of cholesterol, and costs \$1 per serving. The yogurt has 5 grams of fat, 6 mg cholesterol, and costs \$0.80 per serving.

Aaron plans on needing at least 14 servings. He want to have at least 3 servings of both ice cream and yogurt, but no more than 90 grams of fat, and he wants to minimize the cost.

The minimum cost of the ice cream and yogurt is \$__________.

4. Graph the constraints shown by the system of inequalities below. Then indicate by shading the region that corresponds to the feasible region.

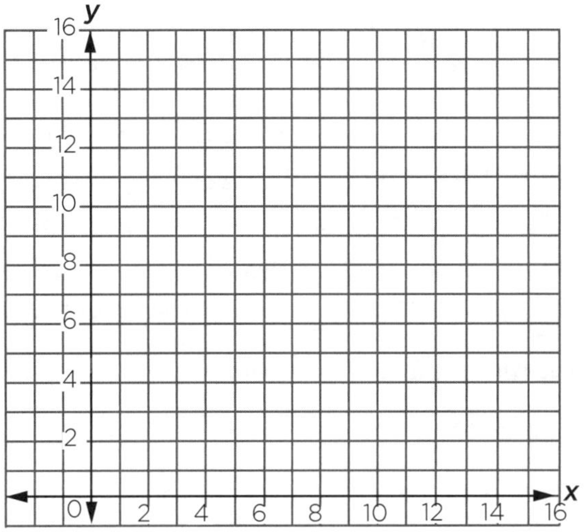

$x + y \leq 8$

$y \geq 2x - 6$

$x \geq 0$

$y \geq 0$

$y \leq 5$

5. Vania is mixing feed to optimize the ratio of oats to beet pulp to feed her horses. She wants a daily mix that includes at least 50 grams of oats and 25 grams of beet pulp.

Brand *A* has 20 grams of oats and 5 grams of beet pulp for \$0.65 per unit.

Brand *B* has 10 grams of oats and 15 grams of beet pulp for \$0.90 per unit.

How much of each type of feed should she use to minimize her cost?

The minimum daily cost per unit is \$__________.

__________ unit(s) of Brand *A* and

__________ unit(s) of Brand *B* should be mixed together.

Practice Exercises *(continued)*

6. The owner of a small business has two employees, Thoma and Marty. The owner needs coverage for 50 hours a week.

   The cost of labor and benefits for Thoma is $14 an hour, and the cost for labor and benefits for Marty is $17 an hour.

   The owner used linear programming to analyze the constraints and came up with the objective function $z = 14x + 17y$, where x is the number of hours Thoma works, y is the number of hours Marty works, and z is the total cost.

   Several feasible points (x, y) are determined: (30, 20), (23, 27), (15, 35), and (20, 30). Use the objective function to minimize the cost of labor.

   The minimum cost of labor is $__________.

   Thoma would work __________ hours.

   Marty would work __________ hours.

7. The director at Camp Cedar Ridge is hiring counselors and instructors for the summer season. There are several constraints that the director must follow in hiring staff.

   Suppose x is the number of counselors and y is the number of instructors that the camp director can hire.

   For each statement given below, complete the inequality to represent the camp director's constraints on hiring.

   The director can hire a maximum of 30 staff: __________ ≤ 30.

   The number of counselors cannot exceed twice the number of instructors: x __________.

   At least 8 counselors must be hired: x __________.

   No more than 15 instructors are needed: y __________ and y __________.

8. Charlie referees soccer games on the weekend. He can referee up to 8 games in one day. He has to allow 30 minutes for the U8 games, and he gets paid $15 to referee that age level. He has to allow 60 minutes for U12 games, and he gets paid $35 for that age level.

   Let x represent the number of U8 games and y represent the number of U12 games. If the games occur within 6 hours, or 360 minutes, determine what number of U8 and U12 games he should sign up to referee in order to make the most money for that day.

   The maximum income for refereeing in one day is $__________.

   Charlie should referee __________ U8 games.

   Charlie should referee __________ U12 games.

9. Nicole decides to make two sizes of bottles of BBQ sauce, 16 ounces and 8 ounces. Her total production is 1,280 ounces for the summer. She has room for up to 100 bottles in her booth at the farmers market and wants to maximize her income from their sale.

   Let x represent the number of 16-ounce bottles sold and y represent the number of 8-ounce bottles. The 16-ounce bottles sell for $25 and the 8-ounce bottles sell for $15.

   The maximum income from the sale of BBQ sauce is $__________.

   Nicole should sell __________ 16-ounce bottles of sauce.

   Nicole should sell __________ 8-ounce bottles of sauce.

Apply

How Much Solar Power Should Your Community Purchase?

A local utility company, HC Electric, wants to reduce its greenhouse gas emissions by purchasing some of its power from a solar power plant.

A solar power plant with a 150-megawatt capacity may have an average power production closer to 50 megawatts.

The chart shows how HC Electric's present power sources compare to the solar power plant.

Comparison of Power Sources		
Measure	Present Sources	Solar Power Plant
Cost per megawatt-hour (Mwh)	$41.97	$55.00
CO_2 equivalent per megawatt-hour	468 kg	46 kg

- CO_2 equivalent is a way of describing a mixture of greenhouse gases. It represents the amount of carbon dioxide (CO_2) that would have the same effect on the atmosphere as the mixture of greenhouse gases.
- A megawatt-hour is the equivalent of one megawatt of electricity used continuously for one hour.

HC Electric must take into account the following information when making its decision:

- The company predicts that it will need to provide its customers with at least 12,800,000 megawatt-hours and no more than 13,440,000 megawatt-hours of energy per year over the next several years.
- It needs to keep its average cost per megawatt-hour (from present sources and the solar power plant) to no more than $45.
- For every 1 megawatt of power capacity that the utility purchases from the solar power plant, it can expect to receive 25,840 megawatt-hours of energy per year.
- As part of its green energy community initiative, it has agreed to purchase at least 1,000,000 megawatt-hours generated from solar energy.

How many megawatts of power capacity should HC Electric purchase from the solar power plant to minimize its greenhouse gas emissions and meet its other constraints?

Show what you've learned by completing the other performance tasks in the online Apply section.

Apply *(continued)*

Your answer to Apply will be assessed on the following criteria:

1. Writing a system of inequalities to model the constraints on the electric utility
2. Using linear programming to minimize the utility's greenhouse gas emissions
3. Deciding how many megawatts of power capacity the utility should purchase from the solar power plant and defending your decision
4. Showing your work and explaining your reasoning

Scale / Criteria	4	3	2	1	0
	Exceeds Criteria	Meets Criteria	Progressing to Criteria	Below Expectations	No Expectation
System of Inequalities	Correctly writes a system of inequalities to model all constraints, and defines all variables used.	Writes a system of inequalities to model all constraints, and defines all variables used, but makes a minor error.	Writes a system of inequalities to model most constraints and defines all variables, but omits one inequality or has errors in two of the inequalities.	Attempts to model the constraints with inequalities, but omits more than one inequality or includes errors in more than two of the inequalities.	Does not write a system of inequalities.
Emissions Function	Correctly writes a function to model the greenhouse gas emissions generated by the utility, and applies linear programming to minimize the function.	Correctly writes a function to model the emissions, but makes minor computation errors when minimizing the function.	Makes an error when writing an emissions function, but minimizes this function correctly; or correctly writes an emissions function, but makes a significant error when minimizing it.	Attempts to write an emissions function and minimize it, but does not apply the correct values when modeling or minimizing the function.	Does not write a function or attempt linear programming.
Number of Megawatts	Correctly determines how many megawatts of power capacity the utility should purchase from the solar power plant, and justifies this decision mathematically.	Determines how many megawatts of solar power capacity the utility should purchase, but bases the decision on work that includes a minor computation error.	Determines how many megawatts of solar power capacity the utility should purchase, but bases the decision on work that includes a significant error in computation or reasoning.	Gives a reasonable value for the amount of power capacity the utility should purchase, but does not base the decision on the results of minimizing the emissions function.	Does not determine the number of megawatts the utility should purchase.
Work and Reasoning	Shows all work and explains all reasoning; explanations are both clear and valid.	Shows all work and explains all reasoning; work or explanations include minor errors.	Shows most work and attempts to explain reasoning; explanations show some of understanding of concepts of linear programming.	Shows very little, if any, work or explanations.	Does not show work or explain reasoning.

UNIT 6: Descriptive Statistics

6.1 Represent and Analyze Data

photo: Getty Images

Lesson Objectives

- Choose a representation that best illustrates data in terms of context.
- Compare the center and spread of data sets using statistical displays appropriate to the shape of the data distributions.
- Interpret differences in shape, center, and spread in the context of data sets.

Essential Question

- What do different displays reveal about data sets?

photo: Getty Images

Investigations

Are Cereals Really That Different?

Use data displays to detail the differences in several brands of cereals.

A Picture Is Worth…

What does it mean? What story do data displays tell?

How Do Cereals Compare?

How much does the data vary? Make comparisons and draw conclusions.

Take Another Look

What can deviation tell you about variance? What does variance tell you about a data set?

Drawing a Conclusion

Analyze the data! Investigate the claim that one store's cereals are more healthful than another's.

Key Vocabulary

box plot (also known as box-and-whisker plot), histogram, mean absolute deviation (MAD), outlier, standard deviation, variability, variance

Discover

As you complete Engage and the investigations, record the most important ideas you've learned.

Engage	
Investigation 1	Investigation 2
Investigation 3	Investigation 4

Name ______________________________ Date ____________

Check for Understanding

Represent and Analyze Data: Investigation 1

The resting and exercising heart rates were taken for 35 people.

Use the displays to answer questions 1–2.

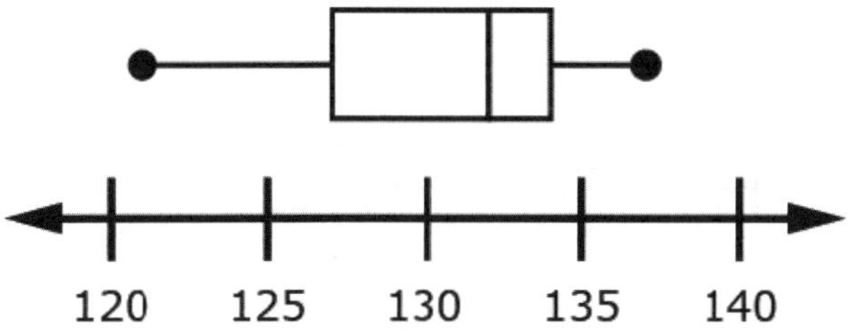

Exercising Heart Rates

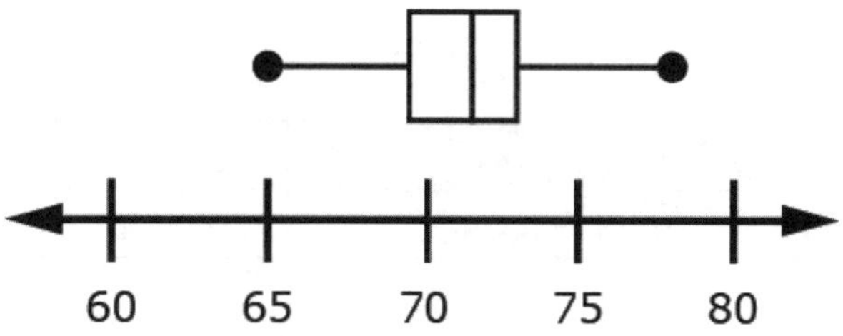

Resting Heart Rates

1. *Circle the terms that best complete the statement.*

   The distribution of the Resting Heart Rates set is [skewed left / skewed right / symmetric] and the Exercising Heart Rates set is [skewed left / skewed right / symmetric].

2. Choose the true statements from the following. *Select all that apply.*

   A. The mean and median for the Resting Heart Rates are very close in value.

   B. The mean and the median are equally good indicators of the typical Exercising Heart Rate.

   C. There is greater variability among the Exercising Heart Rates.

   D. The IQR for Resting Heart Rates as a percentage of its range is less than the IQR for Exercising Heart Rates as a percentage of its range.

   E. If a value of 92 were added to the Resting Heart Rates, the data would be skewed left.

   F. There are twice as many people with exercising heart rates from 121 to 126 as there are people with exercising heart rates from 134 to 137.

   G. The median for the Exercising Heart Rates is greater than the median for the Resting Heart Rates.

3. Explain what effect, if any, removing the extreme data point from a set will have on the measures of center, spread, and variability for the set.

Name ______________________________ Date ____________

Check for Understanding

Represent and Analyze Data: Investigation 2

Melanie is researching summer jobs with a jewelry store and a hobby store at the mall, and she wants to work as near to 20 hours per week as possible. She speaks with the employees at each store, and records their weekly hours in the table shown.

Use the data to answer questions 1–3.

Jewelry Store (hr/wk)	21	28	12	22	18	8	15	24	14
Hobby Store (hr/wk)	28	26	16	27	11	22	21	10	28

1. Compare the measures of center. *Fill in the blanks to complete each statement.*

A. The jewelry store employees work a median of ________ hours and a mean of ________ hours per week.

B. The hobby store employees work a median of ________ hours and a mean of ________ hours per week.

2. Compare the spread for each data set. *Circle the correct terms and fill in the blanks to complete the statements.*

A. The IQR for the jewelry store data is [less than / greater than / equal to] the IQR for the hobby store data.

B. The MAD for the jewelry store data is [less than / greater than / equal to] the MAD for the hobby store data.

C. The IQR as a percentage of the range for the jewelry store is ________, and the MAD as a percentage of the range is approximately ________.

D. The IQR as a percentage of the range for the hobby store is ________, and the MAD as a percentage of the range is approximately ________.

E. The better measure of spread for the jewelry store is the [IQR / MAD], and the better measure for the hobby store is the [IQR / MAD].

F. The store with the greater variability in hours is the [jewelry / hobby] store.

3. Based on the measures of center and spread, make a recommendation to Melanie for which job better meets her goal for working 20 hours per week and explain why.

Name ______________________________ Date ____________

Check for Understanding

Represent and Analyze Data: Investigation 3

Luca and his friends record prices for a bucket of popcorn and a large soda at 8 theaters.

Use their data to answer questions 1–2.

Popcorn ($)	8.25	7.95	8.00	8.20	8.25	8.60	8.30	8.45
Soda ($)	6.35	6.55	6.50	6.45	6.60	6.50	6.40	6.55

1. Match each statistic to its approximate value. *Not all values will be used.*

A. The mean price of a bucket of popcorn — **I.** 8.25

B. The mean price of a large soda — **II.** 0.2

C. The median for the price of popcorn — **III.** 6.50

D. The median for the price of a large soda — **IV.** 0.28

E. The standard deviation for the price of popcorn — **V.** 8.15

F. The standard deviation for the price of a large soda — **VI.** 0.13

G. The IQR for a bucket of popcorn — **VII.** 0.08

H. The IQR for a large soda — **VIII.** 6.49

2. Fill in the blanks or circle the correct term to answer each of the following.

A. What percent of the popcorn prices is within a standard deviation for that data set? ________

B. What percent of the soda prices is included in a standard deviation for that data set? ________

C. A histogram for the popcorn prices would show the distribution for the set to be [skewed left / skewed right / symmetric].

D. A histogram for the soda prices would show the distribution for the set to be [skewed left / skewed right / symmetric].

E. For which set is the mean the better measure of center? ________

3. Explain what the size of the standard deviation reveals about the variation of the data set, and when it is best to use it as a measure of variability.

Name ______________________________ Date ____________

Check for Understanding

Represent and Analyze Data: Investigation 4

Ms. Hardy's engineering class named and tested two spud launcher designs to decide which to enter in the state contest. The results of the test launch distances for each, in meters, are shown below.

Use the data in the table to answer questions 1–4.

Spudnik (m)	298	233	243	263	222	242	225	239
Later Tater (m)	225	234	241	229	253	246	255	265

1. Fill in the statistics for each of the launchers, rounded to the nearest tenth.

	Mean Distance (m)	Median Distance (m)	Standard Deviation	IQR
Spudnik				
Later Tater				

2. Choose the true statements from the following. *Select all that apply.*

A. The distribution of the distances for Spudnik is skewed left because of the outlier.

B. The distribution of the distances for Later Tater is symmetric because the mean and median values are equal.

C. The standard deviation for Spudnik is greater than for Later Tater, so the mean is a better measure of center for the Spudnik distances.

D. The mean and standard deviation are better measures of center and variability for the Later Tater distances.

E. The distances for Spudnik are more variable than those for Later Tater.

3. Circle the term that correctly completes each statement.

A. A box plot for Spudnik data will be [skewed left / skewed right / symmetric] and the median will be [left of / right of / directly on] the center of the IQR.

B. A box plot for Later Tater data will be [skewed left / skewed right / symmetric] and the median will be [left of / right of / directly in] the center of the IQR.

4. The class can enter only one launcher in the qualifying round, in which a minimum distance of 240 meters is required for three of five launches. Based on the statistics for their launch tests, which launcher should they enter for this round and why?

Summary

Before you attempt the Practice Exercises, review what you've learned.

Data may be numerical or categorical. **Numerical data** consist of values that you can measure or count. **Categorical data** consist of characteristics that you can sort into groups or categories. Unlike categorical data, numerical data can be ordered from least to greatest.

Sometimes, it is difficult to picture the overall shape of a data set when it is displayed in a table. Dot plots, histograms, and box plots are ways of plotting a data set on a number line to show its shape.

A **dot plot** is a method of displaying a data set in which each data value is shown by a dot above a number line.

A **histogram** is a type of bar graph in which a data set is grouped into non-overlapping intervals of the same size. The height of each bar represents the frequency of data in a particular interval. There are no spaces between the bars of a histogram.

An **outlier** is a data value that is significantly greater or less than other values in a data set. One way to determine if a data point is a true outlier is by using a calculation involving the first and third quartile and the interquartile range. An outlier is a value x such that $x < Q_1 - 1.5(\text{IQR})$ or $x > Q_3 + 1.5(\text{IQR})$, where Q_1 is the first quartile, Q_3 is the third quartile, and IQR is the interquartile range.

The five-number summary a data set is shown in the table below.

Minimum	Q_1	Median	Q_3	Maximum
26	50.5	63	76	135

The interquartile range is 25.5, so
$Q_1 - 1.5(\text{IQR}) = 50.5 - 1.5(25.5) = 12.25$ which is below the minimum data range. Also,
$Q_3 + 1.5(\text{IQR}) = 76 + 1.5(25.5) = 114.25$,
this means that any value greater than 114.25 is an outlier for the data set.

A **box plot** is a method of displaying a data set on a number line by using the median and the first and third quartiles, and the maximum and minimum values.

- The **median** is the middle value, or the mean of the two middle values, when the data are listed in order.
- For a data set with median M, the **first quartile** (Q_1) is the median of the data values less than M.
- For a data set with median M, the **third quartile** (Q_3) is the median of the data values greater than M.

My Notes

Summary *(continued)*

Measures of Center

One way to compare two data sets is by comparing their centers. Measures of center include the **mean** and the **median**.

Note that because the mean depends on all values in a data set, it may not be the most appropriate measure of center for a data set that includes extreme data points or is heavily skewed.

Measures of Variability

You also can compare two data sets by comparing their variability. The variability of a data set refers to how much the values in a data set vary or how spread out they are. Measures of variability include the following.

- The **interquartile range** (IQR) is the difference between the third and first quartiles of a data set. It indicates the spread of the middle half of the data.
- The **mean absolute deviation** (MAD) is the average of the absolute deviations of each data value from the mean. To find the MAD, subtract the mean of the data set from each data value; these differences are the deviations from the mean. Next, take the absolute value of the deviations. Lastly, find the mean of the absolute values.
- The **variance** is the average of the squared deviations from the mean. To find the variance, first determine the deviations from the mean of the data set. Next, square the deviations. Lastly, find the mean of the squares.
- The population **standard deviation** (σ) is the square root of the variance. It is a measure of the typical distance of a data value from the mean.

In many cases, the data is a sample of a larger population. When the data is a sample, the variance is calculated slightly differently. In the calculation of the variance, the sum of the squared deviations is now divided by the number that is one less than the number of values in the sample. That is, if there are n data points, first calculate the deviation from the mean of the sample data. Next, square the deviations. Lastly, divide the sum of the squares by $(n - 1)$. The **sample standard deviation** (s) is the square root of the sample variance.

Note that because standard deviation depends on the squares of the distances of data values from the mean, it is greatly influenced by extreme data points. Therefore, the standard deviation is more appropriate for symmetric data sets than for data sets that contain extreme data points or that are heavily skewed.

My Notes

Practice Exercises

Review what you've learned using these practice problems. For practice problems with feedback, try the Coach and Play items in the Practice section online.

1. Mariana is taking a survey of the ages of every resident on her block. She tallies how many people are in their teens, twenties, thirties, and so forth, and enters the data into a spreadsheet.

   What is the best graph for Mariana to use to represent this data?

   A. dot plot

   B. histogram

   C. box plot

   D. line graph

2. Felipe and Calista have a summer job capturing frogs, measuring and recording the frogs' lengths, and then releasing them. Felipe and Calista work in different areas, and their data will be used to compare the frog populations for those areas.

   Using the choices below, fill in the blanks to show which display would best represent the data for each question that is asked.

   A. dot plot

   B. histogram

   C. box plot

   How do the shortest, longest, and median lengths compare? ____________

   How do the shapes of the data sets compare, based on defined intervals in length? ____________

   How do different measures of center and spread compare? ____________

3. The results of a test for two different class periods are shown. Using the choices below, complete the sentences comparing the data.

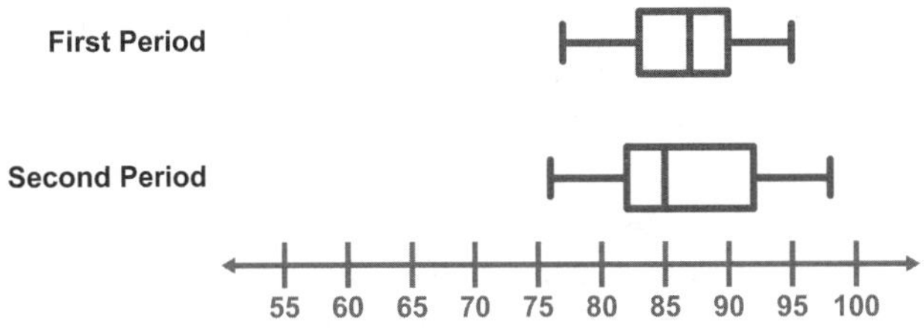

   A. first period

   B. second period

   The __________ class has the smallest spread.

   The __________ class has the highest median.

   The __________ class has the student with the highest grade.

4. For any data set, which of the following are always true? *Select all that apply.*

   A. The mean is greater than the median.

   B. The standard deviation is greater than or equal to the mean absolute deviation.

   C. The interquartile range is greater than the mean absolute deviation.

   D. The greater the spread, the greater the value of the variance.

   E. The mean absolute deviation is the best measure of variability.

   F. There will be a greater number of points within the standard deviation than within the mean absolute deviation.

Practice Exercises *(continued)*

5. Eileen is working with several data sets and notices that one of the sets has 12 values and a standard deviation of 0. Which of the following must be true of the data set? *Select all that apply.*

   A. The mean and median of the set are equal.

   B. The set has no repeated values and pairs of values are equally placed on opposite sides of the mean.

   C. The mean absolute deviation is equal to the standard deviation.

   D. All values in the data set are equal.

   E. The spread of the set is greater than its range.

6. Lisetta is working with a set of data showing the temperature at noon on 10 consecutive days.

   She adds today's temperature to the data set, and after doing so, the standard deviation falls.

   What conclusion can be made?

   A. Today's temperature is lower than on any of the previous 10 days.

   B. Today's temperature is close to the mean for the previous 10 days.

   C. Today's temperature is lower than the mean for the previous 10 days.

   D. Today's temperature is lower than the mean for the 11 days.

   E. Today's temperature is close to the mean for the 11 days.

7. Which of the following statements about the interquartile range (IQR) and the mean absolute deviation (MAD) of a data set are true? *Select all that apply.*

   A. The IQR and the MAD are both measures of center of data.

   B. The IQR and the MAD are both measures of variation of data.

   C. The IQR is the measure of the spread of the middle 50% of data.

   D. The MAD is the measure of the spread of the middle 50% of data.

   E. The MAD is the measure of the amount that data values differ from the mean of the data.

   F. The IQR is the measure of the amount that data values differ from the mean of the data.

8. Karen is analyzing a data set of the number of movies watched by her classmates each month. She wants to determine how much variability the data set has. She calculates the mode, median, mean absolute deviation (MAD), and interquartile range (IQR). Which measures could Karen use to describe the variability? *Select all that apply.*

   A. MAD **B.** IQR

   C. mode **D.** median

9. Terri recorded the number of text messages that she sent each hour, for several hours, as shown below:

   Number of Text Messages Sent: 4, 12, 3, 0, 9, 1, 4, 2, 5, 3, 9, 6, 10, 7, 4

   What is the interquartile range (IQR) of the data?

   A. 4 **B.** 5

   C. 6 **D.** 12

Apply

How Should Data Direct Tornado Research?

Scientists collect data about tornadoes and use the data to create models in order to understand how tornadoes behave. The ability to predict when a tornado might form can potentially save lives.

photo: Pixabay

What kinds of data do scientists gather to study tornadoes, and what do the scientists learn from the data?

Studying Tornadoes: View this video segment about tornado data.

The National Oceanic and Atmospheric Administration (NOAA) is an agency that provides information about weather forecasts, severe storms, and other environmental issues. NOAA's scientists collect and interpret tornado activity to determine how and where to focus future research. As you saw in the video, some areas of the country have more tornadoes than others.

The following table displays the Enhanced Fujita Scale, or EF-Scale, used by the National Weather Service to describe the damage caused by a tornado.

EF Number	0	1	2	3	4	5
3-Second Wind Gust (mph)	65–85	86–110	111–135	136–165	166–200	Over 200

- Gather data on tornado strength, according to the EF–scale, in two states for one year. Create a graphical display to present your data.
- Justify your reasons for choosing this way to display your data.
- Compare and contrast the data, using appropriate statistical measures of center and measures of variability.
- Interpret the differences that exist within the shape, center, and spread of the data you gathered. Which of the two states is more likely to have severe tornadoes?

As you research the tornado activity, think about the factors that might affect the development and severity of tornadoes. What trends or patterns do you observe?

Show what you've learned by completing the other performance tasks in the online Apply section.

Apply *(continued)*

Your answer to Apply will be assessed on the following criteria:

1. Collecting, recording, and citing appropriate data
2. Constructing an appropriate graphical display and explaining your choice
3. Calculating, comparing, and contrasting measures of center and measures of variability
4. Interpreting and analyzing the results

Scale / Criteria	4 Exceeds Criteria	3 Meets Criteria	2 Progressing to Criteria	1 Below Expectations	0 No Expectation
Data Collection	Collects accurate data, records it in a clear and organized way, and cites appropriate source.	Collects accurate data, records it in a mostly clear and organized way, and cites appropriate source.	Collects inaccurate data or records it in an unclear and unorganized way, and does not cite appropriate source.	Collects inaccurate data and records it in an unclear and unorganized way, and does not cite appropriate source.	Does not attempt task.
Graphical Display and Justification	Represents data with appropriate graphical display and gives justification for choosing the display.	Represents data graphically, but some parts are missing; gives justification for choosing the display.	Provides graphical display and justification, but both are inconsistent, or provides correct graphical display but gives no justification for why the display was chosen.	Does not use sufficient graphical display and gives no justification for why the display was chosen.	Does not attempt task.
Measures of Center and Variability	Correctly compares and contrasts the two states using a measure of center and a measure of variability.	Compares and contrasts the two states using a measure of center and a measure of variability, but makes an error with one measure.	Correctly compares and contrasts the two states using only one measure, a measure of center or a measure of variability.	Compares and contrasts the two states using only one measure, a measure of center or a measure of variability, and makes an error.	Does not attempt task.
Interpretation and Observations	States a reasonable, clear conclusion, uses appropriate assumptions and calculations, and explains observations.	States a reasonable conclusion and uses logical assumptions, but does not explain the observations.	States a reasonable conclusion but assumptions and observations are incomplete or unclear.	States a conclusion but does not include assumptions or clear explanations of observations.	Does not attempt task.

UNIT 6: Descriptive Statistics

6.2 Analyze Scatter Plots

photo: Getty Images

Lesson Objectives

- Draw a line of best fit through a scatter plot by hand and using technology.
- Assess the fit of a function by calculating residuals.
- Determine the equation of a line of best fit and interpret the meaning of slope and y-intercept in context.
- Calculate and interpret the correlation of a line using r.
- Understand that correlation does not imply causation.
- Use the line of best fit to solve problems within the constraints of the data set.

Essential Questions

- How can you determine a mathematical model that best describes the relationship between two variables as shown in a scatter plot?
- What can residual plots tell you about mathematical models?

Investigations

What Is the Association?

Do some situations have an effect on others? Is it positive or negative?

Hurricane Rita

How does wind speed correlate to barametric pressure? A coefficient can help you decide.

Running the Mile

Could a line model a scatter plot? Some fit better than others.

Forearm Length vs. Foot Length

Investigate the correlation coefficient of arm and foot lengths. Could their lengths be associated?

Jumping to Conclusions

Guilt by correlation? Maybe so, but maybe not.

Key Vocabulary

bivariate data, causation, correlation, correlation coefficient, line of best fit

Discover

As you complete Engage and the investigations, record the most important ideas you've learned.

Engage	
Investigation 1	**Investigation 2**
Investigation 3	**Investigation 4**

Name ______________________________ Date ______________

Check for Understanding

Analyze Scatter Plots: Investigation 1

1. Which of the following are valid values of the correlation coefficient, r, for a best-fit line? *Select all that apply.*

   A. $r = 9.89$
   B. $r = -0.75$
   C. $r = -1.11$
   D. $r = -0.09$
   E. $r = 0.623$
   F. $r = 0.11$

2. Each of the following equations represents a best-fit line for a set of data. Match each equation and the observed association of the data it represents with a reasonable correlation coefficient.

   A. $y = 3.627x + 8.441$, strong association
   B. $y = -15.34x - 6.31$, no association
   C. $y = 0.132x + 4.65$, weak association
   D. $y = -8.933x - 6.87$, strong association

   I. $r = -0.031$
   II. $r = -0.925$
   III. $r = 0.897$
   IV. $r = 0.316$

3. Circle the correct term to complete each statement.

   A. The coefficient of x in the equation for a best-fit line represents [the slope / the correlation coefficient / both values / neither value].

   B. The rate at which the dependent variable is changing with respect to the independent variable is represented by [the slope / the correlation coefficient / both values / neither value].

   C. The strength of the linear association of the data is represented by [the slope / the correlation coefficient / both values / neither value].

   D. The sign of the slope and the sign of the correlation coefficient are [the same / opposites / not related].

   E. The direction of the best-fit line is represented by [the slope / the correlation coefficient / both values / neither value].

4. A best-fit line for a scatter plot showing an association between regular hours of exercise per week and resting heart rate (BPM) has the equation $y = -4.23x + 78$ and correlation coefficient -0.92. Describe the association between hours of exercise per week and resting heart rate, and explain the meanings of the slope and the value of r.

Name ____________________ Date __________

Check for Understanding

Analyze Scatter Plots: Investigation 2

1. Circle the terms that correctly complete each sentence.

A. A value for a residual for a best-fit line is calculated from the [sum / difference / product / quotient] of the observed [x value / y value] in the plot and the predicted [x value / y value] from the equation.

B. Each residual gives the [vertical / horizontal] distance between the observed and predicted values.

C. The sum of the residuals from a scatter plot and its best-fit line has a [large positive / large negative / near zero] value.

2. Given the best-fit line equation $y = 3.17x - 4.92$, calculate the residuals for the following points from the scatter plot that the line models.

x	2	5	7	8	11	12
y	1.5	9.8	18.9	20.2	30.1	32.7
Residual						

3. Each of the following represents a residual plot created from a scatter plot and its best-fit line equation. Choose the plot(s) indicating that the associated linear equation is a good model for the data.

A.

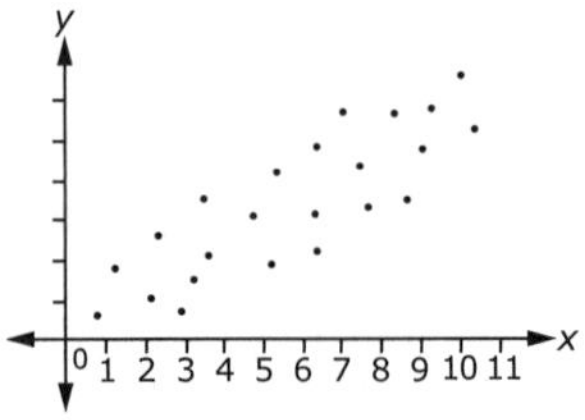

B.

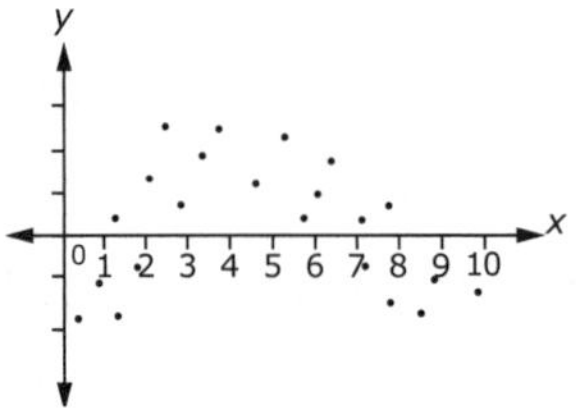

C.

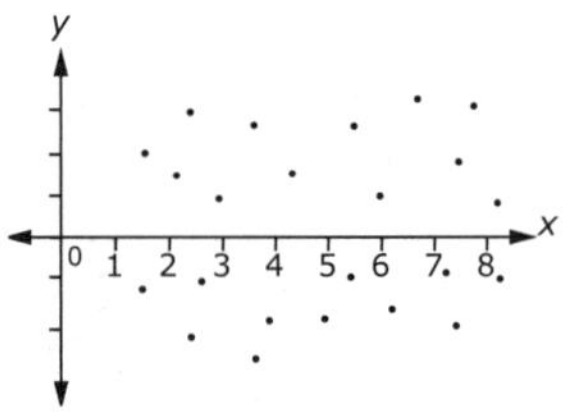

4. International 400m freestyle winning times as a function of years have been decreasing according to a linear model for which the residual plot indicates a good fit. Why might the linear model not be a good predictor of future winning times in this event?

Name ______________________________ Date ______________

Check for Understanding

Analyze Scatter Plots: Investigation 3

Mr. Baird's class is testing the claim that airline fares seem to be independent of distances flown. They check fares to various cities from New York, record the distances, x, and the associated fares, y. Their line of best fit has the equation $y = 0.115x + 90.074$, with a correlation coefficient of $r = 0.81$.

Use the information to answer questions 1–3.

Distance (miles)	572	615	221	784	1,508	790	1,225	980	378
Fare ($)	182	98	142	175	250	182	285	178	132

1. Which of the following statements are true about the residuals? *Select all that apply.*

A. The residuals are all positive.

B. A plot of the residuals indicates that the linear model is the best fit.

C. The data show a strong positive correlation between the distance and the fares.

D. The residual plot shows a definite pattern.

E. The residuals have a sum that is close to 0.

2. Which of the following statements correctly interpret the meanings of the slope and y-intercept? *Select all that apply.*

A. The y-intercept has no meaning in this context.

B. The y-intercept means that the shortest flight is 90 miles.

C. The y-intercept means that each fare has a minimum of about $90.

D. The slope indicates that there is no correlation because it is near to 0.

E. The slope means that fares tend to increase about $0.115 per mile.

F. The slope means that there is an increase of $90 associated with each increase of 115 miles.

3. Many independent sites claim that airfares are not based on miles flown. Why does Mr. Baird's class not have enough evidence to claim that fares are based on distances?

Name ______________________________ Date ______________

Check for Understanding

Analyze Scatter Plots: Investigation 4

1. Data from multiple observations showed a correlation between sleep pattern disruptions in lab rats for several days and subsequent earthquake activity. Which of the following statements can be made? *Select all that apply.*

   A. Impending earthquakes cause lab rats to lose sleep.

   B. Sleep disturbances in lab rats may be connected to upcoming earthquake events.

   C. Lab rats having sleep difficulties can be used to predict earthquakes.

   D. Lab rats tend to experience sleep disturbances prior to an earthquake.

   E. Controlled studies of a link between the sleep patterns of lab rats and impending earthquakes may be warranted.

2. In each pair of statements, circle the one that accurately reflects a correlation between the variables from published results.

   A. Students who play sports have better grades.

   Playing sports is related to higher grades among students.

   B. Maintaining strong social ties is connected to better long-term health.

   People who maintain strong social connections live longer.

   C. People who eat Italian pizza weekly reduce their risk of certain types of cancer.

   A decreased risk for some cancers is linked to eating Italian pizza each week.

3. Which of the following terms can be used to describe a correlation? *Circle all that apply.*

tied	causes	increases
improves	tends	linked
related	benefits	affects
influences	decreases	associated

4. Can a strong correlation ever be used to prove causation? If so, give an example. If not, explain why not.

Summary

Before you attempt the Practice Exercises, review what you've learned.

Association

Bivariate data is a data set that contains data for two variables.

You can display numerical bivariate data by using a scatter plot. A scatter plot may show that there is an association between two variables, which means that their values vary with each other according to some pattern.

- If the value of one variable tends to increase as the value of the other variable increases, then there is a positive association between the variables.
- If the value of one variable tends to decrease as the value of the other variable increases, then there is a negative association between the variables.
- If there is no clear pattern between the value of one variable and the value of the other, then there is no association between the variables.

Lines of Best Fit

A **line of best fit** is a line that best fits the data on a scatter plot. The line that best fits the data is the one that comes as close as possible to each of the data points.

EXAMPLE: Approximating a Line of Best Fit

A website lists the manufacturer's suggested price and the lowest likely sale price for a variety of new vehicles. The table shows this data set for several cars. Make a scatter plot of the data and approximate the line of best fit.

Car	Manufacturer's Suggested Price ($), x	Lowest Likely Sale Price ($), y
A	22,900	21,800
B	14,900	14,100
C	22,700	20,600
D	35,900	32,900
E	17,700	16,800
F	32,200	29,100
G	31,800	29,600
H	15,500	14,500

SOLUTION:

Plot the data. Draw a line that has about the same number of data points above it as below it. Try to draw the line so that it comes as close to the data points as possible.

My Notes

Summary *(continued)*

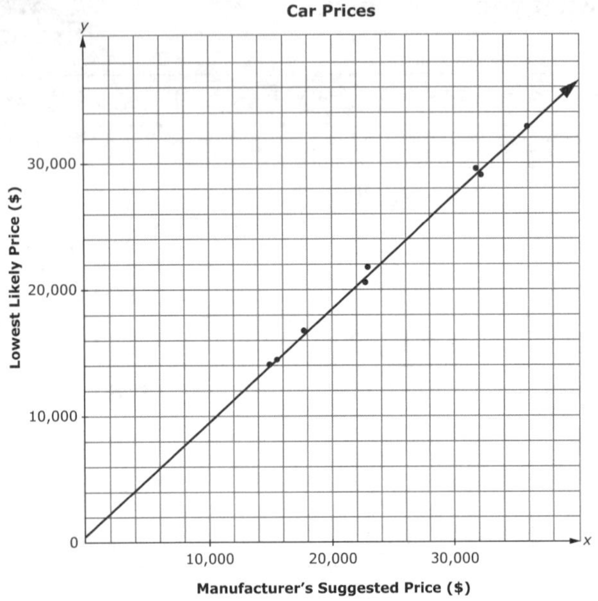

Now write the equation of the line in slope-intercept form, $y = mx + b$.

The y-intercept of the line is approximately 800, so $b = 800$.

Estimate the coordinates of any two points on the line to determine the slope. Use (0, 800) and (30,000, 27,500).

$$m = \frac{27,500 - 800}{30,000 - 0} = 0.89$$

So, the line of best fit is approximately $y = 0.89x + 800$.

My Notes

Note that the slope of the line indicates that the lowest likely sale price of a car increases by about $0.89 for every $1 that the manufacturer's suggested price increases.

You can also use technology to help you determine the line of best fit for a set of bivariate data. Graphing calculators generate a line of best fit by minimizing the sum of the squares of the residuals.

EXAMPLE: Using Technology to Find a Line of Best Fit

Use technology to find the line of best fit for the car price data.

SOLUTION:

Enter the data set into the Graphing Calculator. Next, use the Best Fit feature to generate a line of best fit.

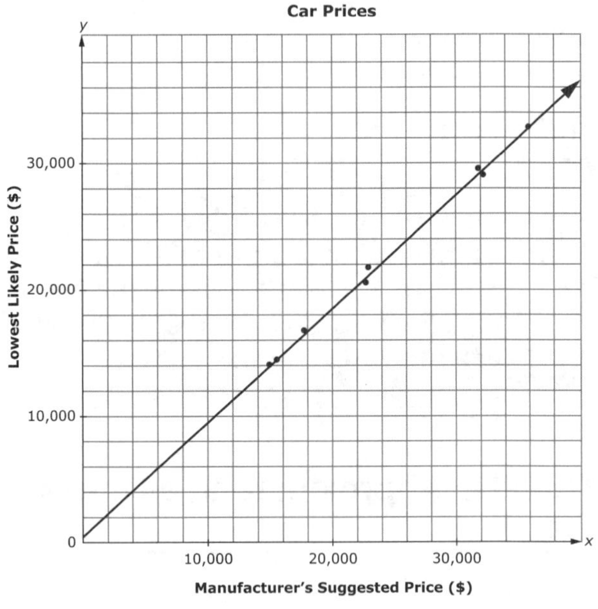

The line of best fit is given by the equation $y = 0.893x + 818.486$.

Notice that the equation we generated earlier is very close to this equation.

Summary *(continued)*

You can use residuals to help you analyze a line on a scatter plot. A residual is a measure of the vertical distance between a data point and the line. If a linear model is a line of best fit for a data set, the sum of the residuals should be 0.

EXAMPLE: Calculating Residuals

Analyze the residuals of the line from the car price example to confirm that it is the line of best fit for the data.

SOLUTION:

Let the function f represent the line of best fit. So, $f(x) = 0.893x + 818.486$.

For each value of x in the data table, calculate the corresponding value of $f(x)$. Find the difference between the value of y from the table and the value of $f(x)$.

Car	Manufacturer's Suggested Price (\$), x	Lowest Likely Sale Price (\$), y	$f(x)$	Residual $y - f(x)$
A	22,900	21,800	21,268	532
B	14,900	14,100	14,124	–24
C	22,700	20,600	21,090	–490
D	35,900	32,900	32,877	23
E	17,700	16,800	16,625	175
F	32,200	29,100	29,573	–473
G	31,800	29,600	29,216	384
H	15,500	14,500	14,660	–160

Next, find the sum of the residuals.

Add 532, –24, –490, 23, 175, –473, 384, and –160 together.

The sum of the residuals is –33, which is close to 0 when compared to the prices of the cars. (The sum is not exactly equal to 0 because of rounding in our calculations.)

The sum of the residuals confirms that the line given by $f(x) = 0.893x + 818.486$ is the line of best fit.

We can also make and analyze a plot of the residuals from a linear regression. If the points on the graph of the residuals appear to be randomly dispersed above and below the x-axis, then you can conclude that a linear model is appropriate for the data. If the points exhibit a distinctly nonrandom pattern, a linear model may not be appropriate for the data.

Correlation Coefficients

A **Correlation** is a measure of the strength and direction of an association between two variables that have a linear relationship.

A A **correlation coefficient** r is a number between –1 and 1 that indicates how closely a set of data points is to the line of best fit. The correlation coefficient also indicates whether the association between two variables is positive or negative.

My Notes

Summary *(continued)*

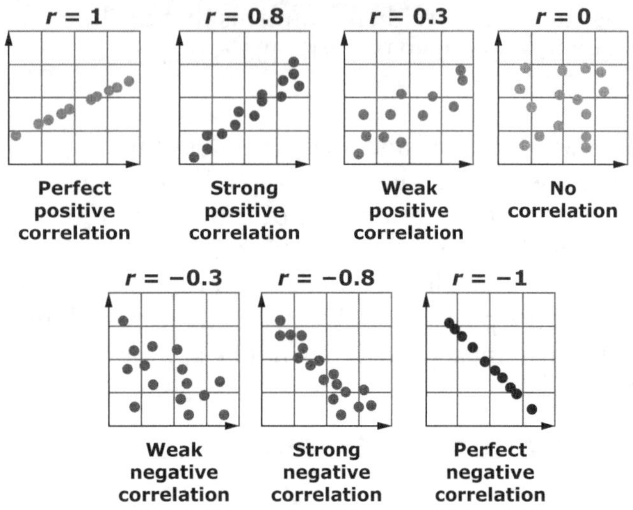

- A positive value of r indicates a line of best fit with a positive slope. A negative value of r indicates a line of best fit with a negative slope.
- The closer the absolute value of r is to 1, the stronger the linear correlation between the variables.
- If $r = 1$ or $r = -1$, then all data points lie on the line of best fit.

You can use a graphing calculator or other technology to display the correlation coefficient when you generate a line of best fit for a data set.

If the line of best fit is a good model for the data, you can use the line to make predictions

Correlation vs. Causation

If two variables are related by causation, then a change in one variable causes a change in the other variable. If there is a strong correlation between two variables, they may or may not be related by causation. Correlation does not imply causation.

In the car price example, it seems likely that an increase in the manufacturer's suggested price of a car would cause an increase in the lowest likely sale price of the car. For these two variables, there is both a strong correlation and a likelihood of causation.

My Notes

Practice Exercises

Review what you've learned using these practice problems. For practice problems with feedback, try the Coach and Play items in the Practice section online.

1. The equation of a line of best fit for a given scatter plot is $y = 2.2x + 10$.

   Point A (5, 25) is a data point on the scatter plot.

   What is the residual of Point A? ___________

2. Mr. Valentine made a scatter plot with his students' homework averages on the x-axis and their test score averages on the y-axis. He calculated the equation of the line of best fit as $y = 0.9x + 12$, with x and y as scores from 0 to 100. Joseph, one of Mr. Valentine's students, had a homework average of 80 and a residual of 3.

   What was Joseph's average test score? ________%

3. Which of the following measures would indicate that a linear model is a good fit for a data set? *Select all that apply.*

   A. distinct pattern in the distribution of the residual points

   B. each residual has a value of 0

   C. mostly positive residuals

   D. random pattern in distribution of the residual points

   E. balance of positive and negative residuals

   F. positive sum of residuals

4. Trè knows that the correlation coefficient for a best-fit line is $r = -0.837$. What conclusions can he make? *Select all that apply.*

   A. The best-fit line is decreasing.

   B. There is no association between x and y.

   C. Each increase in x causes a decrease in y.

   D. The best-fit line's equation is of the form $y = -0.837x + b$.

   E. There is a strong negative correlation.

   F. The best-fit line has a negative slope.

5. Which of the following statements below best complete this sentence? *Select all that apply.*

   To establish a causal relationship between two variables, ___________.

   A. a controlled scientific experiment must be done

   B. a line of best fit modeling the relationship between the variables must show a positive correlation

   C. there must be no residuals for the line of best fit

   D. the correlation coefficient for the line of best fit modeling the relationship between the variables must be greater than 0.9

   E. all other possible causes must be accounted for

6. Which of the following is the best prediction equation for a line of best fit representing the following data set?

   (0, 2), (−2,0), (1, 4), (−1, 2), (−2, 2), (1, 8), (2, 6), (3, 10)

   A. $y = 2x + 3$ **B.** $y = x + 3$

   C. $y = x + 1$ **D.** $y = 1$

7. A newspaper reported that four nearby neighborhoods have had cars broken into recently. The report stated that most of the cars that had items stolen out of them had been left unlocked by the owners.

   Based on this information, answer the questions below.

   Write Yes or No in each blank.

   Would this indicate a correlation between car break-ins and the owners leaving the cars unlocked? ________

   Would this indicate a causal relationship between car break-ins and the owners leaving the cars unlocked? ______

Practice Exercises *(continued)*

8. The table includes the data relating the age of a person and a corresponding systolic blood pressure measurement.

Age, x	Systolic Blood Pressure, y
25	128
39	120
56	135
45	143
49	136
64	170
29	141
57	132
43	144
22	122

Create a scatter plot and best-fit line. Use the equation of the best-fit line to predict the systolic blood pressure for a person of age 60.

The approximate systolic blood pressure for a person age 60 is ____________.

9. Which of the following demonstrates a correlated relationship that is not causal?

A. the number of miles driven and the amount of gas used

B. the number of hours worked and the amount of pay earned

C. the number of calories burned and the amount of exercise done

D. the age of a child and the child's math skills

10. Chelsea asked each of her classmates their month of birth and their favorite type of music. She plans on putting the data in a scatter plot.

What type of correlation would be expected with this data?

A. There will be a negative correlation.

B. There will be a positive correlation.

C. There will be no correlation.

D. The correlation will be even.

11. The following four data points are on a scatter plot.

(0, 4), (1, 4), (2, 11), (3, 15)

Find the sum of the residuals to determine which of the following is the line of best fit for this data. Identify the line and the sum of the residuals.

A. Line of best fit: $f(x) = 2.5x + 4$

B. Line of best fit: $f(x) = 4x + 2.5$

C. Sum of residuals: –5

D. Sum of residuals: –3

E. Sum of residuals: –0.5

F. Sum of residuals: 0

G. Sum of residuals: 0.5

H. Sum of residuals: 3

I. Sum of residuals: 5

12. The newspaper will print an article about a study that reported a correlation between texting and increased language skills. Which headline would accurately reflect the results of the study?

A. Texting Boosts Language Skills

B. Texting Raises Language Scores

C. Texting Linked to Improved Language Skills

D. Text More for Better Grades!

Apply

How Do Populations Change over Time?

How do populations change over time? How can you use math to help you predict how much populations will increase or decrease? Watch the video to learn more.

How can you use statistics to predict changes in populations?

Population Growth: View this video segment about population changes.

Over long periods of time, human population growth has followed a certain trend. One way to learn about that trend is to find a line of best fit and study the residuals. If there is a pattern in the residuals, the pattern tells you something about the trend and the type of function that models the data.

Population Table: View this table of data (in millions) from the United Nations, World Population Prospects.

- Choose a continent that is of interest to you, and create a scatter plot of the data.
- Find the line of best fit that models the population over time, and graph it on your scatter plot.
- Calculate and plot the residuals. Do they show a pattern? If so, explain what you think the pattern means. If not, what do you think the absence of a pattern means?
- Calculate the correlation coefficient. Is the linear model a good fit? If not, describe another model that better fits the data. Justify your results and conclusions.
- Based on your observations, what do you predict about the population growth of your chosen continent in 10 years?

Show what you've learned by completing the other performance tasks in the online Apply section.

Apply *(continued)*

Your answer to Apply will be assessed on the following criteria:

1. Creating a scatter plot, finding the equation of the line of best fit, and graphing it
2. Calculating and plotting the residuals and analyzing the pattern
3. Calculating the correlation coefficient and discussing whether the model is a good fit
4. Predicting population growth of the continent in 10 years

Criteria \ Scale	4 Exceeds Criteria	3 Meets Criteria	2 Progressing to Criteria	1 Below Expectations	0 No Expectation
Scatter Plot and Line of Best Fit	Graphs the data accurately; finds the equation of the line of best fit and graphs it correctly.	Graphs the data accurately; makes minor errors in finding or graphing the equation of the line of best fit.	Graphs the data accurately; makes minor errors in finding and graphing the equation of the line of best fit.	Graphs the data inaccurately; makes errors in finding and graphing the equation of the line of best fit.	Does not graph data or find the line of best fit.
Residuals and Analysis	Calculates and plots the residuals accurately and correctly analyzes the pattern.	Calculates and plots the residuals and analyzes the pattern but makes minor errors.	Calculates and plots the residuals incorrectly, or analyzes the pattern with faulty logic.	Calculates and plots the residuals incorrectly and analyzes the pattern with faulty logic.	Does not calculate or plot the residuals.
Correlation and Discussion	Calculates the correlation coefficient accurately and correctly describes how closely the model fits the data.	Calculates the correlation coefficient and describes how closely the model fits the data but makes minor errors.	Calculates the correlation coefficient incorrectly or does not describe how closely the model fits the data.	Calculates the correlation coefficient incorrectly and does not describe how closely the model fits the data.	Does not calculate the correlation coefficient or describe how the model fits the data.
Prediction and Justification	Correctly justifies whether the line of best fit is an accurate model and makes a reasonable prediction.	Justifies whether the line of best fit is an accurate model with minor errors and makes a reasonable prediction.	Incorrectly determines whether the line of best fit is an accurate model or does not make a reasonable prediction.	Incorrectly determines whether the line of best fit is an accurate model and does not make a reasonable prediction.	Does not analyze the model or make a prediction.

UNIT 6: Descriptive Statistics

6.3 Interpret Two-Way Frequency Tables

Photo: Getty Images

Lesson Objectives

- Understand how data are organized in a two-way table.
- Construct a two-way table and interpret the table to draw conclusions.
- Calculate joint, marginal, and conditional relative frequencies.

Essential Question

- How can a two-way table be used to draw conclusions about data?

photo: iStock

Investigations

Teenage Jobs

What can a two-way table tell you about teenage employment? Make a claim and justify it with data.

Business or Education?

Percentages describe labor statistics, but it's all relative. Which percentage makes the better argument?

Surveys

Survey results are used to make recommendations. Conduct a survey and make a convincing argument.

Curfew: Fair or Unfair?

What's a fair curfew? Use data and relative frequencies to construct a convincing argument.

Key Vocabulary

bivariate data, univariate data

Discover

As you complete Engage and the investigations, record the most important ideas you've learned.

Engage	**Investigation 1**
Investigation 2	**Investigation 3**

Name ______________________________ Date ____________

Check for Understanding

Interpret Two-Way Frequency Tables: Investigation 1

To analyze traffic patterns during school hours, juniors and seniors were asked if they regularly drove themselves to school or arrived to school using some other transportation.

1. Complete the following frequency table and use it to answer questions 2–3.

	Drive Themselves	Use Other Transportation	Total
Juniors	48		128
Seniors			112
Total	138		240

2. Calculate the relative and marginal frequencies to hundredths for each cell in the table.

	Drive Themselves	Use Other Transportation	Total
Juniors			
Seniors			
Total			

3. Match the given frequency descriptions to their approximate percentages.

A. the marginal relative frequency of students using other transportation

B. the joint relative frequency of seniors who drive themselves

C. the conditional relative frequency of juniors among students who drive themselves

D. among seniors, the conditional relative frequency of students who don't regularly drive themselves to school

I. 0.35

II. 0.20

III. 0.38

IV. 0.43

4. Jonathan used the frequency table to calculate the conditional relative frequencies, while Deidre used the relative frequency table. They were surprised to see that they had the same answers. Explain why their percentages were the same.

Name ______________________________ Date ______________

Check for Understanding

Interpret Two-Way Frequency Tables: Investigation 2

Students at the high school were asked to vote yes or no to put on a Fall Festival and a Spring Fling with the available funds.

Use the results in the table to answer questions 1–3.

		Fall Festival		
		Yes	No	Total
Spring Fling	Yes	120	160	280
	No	140	60	200
	Total	260	220	480

1. Complete the relative frequency table for the survey. *Enter each value as a decimal rounded to the nearest hundredth.*

		Fall Festival		
		Yes	No	Total
Spring Fling	Yes			
	No			
	Total			

2. Use the frequencies you calculated to match the percentage of students to their approximate values for each category.

   A. Percent who voted yes to both Fall Festival and Spring Fling — **I.** 0.72

   B. Percent who voted no to Fall Festival and yes to Spring Fling — **II.** 0.46

   C. Percent who voted no to Fall Festival — **III.** 0.43

   D. Percent who voted yes to Spring Fling — **IV.** 0.25

   E. Among those voting no to Fall Festival, percent who voted yes to Spring Fling — **V.** 0.58

   F. Among those voting yes to Spring Fling, percent who voted yes to Fall Festival — **VI.** 0.33

3. Given that only one event will be approved, make a recommendation to the school administration based on the data in the table. Use the data to support your argument.

Name ______________________________ Date ______________

Check for Understanding

Interpret Two-Way Frequency Tables: Investigation 3

The table below shows incomplete results from a survey of high school students and their reported curfew times on the night of the spring formal.

1. Complete the frequency table for the survey.

	11 p.m.	12 a.m.	1 a.m.	Later or No Curfew	Total
9th Grade	26		46	4	128
10th Grade		62	42	14	
11th Grade	6	28		32	96
12th Grade	0	8	28		88
Total		150		102	448

2. Circle the term that correctly completes each of the following statements. *Consider frequencies to the nearest whole percent.*

   A. The percentage of surveyed 10th graders who have no curfew is a [joint relative / conditional relative / marginal relative] frequency.

   B. The percentage of surveyed students having a curfew of 12 a.m. is a [joint relative / conditional relative / marginal relative] frequency.

   C. The percentage of students surveyed who are in 12th grade and who have a curfew of 12 a.m. is a [joint relative / conditional relative / marginal relative] frequency.

   D. The percent of students in 11th grade surveyed is [20% / 21% / 25% / 30%].

   E. The percent of students surveyed with curfews no later than 1 a.m. is [11% / 21% / 45% / 77%].

   F. The percent of [9th / 10th / 11th / 12th] graders with 12 a.m. curfews is 29%.

3. What might an 11th grader say to support his argument that he should be allowed a curfew of 2 a.m. for the spring formal? Use the data to support your argument.

Summary

Before you attempt the Practice Exercises, review what you've learned.

A two-way table, or two-way frequency table, can be used to summarize data about two categorical variables. The cells in the table represent the number of observations that occur in the intersection of two categories.

EXAMPLE: Analyze a Two-Way Table

The two-way table shows data for the categories of gender and age group.

Compare the number of females who have no vision difficulty to the number of females who have some type of vision difficulty. What conclusions can you draw about randomly-selected females from the population?

Vision Difficulty for Ages 5–17

	No Vision Difficulty	With Vision Difficulty	Total
Male	27,230,890	214,675	27,445,565
Female	26,053,579	201,014	26,254,593
Total	53,284,469	415,689	53,700,158

My Notes

SOLUTION:

The total number of females is 26,254,593. Of these, 26,053,579 have no vision difficulty. This represents about 99.2% of the total number of females.

The remaining 201,014 females have some type of vision difficulty. This represents about 0.8% of females.

The data indicate that it is very likely that a randomly-selected female will have no vision difficulty.

The data in a two-way frequency table can be used to make a two-way relative frequency table. In a **two-way relative frequency table**, each cell in the table is compared to the total number of observations that are recorded in the table. The relative frequency of each observation can be stated as a decimal or percent.

A **joint relative frequency** gives the relative frequency that two conditions occur at the same time. The relative frequency of people who are male and have no vision difficulty is an example of joint relative frequency. A joint relative frequency is the ratio of the value of an interior cell of the table to the total number of observations in the table.

A **marginal relative frequency** gives the relative frequency at which a particular category occurs. The relative frequency of people with vision difficulty is an example of marginal relative frequency. A marginal relative frequency is the ratio of the total for one column or row to the total number of observations in the table.

Summary *(continued)*

EXAMPLE: Find Joint and Marginal Relative Frequencies

Make a relative frequency table to find the following joint or marginal relative frequencies.

A. What percent are female?

B. What percent have no vision difficulty?

C. What percent are female and have vision difficulty?

SOLUTION:

To make the relative frequency table, divide each value in the table by the total number of people represented by the data, 53,700,158.

Vision Difficulty for Ages 5–17

	No Vision Difficulty	With Vision Difficulty	Total
Male	0.507	0.004	0.511
Female	0.485	0.004	0.489
Total	0.992	0.008	1

Marginal relative frequencies are the totals for each row and column. Joint relative frequencies are in the body of the table.

A. The marginal relative frequency for females is 0.489, or about 48.9%.

B. The marginal relative frequency for people with no vision difficulty is 0.992, or about 99.2%.

C. The joint relative frequency for all participants who are female and have vision difficulty is 0.004, or about 0.4%.

A **conditional relative frequency** gives the relative frequency at which a certain condition occurs, given that the observation is in a particular category. The relative frequency of being male, given that the person has no vision difficulty, is an example of conditional relative frequency. A conditional relative frequency is the ratio of a joint frequency to a marginal frequency. You need to determine which category the given condition represents and which category the question wants you to consider.

In the example of being male, given that the person has no vision difficulty, the category we are interested in is being male. The given condition is having no vision difficulty.

My Notes

Summary *(continued)*

EXAMPLE: Conditional Relative Frequency

Identify the category and the condition that is given. Then find the conditional relative frequency.

A. What percent of people who have no vision difficulty are female?

B. What percent of males have vision difficulty?

SOLUTION:

A. Write the question as "What percent is female, given that they have no vision difficulty?" The given condition is having no vision difficulty, and the category is being female. Identify the column "having no vision difficulty." Calculate the portion of this column total that is female. This is the ratio of a joint relative frequency to a marginal relative frequency.

$$\frac{\text{female and no difficulty}}{\text{no difficulty}} = \frac{0.485}{0.992} \approx 48.9\%$$

B. Write the question as "What percent have vision difficulty, given that they are male?" The given condition is being male, and the category is having vision difficulty.

Identify the row "male," and calculate the ratio of this row total who have vision difficulty. This is the ratio of a joint relative frequency to a marginal relative frequency.

$$\frac{\text{with difficulty and male}}{\text{male}} = \frac{0.004}{0.511} \approx 0.8\%$$

You can use joint, marginal, and conditional relative frequencies to answer questions about the population or to make a convincing argument.

EXAMPLE: Draw a Conclusion

Are 5–17-year-old females more likely to have no vision difficulty than 5–17-year-old males?

SOLUTION:

First, determine whether males and females are approximately equally represented among people who have no vision difficulty.

Find the marginal relative frequencies for males and females.

Male: 0.511, or 51.1%

Female: 0.489, or 48.9%

Find the conditional frequencies for people with no vision difficulty who are males and females.

Male: $\frac{0.507}{0.992} \approx 0.511$, or about 51.1%

Female: $\frac{0.485}{0.992} \approx 0.489$, or about 48.9%

The representation of females among those with no vision difficulty is consistent with the representation of females in the total population. This means that females are proportionally represented among those with no vision difficulty.

Now you know that it is valid to compare the joint relative frequencies of males with no vision difficulty to females with no vision difficulty.

In the category of those with no vision difficulty, the frequency for females is a bit lower than the frequency for males. Females are slightly less likely to have no vision difficulty than males.

My Notes

Practice Exercises

Review what you've learned using these practice problems. For practice problems with feedback, try the Coach and Play items in the Practice section online.

1. A survey of students in grades 11 and 12 was taken to find out their primary plans after high school graduation. The response options were four-year college, community college, technical college, employment, or other.

   Which headings are relevant for creating a two-way frequency chart to report the results? *Select all that apply.*

   A. College
   B. 12th Graders
   C. Males
   D. Community College
   E. Travel
   F. 11th Graders

2. A survey of high school students was conducted in order to determine if there was an association between gender and the type of pet the students own. The pet choices were dog, cat, both, or neither. Other pets were not considered for this survey. The results are shown in the table.

	Male	Female	Total
Dog	50	26	
Cat	27	35	
Both	10	22	
Neither	13	17	
Total			

   What are the joint and relative marginal frequencies for the following?

   The marginal relative frequency for those who own cats is ___________.

   The marginal relative frequency for those who own neither a cat nor a dog is ___________.

   The joint relative frequency for males who own dogs is ___________.

   The joint relative frequency for females who own both cats and dogs is ___________.

3. A survey was taken of 12th graders to determine if time studying per week is associated with better grades. The results were recorded in a table.

	A	B	C	D	E	Total
0–2 hr	2	5	9	12	6	34
3–5 hr	3	8	7	9	4	31
6–8 hr	5	10	12	8	3	38
9–11 hr	9	13	10	6	2	40
12–14 hr	7	13	9	5	2	36
Total	26	49	47	40	17	

   Which study time group nets the most As? ___________

   Which grade occurs most often? ___________

   How many total students were polled? _________

4. Boys and girls were polled to see who bought school lunches versus those who packed their lunches. There are more girls in the school than boys, and 5 more boys said they packed lunches than said they bought lunches.

   Choose the correct labels for each column and row.

	_______	_______	Total
_______	15	20	35
_______	31	10	41
Total	46	30	

   A. Boys
   B. Packed
   C. Girls
   D. Bought

Practice Exercises *(continued)*

5. A survey was taken of 10th and 11th graders to see if they were going to participate in homecoming events. Their response options were dance only, game only, both, and neither. The table below represents the results of the survey.

	Dance Only	Game Only	Both	Neither	Total
10th	50	30	25	15	120
11th	65	20	30	10	125
Total	115	50	55	25	

Calculate the following in decimal form.

What is the relative frequency of 11th graders who responded that they are going to the dance only? ________

What is the relative frequency of students who responded that they are going to both the dance and the game? ________

What is the relative frequency of 10th graders who responded that they are going to the dance?

____

6. A survey was taken to learn the associations of people with black, blonde, red, and brown hair and their eye color (brown, blue, green, or other).

Categorize these descriptions into marginal, joint, or conditional relative frequency data.

A. all brown haired people

B. brown haired people who have blue eyes

C. the ratio of people with blonde hair who also have green eyes

D. blonde haired people with brown eyes

E. the ratio of people with black hair people who also have brown eyes

F. all green eyed people

Marginal: ________

Joint: ________

Conditional relative frequency: ________

7. Happyvale High School is considering adding two new fruit drinks to their menu. Selected 9th and 10th graders were asked to participate in a taste test to choose which drink they preferred. The table shows the results.

	Berry Blend	Yellow Energy	Total
9th	7	11	18
10th	9	12	21
Total	16	23	39

Which statement is a conditional relative frequency shown with the correct supporting ratio?

A. the ratio of students who prefer Berry Blend: $\frac{16}{39}$

B. the ratio of 9th graders who prefer Yellow Energy: $\frac{11}{39}$

C. the ratio of those surveyed who are 10th graders: $\frac{21}{39}$

D. of those who prefer Yellow Energy, the ratio who are 10th graders: $\frac{12}{23}$

8. The frequency table below shows how many baseball or football championships were won by teams in Pittsburgh and San Francisco.

	Baseball	Football	Total
Pittsburgh	5	6	11
San Francisco	7	5	12
Total	12	11	23

Calculate the following in decimal form.

What is the relative frequency of football championships won when the champion was San Francisco? ________

What is the relative frequency that, when the winner was Pittsburgh, the championship was in baseball? ________

Apply

Don't You Just Love the Winter Olympics?

The 2014 Winter Olympics were held in Sochi, Russia. A total of 2,873 athletes from 88 countries competed in 98 events.

Math and Athletics: View this video segment to see how to use math to analyze athletic competitions.

The International Olympic Committee (IOC) maintains a tremendous amount of data on the games. Some data about the Winter Olympic medals won by the United States and Canada are displayed in a two-way frequency table that shows the sport and the type of medal (gold, silver, or bronze).

Winter Olympics Data: Use the data in the Winter Olympics activity to write three questions: one that uses joint relative frequencies, one that uses marginal relative frequencies, and one that uses conditional relative frequencies.

- Provide complete solutions to your questions, including any computation and explanation.
- Research the total number of Winter Olympics medals earned by each of the two countries from 1924 to 2014. With this additional information, explain what else you could find out. Give at least one example.
- Explain one advantage and one disadvantage of using a two-way relative frequency table vs. a two-way table of actual data.

Show what you've learned by completing the other performance tasks in the online Apply section.

Apply *(continued)*

Your answer to Apply will be assessed on the following criteria:

1. Writing three questions, one each to be answered by comparing joint, marginal, and conditional frequencies
2. Providing complete answers to your questions, supported by your comparisons, relevant computation, and explanation
3. Researching actual medal counts and explaining what they could tell you; giving an example
4. Explaining one advantage and one disadvantage of a two-way relative frequency table

Criteria \ Scale	4 Exceeds Criteria	3 Meets Criteria	2 Progressing to Criteria	1 Below Expectations	0 No Expectation
Three Questions	Writes three questions that meet the stated criteria.	Writes two questions that meet the stated criteria and a third that repeats it or doesn't meet any of the criteria.	Writes one question that meets the stated criteria and two others that repeat it or don't meet any of the criteria.	Writes three questions, none of which meet the stated criteria.	Does not attempt task.
Three Answers	Writes three answers that fully address the questions.	Writes two answers that fully address the questions and a third that does not address the question or is incorrect.	Writes one answer that fully addresses the question and two that do not address the questions or are incorrect.	Writes three answers that do not address the questions or are incorrect.	Does not attempt task.
Finding Data; Explaining Its Use	Finds accurate data and explains what the data can tell you; gives at least one example.	Finds accurate data and explains what the data can tell you; gives at least one example.	Finds accurate data and gives at least one example.	Finds inaccurate data and does not explain or give an example.	Does not attempt task.
Advantages and Disadvantages	Clearly explains one advantage and one disadvantage.	Explains one advantage and one disadvantage, but not clearly.	Explains one advantage or one disadvantage.	Explains one advantage or one disadvantage, but not clearly.	Does not attempt task.

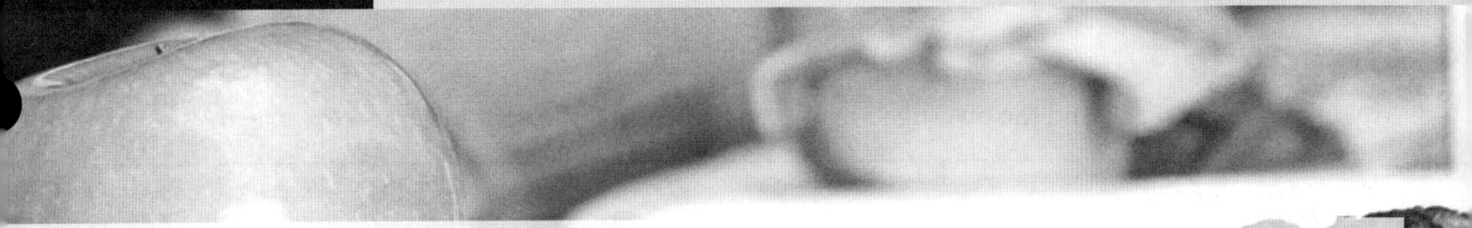

photo: Getty Images

UNIT 7: Data Modeling

7.1 Explore Normal Distributions

Lesson Objectives

- Describe the characteristics of a normal distribution.
- Use a variety of methods for estimating population percentages by finding areas under the normal curve.
- Explain how to use the mean and standard deviation of a data set to fit it to a normal distribution.

Essential Questions

- What is the importance of the mean and standard deviation in relation to the normal distribution?
- What does the empirical rule tell you about the data in a normal distribution?
- What do *z*-scores reveal about a data point?

photo: Getty Images

Investigations

Healthy Choices

When is data "normal"? Roll the die, analyze and graph the data. Watch that curve!

Cookie Packaging

Get empirically precise with your definition of normal.

Interpreting Test Scores

When data is normal, classify the deviations with *z*-scores.

Key Vocabulary

center, dot plot (line plot), mean, median (statistics), mode, normal distribution, outlier, standard deviation

Discover

As you complete Engage and the investigations, record the most important ideas you've learned.

Engage

Investigation 1

Investigation 2

Name ______________________________ Date ______________

Check for Understanding

Explore Normal Distributions: Investigation 1

1. Consider what the empirical rule says about a normal distribution. Fill in the blanks in the sentence below. *Give answers to the nearest tenth of a percent, when necessary.*

   Approximately ______% of the data fall within one standard deviation of the mean;

   approximately ______% of the data fall within two standard deviations; and

   approximately ______% of the data fall within three standard deviations.

2. On a national fitness test, the times of 47,000 high school students who ran the 100-meter dash were normally distributed, with a mean of 16 seconds, and a standard deviation of 2 seconds. Which statements are correct? *Select all that apply.*

   A. About 42,300 students had times within two standard deviations of the mean.

   B. About 44,650 students had times ranging from 12 seconds to 20 seconds.

   C. About 31,960 students had times within one standard deviation of the mean.

   D. About 42,300 students had times ranging from 14 seconds to 18 seconds.

   E. About 44,650 students had times within three standard deviations of the mean.

3. An entomologist caught 50 imperial moths at random and measured each of their wingspans in centimeters. Her results are given in the table shown below.

Wingspans of 50 Imperial Moths (cm)									
14	15	12	11	15	10	12	12	14	14
11	13	12	13	14	12	11	14	11	12
9	11	12	12	12	16	13	11	14	13
10	12	12	14	12	13	12	13	13	11
13	10	12	12	11	10	13	13	13	14

Use a calculator to determine the mean, median, mode, and standard deviation of the data. Does the data appear to approximate a normal distribution? Explain your answer.

Name ______________________________ Date ____________

Check for Understanding

Explore Normal Distributions: Investigation 2

1. On a particular day, the prices charged at all gas stations across a state for a gallon of gas were normally distributed, with a mean price of \$2.43 and a standard deviation of \$0.16. Which of the following statements are correct? *Select all that apply.*

   A. The z-score for a price of \$2.43 is $z = 0$.

   B. The z-score for a price of \$2.29 is positive.

   C. The z-score for a price of \$2.72 is negative.

   D. The z-score for a price of \$2.35 is the same as the z-score for a price of \$2.51.

   E. The z-score for a price of \$2.65 is greater than the z-score for a price of \$2.56.

2. Consider the positions on a normal curve of the data points with z-scores of $z = -1.39$ and $z = -1.11$, respectively. *Circle the correct words in the following sentence.*

   The data point with a z-score of $z = -1.11$ is to the [right / left] of the data point with a z-score of $z = -1.39$ and to the [right / left] of the mean.

3. The lengths of the songs available for download on a website are normally distributed, with a mean length of 3 minutes and 8 seconds, and a standard deviation of 24 seconds. Xavier downloaded a song with a length of 2 minutes and 50 seconds. *Fill in the blanks.*

   The z-score for Xavier's song is $z =$ ______, so to the nearest tenth of a percent, the song is longer than ______% of the songs available for download on the website.

4. All students in Penelope's school district had their resting heart rates measured in beats per minute (bpm). The heart rates were normally distributed, with a mean rate of 79 bpm, and a standard deviation of 8 bpm. Penelope's heart rate was faster than 6.68% of the students in the district. About how many beats per minute was her heart rate?

   A. 64 **B.** 67 **C.** 71 **D.** 74

5. The weights of the pumpkins in a pumpkin patch are normally distributed, with a mean weight of 18 pounds, and a standard deviation of 3 pounds. Approximately what percentage of the pumpkins weigh between 13 and 20 pounds? Explain your answer.

Summary

Before you attempt the Practice Exercises, review what you've learned.

When data are graphed, a common pattern often occurs. Numerical data values often cluster around the mean. This is called a **normal distribution**. When these data values are graphed in a **dot plot**, a shape called the **bell curve** appears. Since there are variations in data, the curves may look slightly different, but they are the same general shape.

The Bell Curve

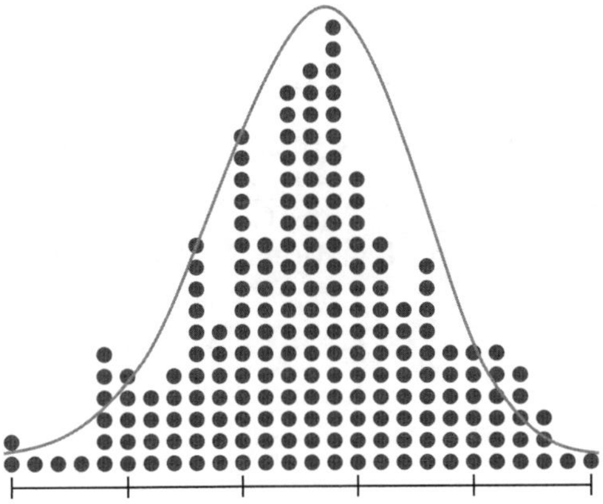

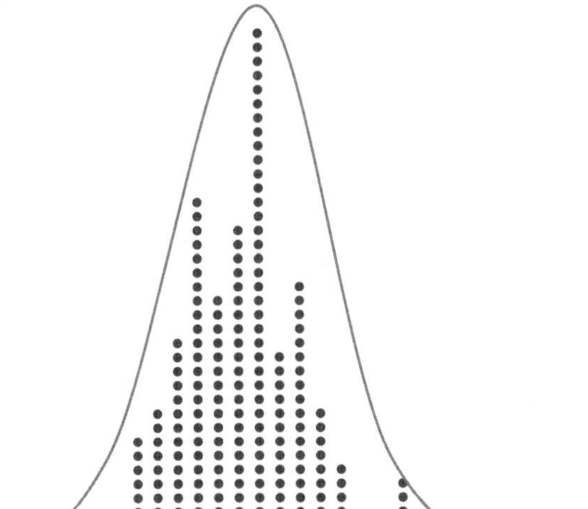

Characteristics of a Normal Distribution

In a normal distribution, the mean, median, and mode are the same and at the peak of the curve. The curve is symmetrical. Half of the data lie above the mean, and half below the mean.

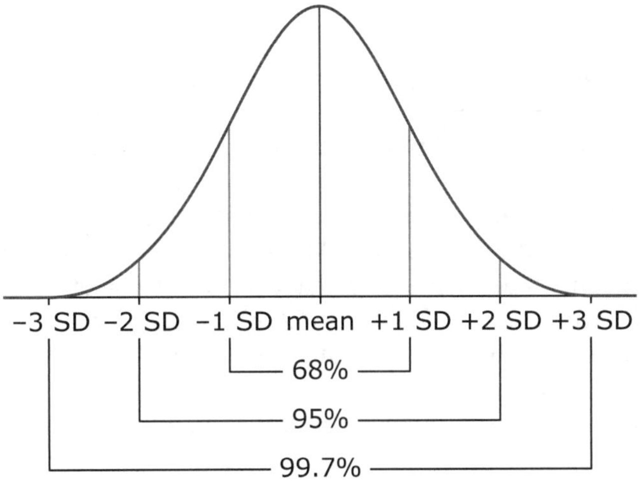

My Notes

Summary *(continued)*

Looking at the standard deviations above and below the mean, the data are separated into intervals. About 68 percent of the data lie within one standard deviation of the mean, 95 percent within two standard deviations, and 99.7 percent within three standard deviations of the mean. While data can vary, these percentages are not always exact, especially with smaller sample sizes. This is called the empirical rule, 68-95-99.7.

EXAMPLE: Normal Distribution

Cholesterol level in men is normally distributed with a mean of 4.8 and a standard deviation of 0.6. Since 4.2 is one standard deviation below the mean and 5.4 is one standard deviation above the mean, 68 percent of men have a cholesterol level between 4.2 and 5.4. Since 3.6 is two standard deviations below the mean and 6.0 is two standard deviations above the mean, 95 percent of men have a cholesterol level between 3.6 and 6.0. Since 3.0 is three standard deviations below the mean and 6.6 is three standard deviations above the mean, 99.7 percent of men have a cholesterol level between 3.0 and 6.6.

A z-score is a measure of the position of a data value. It tells you how many standard deviations the value is above or below the mean.

The formula for the z-score is $\frac{\text{data value} - \text{mean}}{\text{standard deviation}}$.

EXAMPLE: *Z*-Score

Cholesterol level in men is normally distributed with a mean of 4.8 and a standard deviation of 0.6. If a man has a cholesterol level of 5.2, the z-score is $\frac{5.2 - 4.8}{0.6} = 0.667$. A cholesterol level of 5.2 is 0.667 standard deviations above the mean. A cholesterol level of 4.5 has a z-score of $\frac{4.5 - 4.8}{0.6} = -0.5$. A level of 4.5 is 0.5 standard deviations below the mean.

My Notes

Practice Exercises

Review what you've learned using these practice problems. For practice problems with feedback, try the Coach and Play items in the Practice section online.

1. Ryan is doing some statistical analysis on the members of his rugby club. He graphs their weights and finds that the weights are normally distributed, with a mean of 220 pounds. He also finds that 25% of the team members are below 216 pounds.

   What percentage of the members would you expect to weigh more than 224 pounds?

   A. 25%

   B. 34%

   C. 50%

   D. 58%

   E. 75%

2. Emily is looking at a population of starfish. She finds that their size is normally distributed, with a mean of 12 centimeters. She determines there is a 34% probability that a randomly selected starfish is between 11 and 12 centimeters.

   What is the probability that a randomly selected starfish is between 12 and 13 centimeters?

   Probability = ______%

3. Marty measures the lengths of the stems of a certain type of flower and finds that the lengths are normally distributed with a mean of 26 centimeters and a standard deviation of 2 centimeters.

   Marty then picks one flower at random. What is the probability that the length of the stem is between 24 and 26 centimeters?

   A. 5%

   B. 15%

   C. 34%

   D. 68%

   E. 95%

4. Suppose there is a city where the mean cost per month of a one-bedroom apartment is $1,500, with a standard deviation of $250.

   Approximately what percentage of one-bedroom apartments in the city cost between $750 and $1,000?

   A. 49.85%

   B. 99.7%

   C. 2.35%

   D. 0.15%

   E. 0.3%

5. The Spanish class is testing how many vocabulary words they can recite in one minute. The count for the population is normally distributed, with a mean of 15 and a standard deviation of 2. Lisa is able to recite 3 more words than Matt. What are possible *z*-scores for Lisa and Matt?

   Lisa's *z*-score: ______

   Matt's *z*-score: ______

6. While practicing for a competition, the members of the academic bowl team take a total of 4 tests. To be eligible for the competition, a member must get an average score of 90 on the 4 tests. This equation shows the formula used to determine the average based on the first three scores (x, y, and z) and the last test score, L.

   $$90 = \frac{x + y + z + L}{4}$$

   As the last test approaches, several members want to calculate the score they must get to be eligible. What is the equation they can use to determine this value?

   ☐ = ☐ − ☐

Practice Exercises *(continued)*

7. Suppose that the number of sit-ups that students can do in one minute is normally distributed, with a mean of 30 and a standard deviation of 5.

Paul was able to do 38 in one minute. What is his *z*-score? *Express your answer to the nearest tenth.*

Paul's *z*-score is ______.

8. The gym class is taking a fitness test. Kayla and Sabrina are seeing how many one-legged hops they can do in 30 seconds. The count for the population is normally distributed, with a mean of 22 hops and a standard deviation of 3 hops.

Kayla's *z*-score is 2 and Sabrina's is (−1).

How many more one-legged hops did Kayla do than Sabrina?

Kayla did ______ more one-legged hops than Sabrina.

9. Natalie's bowling league tracks its members' bowling averages, which are normally distributed, with a mean of 160 pins per game. Natalie's bowling average is 143.

If her *z*-score is −1.7, what is the standard deviation (SD) of the members' bowling averages?

SD = ______

10. The scores on your last Statistics test were normally distributed, with a mean of 84% and a standard deviation of 3%. To show you how you compared with your classmates, the teacher did not record your grades as percentages; instead, she wrote them as *z*-scores. You look at your paper and see a zero (0) at the top.

What was your score as a percentage?

Score = ______%

11. Samuel and Carlos are competitive brothers who ran in the same 10K race. The race results showed that the runners' times followed a normal distribution. The data analysis showed that Samuel had a *z*-score of −1.4 and Carlos had a *z*-score of −0.7.

Using the partial *z*-score table below, find the percentage of runners who finished between Samuel and Carlos.

z	0.00	0.01	0.02	0.03	0.04	0.05	0.06	0.07	0.08	0.09
−1.6	0.0548	0.0537	0.0526	0.0516	0.0505	0.0495	0.0485	0.0475	0.0465	0.0455
−1.5	0.0668	0.0655	0.0643	0.0630	0.0618	0.0606	0.0594	0.0582	0.0571	0.0559
−1.4	0.0808	0.0793	0.0778	0.0764	0.0749	0.0735	0.0721	0.0708	0.0694	0.0681
−1.3	0.0968	0.0951	0.0934	0.0918	0.0901	0.0885	0.0869	0.0853	0.0838	0.0823
−1.2	0.1151	0.1131	0.1112	0.1093	0.1075	0.1056	0.1038	0.1020	0.1003	0.0985
−1.1	0.1357	0.1335	0.1314	0.1292	0.1271	0.1251	0.1230	0.1210	0.1190	0.1170
−1.0	0.1587	0.1562	0.1539	0.1515	0.1492	0.1469	0.1446	0.1423	0.1401	0.1379
−0.9	0.1841	0.1814	0.1788	0.1762	0.1736	0.1711	0.1685	0.1660	0.1635	0.1611
−0.8	0.2119	0.2090	0.2061	0.2033	0.2005	0.1977	0.1949	0.1922	0.1894	0.1867
−0.7	0.2420	0.2389	0.2358	0.2327	0.2296	0.2266	0.2236	0.2206	0.2177	0.2148
−0.6	0.2743	0.2709	0.2676	0.2643	0.2611	0.2578	0.2546	0.2514	0.2483	0.2451
−0.5	0.3085	0.3050	0.3015	0.2981	0.2946	0.2912	0.2877	0.2843	0.2810	0.2776
−0.4	0.3446	0.3409	0.3372	0.3336	0.3300	0.3264	0.3228	0.3192	0.3156	0.3121

A. 25.8% **B.** 41.92% **C.** 16.12% **D.** 24.2% **E.** 34%

Apply

Which Score Is Better?

photo: Getty Images

Many students take standardized tests for college applications. They are called standardized tests because they are scored so the population of student scores for any one particular test follows a normal distribution. The most common ones are the SAT and the ACT.

Research the mean and standard deviation for the ACT composite score, the SAT critical reading score, and the SAT mathematics score for the same year (for example, 2010). Make a bell curve for each set of data with markings on the horizontal axis for ±1 standard deviation, ±2 standard deviations, and ±3 standard deviations.

Use the results of your research to determine how unusual it is for each of the following to happen in that year:

- an ACT composite score of at least 32
- an SAT mathematics score of at least 760
- an SAT critical writing score of at least 780

Report your answers as percentiles. Mark each of these scores on your graph and shade in their corresponding areas.

Suppose a student wants to submit scores at the 90th percentile or higher for each test. What are the cutoff scores the student must earn to be able to submit all three scores?

Show what you've learned by completing the other performance tasks in the online Apply section.

Apply *(continued)*

Your answer to Apply will be assessed on the following criteria:

1. Researching to find the mean and standard deviation for the three tests and graphing a normal curve for each
2. Creating and marking the intervals for ±1 standard deviation, ±2 standard deviations, and ±3 standard deviations for each graph
3. Determining the probabilities of earning the three given scores, reporting the results as percentiles, and graphing the areas associated with the probabilities
4. Determining the correct score on each of the tests to meet the student's goal

Scale / Criteria	4	3	2	1	0
	Exceeds Criteria	**Meets Criteria**	**Progressing to Criteria**	**Below Expectations**	**No Expectation**
Research	Correctly completes the following six items: determines the mean and standard deviation for each of the three tests, and graphs a normal curve for each of the three tests.	Correctly completes five of the following: determines the mean and standard deviation for each of the three tests, and graphs a normal curve for each of the three tests.	Correctly completes three or four of the following: determines the mean and standard deviation for each of the three tests, and graphs a normal curve for each of the three tests.	Correctly completes one or two of the following: determines the mean and standard deviation for each of the three tests, and graphs a normal curve for each of the three tests.	Does not attempt task.
Standard Deviation Graph	Correctly completes the following nine items: marks the intervals for ±1 standard deviation, ±2 standard deviations, and ±3 standard deviations for each of the three graphs.	Correctly completes seven or eight of the following: marks the intervals for ±1 standard deviation, ±2 standard deviations, and ±3 standard deviations for each of the three graphs.	Correctly completes four, five, or six of the following: marks the intervals for ±1 standard deviation, ±2 standard deviations, and ±3 ±3 standard deviations for each of the three graphs.	Correctly completes one, two, or three of the following: marks the intervals for ±1 standard deviation, ±2 standard deviations, and standard deviations for each of the three graphs.	Does not attempt task.
Probabilities	Correctly completes the following six items: reports probabilities for the three scores as percentiles and graphs the area associated with each of the three probabilities.	Correctly completes five of the following: reports probabilities for the three scores as percentiles and graphs the area associated with each of the three probabilities.	Correctly completes three or four of the following: reports probabilities for the three scores as percentiles and graphs the area associated with each of the three probabilities.	Correctly completes one or two of the following: reports probabilities for the three scores as percentiles and graphs the area associated with each of the three probabilities.	Does not attempt task.
Scores to Meet Student's Goal	Correctly determines the three scores that the student must have in order to be considered in the 90th percentile or higher.	Correctly determines the three scores that the student must have in order to be considered in the 90th percentile or higher with minor errors.	Correctly determines two scores that the student must have in order to be considered in the 90th percentile or higher.	Correctly determines one score that the student must have in order to be considered in the 90th percentile or higher.	Does not attempt task.

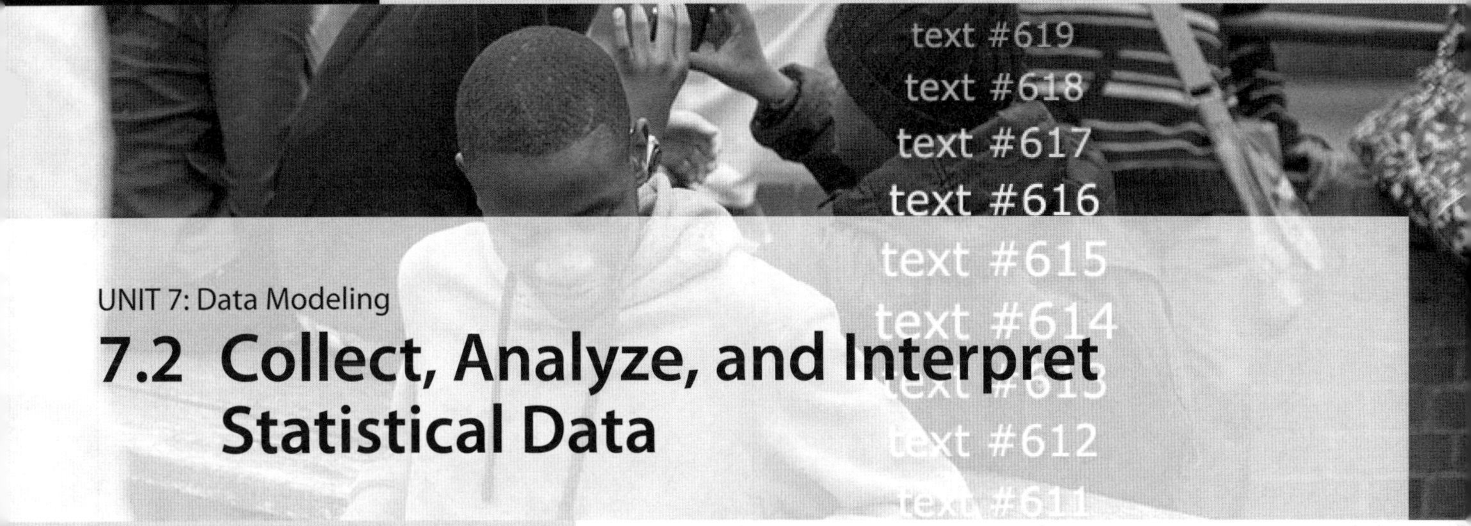

UNIT 7: Data Modeling

7.2 Collect, Analyze, and Interpret Statistical Data

photo: Getty Images

Lesson Objectives

- Describe the characteristics of sample surveys, experiments, and observational studies.
- Explain how randomization relates to a given sample survey, experiment, or observational study.
- Understand the relationship between a statistic and a population parameter.
- Estimate a population mean or proportion based on a sample survey and approximate a margin of error.
- Compare a statistical model to a particular experimental result, and decide whether the model is reasonable.
- Gather and interpret information from experiments or simulation.
- Evaluate reports based on data.

Essential Questions

- When and why is it necessary to use a sample statistic to estimate a population parameter?
- Why are random rather than non-random samples preferable for estimating population parameters?

Investigations

Phone Contest

What data analysis could you do to have a better chance at winning a contest?

Statistical Sampling Methods

Which is the better sample? Make sure your your sample is free of bias.

Modeling with Statistics

Conduct a skeeball simulation to help develop a mobile app. Is your simulation realistic?

Sample Proportions

How do sample size and method affect research results? Watch your margin of error!

Sample Mean

How high does your confidence level need to be? It depends on your tolerance for error.

Observational Study vs. Experiment

Not all conclusions are valid. How can you tell?

Key Vocabulary

bias, descriptive statistics, inferential statistics, margin of error, population, sample

Discover

As you complete Engage and the investigations, record the most important ideas you've learned.

Engage	**Investigation 1**
Investigation 2	**Investigation 3**
Investigation 4	**Investigation 5**

Name ______________________________ Date ______________

Check for Understanding

Collect, Analyze, and Interpret Statistical Data: Investigation 1

1. Researchers correctly used simple random sampling to obtain a sample of a magazine's subscribers. Two of the magazine's subscribers are Desmond and Ava. Desmond got selected for the sample, but Ava did not. Before the researcher took the sample, the chance of Desmond being selected was $\frac{1}{70}$. *Fill in the blank in the sentence below.*

   The chance of Ava being selected before the researcher took the sample was ____________

2. A bank with many branches has decided to obtain a sample of its customers and ask for their opinions about some new account types. *Circle the correct words in the sentence.*

   If the bank uses clustered sampling, it will likely choose customers from [homogeneous / heterogeneous] group(s), and if it uses stratified sampling, it will likely choose customers from [homogeneous / heterogeneous] group(s).

3. After a marathon, the organizers selected a random sample of 50 runners that was representative of all those who ran the race. They asked the runners to fill out a survey rating the race on a scale of 1 to 10. However, they only got 30 surveys back. Also, there were many 1's and 10's, but not a lot of ratings in-between the two extremes. From which type(s) of bias did the results of the marathon survey likely suffer?

   A. only under-coverage

   B. both non-response bias and under-coverage

   C. both under-coverage and voluntary-response bias

   D. both non-response bias and voluntary-response bias

4. Jamal works in the office of an association of college and university professors. He has a list of the association's members that is thousands of pages long, and each page has hundreds of names on it. He wants to use systematic sampling to obtain a sample of the members and then survey those in the sample to find their mean annual salary. He randomly selected a page from the list, and then chose every 5th member on that page to survey. Has Jamal used systematic sampling correctly? Is the mean salary he finds likely to be representative of the population of the association's members? Explain.

Name ______________________________ Date ____________

Check for Understanding

Collect, Analyze, and Interpret Statistical Data: Investigation 2

1. Jill wants to simulate an event with a 50% probability. She will number pieces of paper from 1 to 10, put them in a bowl, and randomly pick one. Picking one with which of these numbers on it could represent the event occurring, assuming that picking one with any other number on it represents the event not occurring? *Select all that apply.*

   A. an even number
   B. a number from 2 to 7
   C. a number from 5 to 10
   D. a number from 4 to 9
   E. a number from 1 to 5
   F. an odd number

2. Alberto will flip multiple coins at once to simulate an event that has a probability of 12.5%. A certain outcome will represent the event happening, and the rest of the possible outcomes will represent the event not happening. Fill in the blank.

   The fewest number of coins that Alberto can use in his simulation is ______.

3. A game of bowling consists of 10 frames. In all but the last frame, a bowler has a maximum of 2 throws of her ball to knock down 10 pins, which are numbered from 1 to 10. If the bowler doesn't knock down all 10 pins on her first throw, then she attempts to "pick up the spare" by knocking down the remaining pins on her second throw. The table shown below gives the number of spares attempted and picked up in a professional bowling league for some common splits, which are combinations of non-adjacent pins.

Split	Number of Spares Attempted	Number of Spares Picked Up
6-8	174	29
6-8-10	342	38
8-10	252	7

   Suppose you want to simulate each of these splits individually by rolling two number cubes, each with faces that are labeled with the integers 1 to 6, and finding the sum of the two numbers that come up. For each split, a certain sum will represent picking up the spare, and the rest of the possible sums will represent not picking up the spare. Which sum(s) could appropriately represent picking up the spare for each split? Explain.

Name ______________________________ Date ______________

Check for Understanding

Collect, Analyze, and Interpret Statistical Data: Investigation 3

1. A spinner is equally divided into the colors blue, green, red, and yellow. Tim and Sophie each spun the arrow 500 times. Tim counted it landing on blue in 100 of his spins, while Sophie counted it landing on blue in 125 of her spins. Who must have counted wrong?

   A. only Tim
   B. only Sophie
   C. both Tim and Sophie
   D. neither Tim nor Sophie

2. To see if they would make use of expanded hours, a museum surveyed a random sample of its patrons. After analyzing the results, the museum concluded that for 11% of its patrons, expanded hours would be useful. The margin of error of the survey was 3.5%. Which percentages fall within the margin of error? *Select all that apply.*

   A. 11%
   B. 3%
   C. 16%
   D. 8%
   E. 7%
   F. 13%
   G. 1%
   H. 15%

3. Of all licensed drivers in the United States, 38% are registered to be organ donors. Suppose many random samples of 1,000 licensed drivers were taken, and the proportion of those in each sample who are registered to be organ donors was calculated. *Fill in the blank.*

   As a decimal rounded to four places, the standard deviation of the distribution of the sample proportion of those registered to be organ donors is ____________.

4. A random sample of a movie theater's customers was taken, and the proportion of those in the sample who said they like the concession stand's popcorn was calculated. Now, a 95% confidence interval will be determined for the proportion of all the movie theater's customers who like the popcorn. Circle the correct word in the following sentence.

   If the sample size was 50, then the 95% confidence interval will be [narrower / wider] than if the sample size was 75.

5. When a public radio station surveyed a random sample of its listeners, it found that 165 of the 250 respondents preferred jazz music to classical music. At a 95% confidence level, what percentage of all of the station's listeners prefers jazz to classical? Explain.

Name ____________________ Date __________

Check for Understanding

Collect, Analyze, and Interpret Statistical Data: Investigation 4

1. Each of the 1,800 students in Malcom's high school rolled a number cube with faces labeled with the integers 1 to 6 a total of 225 times and calculated the mean of numbers that came up. The students then added all of their individual means together and divided the sum by 1,800. Which of these is most likely closest to the result they got?

   A. 2.5 **B.** 2.9 **C.** 3.5 **D.** 3.9

2. A large corporation randomly chose 100 of the employees in its offices and measured their typing speeds. The mean speed was 45 words per minute, and the standard deviation was 10 words per minute. Suppose the corporation were to choose 15,000 more random samples of 100 of its employees and find the mean typing speed of those in each sample. Which of these statements are correct? *Select all that apply.*

   A. About 14,250 of the means would be between 43 and 47 words per minute.

   B. About 10,200 of the means would be between 35 and 55 words per minute.

   C. About 14,955 of the means would be between 42 and 48 words per minute.

   D. About 13,500 of the means would be between 25 and 65 words per minute.

   E. About 12,000 of the means would be between 44 and 46 words per minute.

3. The visitor's bureau on a tropical island surveyed 50 tourists at random and asked them how long their vacations would last. The results are given in the table shown below.

Durations of 50 Tourists' Vacations (days)									
10	10	7	9	7	10	10	8	9	9
8	7	8	10	10	7	8	6	10	10
9	6	10	9	9	8	8	12	11	13
11	7	5	9	8	9	12	9	13	9
9	9	9	10	11	7	9	8	11	12

Use a calculator to help determine the mean length in days of the vacations of all tourists who visit the island at a 95% confidence level. Explain your answer.

Name ______________________________ Date ____________

Check for Understanding

Collect, Analyze, and Interpret Statistical Data: Investigation 5

1. Consider the difference between an experiment and an observational study. *Then circle the correct words in the sentence shown below.*

   A researcher does not try to manipulate the independent variable in an [experiment / observational study], while a researcher does try to manipulate the independent variable in an [experiment / observational study].

2. Lindsay attempted to do an experiment to investigate the effectiveness of bee venom in relieving tendonitis. She randomly assigned volunteers reporting the condition into one of two groups. To those in group A, she applied bee venom cream, and to those in the group B, she applied a placebo cream. Then, she asked those in both groups to fill out a survey reporting how they felt afterwards. Which statement is correct?

   A. Both group A and group B were control groups.

   B. Both group A and group B were treatment groups.

   C. Group A was a control group, and group B was a treatment group.

   D. Group A was a treatment group, and group B was a control group.

3. An experiment's results seem to indicate that exercise helps reduce insomnia. The mean number of hours of sleep per night for those in the control group was 5.4. Which of these could be the mean number of hours of sleep per night for those in the treatment group? *Select all that apply.*

   A. 4.1 **B.** 4.4 **C.** 4.8 **D.** 5.1

   E. 5.4 **F.** 5.9 **G.** 6.3 **H.** 6.6

4. In an experiment on a company's e-commerce website, users who clicked on a link to purchase a particular pair of shoes were randomly shown either the regular price of $49.99, or a higher price of $69.99. The company found that the ratio of those shown a price of $49.99 who purchased the shoes was $\frac{17}{100}$, while the ratio of those shown a price of $69.99 who purchased the shoes was $\frac{13}{100}$. Based on the results of the experiment, should the company charge $49.99 or $69.99 for the shoes? Explain.

Summary

Before you attempt the Practice Exercises, review what you've learned.

Introduction to Statistics

There are many terms used to describe statistics.

> A **population** is the entire group that is being studied. The population can be people, animals, or objects. A **sample** is a portion of the population.

A population can be all of the people in the world, tenth-grade girls in the United States, giraffes in Africa, or trees in Oregon. A researcher must clearly define a population that he or she wants to study. If the population is too large, a sample is studied. If the sample is chosen correctly, it will be representative of the population, so the information gathered from the sample will be similar to information about the entire population.

> A **parameter** is a characteristic of a population. A **statistic** is a characteristic of a sample.

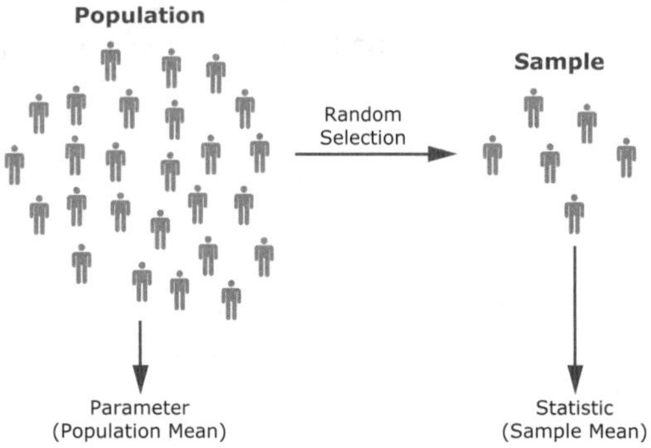

A parameter is often unknown, since information from the entire population can be impossible to gather. For example, imagine trying to find the mean height of every person in the world. Statisticians use information from samples to estimate a parameter, and different samples may produce different statistics. The mean height of different samples is variable; it can change. The parameter does not change. Often, statisticians use many statistics from different samples to estimate the parameter.

> **Descriptive statistics** summarizes data. This includes mean, median, mode, range, and standard deviation. Summaries are often represented in tables and graphs. **Inferential statistics** draws conclusions about a population based on data from a sample.

My Notes

Sampling Techniques

Information is often needed about a population, but it is usually too difficult and expensive to study the whole population. Samples can be used to make estimates about a population, as long as the sample is chosen correctly. If a sample is not chosen correctly, then the results of the study cannot be generalized to the population. For example, using a sample of the heights of the tenth-graders at only one school will give results that cannot be generalized to the heights of all the tenth-graders in the United States.

Simple Random Sampling

Each member of the population has an equally-likely chance of being selected.

Example: randomly choosing a set of students from the entire population of tenth graders.

Summary *(continued)*

Systematic Sampling

Every *n*th person of a random list of the population is selected.

Example: choosing every 100th student.

Stratified Sampling

The population is separated into homogenous groups, and a random sample from each group is selected.

Example: separating students by region, heritage, demographics, or other factors and randomly selecting students from each group.

Cluster Sampling

The population is separated into heterogeneous groups that mirror the characteristics of the population, and then a sample of the groups is selected.

Example: separating the students into groups of 1,000 and randomly selecting one student from each group.

Bias

During a study, bias can occur when the sample has not been chosen fairly. Bias can lead to results that are not accurate, which can cause incorrect inferences about the population.

Undercoverage

A certain group in the population is underrepresented in the sample. For example, if a sample consists of 70% girls and 30% boys, and the population is about half of each gender, boys are underrepresented.

Non-response Bias

Individuals chosen to be in the sample are not willing or able to participate. For example, if a sample is selected for a phone survey and some people do not answer the phone or refuse to participate in the survey, data is not collected from the entire sample.

Voluntary Response Bias

Individuals who volunteer for a study often have strong opinions and are overrepresented in the sample.

Descriptive Statistics and Inferential Statistics

There are two branches of statistics that have different purposes. Descriptive statistics summarizes data collected from a sample. Inferential statistics draws conclusions from the data collected.

Sample Mean – the arithmetic mean of the data in a sample.

Population Mean – the actual, fixed mean of the population.

Margin of Error – the amount above or below the mean that the sample mean may be different from the population mean.

Standard Deviation – a numerical value that describes how spread out the data are.

Standard Error – the standard deviation of the sample that is used to estimate the standard deviation of the population.

The larger the amount of data used to calculate the sample mean, the more accurately the sample mean represents the population mean and the smaller the standard error.

My Notes

Summary *(continued)*

Causation

Simple experiments are used to test hypotheses.

The **independent variable** is the variable that the experimenter manipulates.

The **dependent variable** is the outcome variable, the variable that is measured.

EXAMPLE: Causation

Does a test printed on yellow paper increase test performance? The independent variable is the color of the paper the test is printed on, and the dependent variable is test performance.

A simple experiment can be used to determine whether the cause—the color of the paper a test is printed on—has an effect on test performance.

A **control group** is the group that receives no treatment in an experiment (the test is on white paper).

A **treatment group** is the group that experiences a treatment in an experiment (the test is on colored paper).

In an experiment, participants must be randomly assigned into the two groups, so that the groups are as identical as possible. It is important that only one variable is changed in the experiment, so that the changes measured can be attributed to one variable.

EXAMPLE: Causation

The students who complete the test on white paper are in the control group, and the students who complete the test on yellow paper are in the treatment group.

Simulations

Simulations are used in a controlled setting to imitate behaviors that occur due to chance. Random numbers are often used to represent the events of interest. After performing numerous repetitions of the simulation, it can be evaluated to see how well the simulation aligns with actual behaviors, in order to see how useful it can be in the future.

In the practice of statistics, **random** selection indicates that each item has an equal chance of being chosen. There is no way to predict the outcome, such as when you roll dice or deal properly-shuffled cards. "Random" number generators actually produce **pseudo-random** numbers. They seem random, but they are predictable given the algorithm.

EXAMPLE: Creating a Simulation

Polygraph machines (commonly called lie detectors) are used by the police to aid in determining if people who are suspected of committing crimes are telling the truth. While it is said that these machines are good at detecting lies, there is some controversy over how well they can detect honesty. Some experts say that the rate for false positives (detecting a lie where there is not one) is as high as 35% or 0.35. Create a simulation to determine the amount of false positives one might find in 5 questions.

Use a random number generator for whole numbers 1 – 100; 1 through 35 could mean false positive. Generate 5 of these to simulate the polygraph problem.

My Notes

Practice Exercises

Review what you've learned using these practice problems. For practice problems with feedback, try the Coach and Play items in the Practice section online.

1. Bethany wants to find out what theme the seniors would like to have for their prom this year. She decides to ask every third person who bought a ticket in advance about his or her preference.

   What type of data collection design did Bethany use?

   A. survey

   B. experiment

   C. observational study

   D. census

2. Brandy is reading about several investigations that tried to determine how much material households recycle each week.

   In investigation A, researchers used property-tax records in one town to randomly select 1,000 households to fill out a survey.

   In investigation B, the researchers accompanied a truck for one day as it picked up recycling from households on its route. The researchers examined the recycling bins at each stop.

   Of the two experiments, which one is an observational study?

   Investigation ______ is an observational study.

3. Several big media companies and Internet service providers are tussling over "net neutrality," the idea that any website should be available to users and those providers should not block any particular sites, pages, or products.

   To gauge support for its position, one media company uses a survey on its web page where customers go to pay their bills online. It reports that 80% of its customers are in favor of the company's position.

   Is this data-collection method biased or unbiased?

   This data-collection method is ____________.

4. Dr. Flores wants to study whether liver tissue is damaged in people who smoke.

   Which of these best describes how the study should be conducted?

   A. Create an experiment on non-smokers that are 25 to 30 years old. Two groups are randomly selected where subjects in one group continue not to smoke and the subjects in the other group smoke for the remainder of the study. Dr. Flores observes the livers of all the subjects at the beginning and end of the study.

   B. Conduct an observational study. Dr. Flores compares the livers of randomly chosen non-smokers to the livers of randomly chosen people who have been smoking for at least 5 years previous to the observations.

   C. Design a survey by choosing doctors at random and having them ask their patients suffering from liver problems if they smoke.

   D. Design a survey by choosing doctors at random and having them ask their smoking patients whether they have developed liver problems since they started smoking.

5. Margaret is trying to determine the mean amount of time her 200 classmates spent studying for an exam. She randomly selects a sample of 40 classmates and finds that they spent an average of 1.32 hours studying for the exam.

   Approximate the average time that all 200 students spent studying for the exam.

   The average amount of time that all students spent studying for the exam was about ____________ hours.

Practice Exercises *(continued)*

6. Rosita rolls a number cube 6,000 times and records how many times the cube shows each of the numbers 1 through 6. She finds that the number 6 came up 995 times.

What can Rosita conclude about the number cube?

A. The number cube is weighted unfairly because 6 does not come up as often as it should.

B. The number cube is weighted unfairly because 6 comes up more often than it should.

C. There is more than one 6 on the number cube.

D. Based on these results, Rosita cannot conclude that the number cube is unfair.

7. Tony's summer internship is at a lab that studies Lyme disease. One of his duties is to trap mice and to check any ticks on the mice for the presence of Lyme disease.

Tony cannot catch every mouse, but he does trap 642 over the course of the summer. He finds that 282 of the mice carry ticks with Lyme disease.

What is Tony's best estimate of the population proportion of mice with ticks carrying Lyme disease?

Tony's best estimate of the population proportion of mice with ticks carrying Lyme disease is ____________.

8. Paul runs a computer simulation on a spinner with 8 equally likely colored sections. After several thousand virtual spins, he has the results shown in the table.

Red	2,487
Blue	1,258
Green	1,240
Orange	2,512
Yellow	2,503

Based on Paul's simulation, what is the likely number of yellow sections on Paul's spinner?

There are likely ____________ yellow sections on the spinner.

9. Maria wants to see whether the head of the household income is related to the number of people in the household in the United States.

Which of these would be a good method of collecting data for a report?

A. Conduct a survey of the other families in her school.

B. Look up data on the U.S. Census Bureau's website.

C. Use a library book published in 1980 that contains information on household size and income of the head of the household.

D. Look up information from a reliable study that includes the number of heads of household who report they are satisfied with their income, separated into categories of low, medium, and high income.

Apply

How Many Letters Are in an Average Word?

photo: Discovery Education

Have you ever thought about words that are very short or very long? Have you ever wondered about the average length of English words?

Consider your own experiences and make an estimate of what you think the typical number of letters in a word are. Write a concise rationale for your estimate.

Find a passage from a classic text or from one of these resources to use as a data source.

- A piece of current fiction
- An article from the Internet that contains informational text
- A page from a science or history textbook
- A page of dialogue from a play or a television show

Consider the different types of sampling procedures you have learned. Devise two different plans that you will use on the data. Each one should employ a different sampling technique. Repeated words can be used for your data since each occurrence is a new use of the word.

Carry out your plans and compare the results of each technique. Make a decision about which method gives the best approximation of the average number of letters in a word, and write a brief explanation for your choice.

Finally, show the results of the word counts for each of your two samples as a dot plot.

Show what you've learned by completing the other performance tasks in the online Apply section.

Apply *(continued)*

Your answer to Apply will be assessed on the following criteria:

1. Making an estimate of the average number of letters in a word and providing a clear rationale for your estimate
2. Devising, implementing, and documenting two plans using any two of the following: cluster sample, simple random sample, stratified random sample, or systematic sample
3. Deciding which sampling technique is more appropriate and justifying why that method is the better choice
4. Displaying dot plots from each of the two samples

Scale / Criteria	4 Exceeds Criteria	3 Meets Criteria	2 Progressing to Criteria	1 Below Expectations	0 No Expectation
Estimate	Gives a distinct estimate of the average number of letters in a word with a clear rationale that supports that average.	Gives a distinct estimate of the average number of letters in a word with a rationale that weakly supports that average.	Gives a distinct estimate of the average number of letters in a word with a rationale that does not support that average.	Does not give a distinct estimate of the average number of letters in a word or a rationale that supports the average.	Does not attempt task.
Plan	Correctly uses two sampling techniques to determine the average length of a word.	Uses two sampling techniques to determine the average length of a word with minor errors or lack of clarity.	Correctly uses one sampling technique to determine the average length of a word.	Incorrectly uses two sampling techniques to determine the average length of a word.	Does not attempt task.
Justification	Makes a clear decision about which of the chosen sampling techniques is more appropriate and justifies that choice.	Makes a clear decision about which of the chosen sampling techniques is more appropriate but does not fully justify that choice.	Makes a decision about which of the chosen sampling techniques is more appropriate but does not justify that choice in any way.	Does not make a clear decision about which of the chosen sampling techniques is more appropriate; includes no justification.	Does not attempt task.
Data Display	Correctly displays the data from both of the sampling methods.	Displays the data from both of the sampling methods with minor errors.	Correctly displays the data from one of the two sampling methods.	Attempts to, but does not correctly display any of the data.	Does not attempt task.

UNIT 8: Nonlinear Functions

8.1 Create and Analyze Piecewise Functions

photo: Getty Images

Lesson Objectives

- Analyze piecewise functions.
- Create and graph piecewise-defined functions.
- Interpret piecewise functions in real-world contexts.

Essential Question

- How are algebraic, numeric, and graphical representations of piecewise functions related?

photo: Getty Images

Investigations

Speed of a Racecar

How does a racecar driver's speed change throughout a race? It's certainly not linear!

Graphing a Racecar's Changing Speed

Take a closer look at each piece of the speed graph. Get detailed. Set restrictions.

Mount Everest Expedition

Can a function describe distance? Absolutely! The position of the graph is relative to the parent.

Rube Goldberg Machine

Piecewise functions do not have to be continuous. How does this affect the the domain?

Key Vocabulary

absolute value function, discontinuous function, parent function, piecewise function, step function

Discover

As you complete Engage and the investigations, record the most important ideas you've learned.

Engage	Investigation 1
Investigation 2	Investigation 3

Name ______________________________ Date ______________

Check for Understanding

Create and Analyze Piecewise Functions: Investigation 1

An airplane is cruising at an altitude of 30, 000 feet. The graph shows the remainder of the flight as the pilot makes adjustments to avoid turbulence and other air traffic.

Use the graph of the airplane's altitude over time to answer questions 1–3.

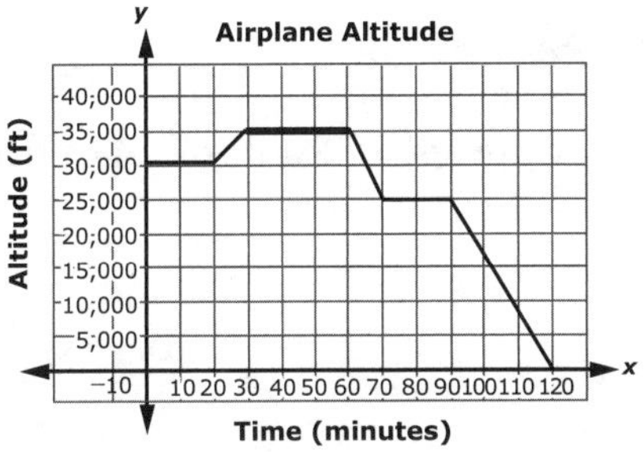

1. Use inequalities to identify the domain and range of the function that describes all the recorded altitudes.

   A. Domain: ______________

   B. Range: ______________

2. Match each equation to the interval over which it is defined on the graph.

A. $y = -1{,}000x + 95{,}000$	I. $0 \le x < 20$
B. $y = 25{,}000$	II. $20 \le x < 30$
C. $y = 30{,}000$	III. $30 \le x < 60$
D. $y = -\frac{2{,}500}{3}x + 100{,}000$	IV. $60 \le x < 70$
E. $y = 500x + 20{,}000$	V. $70 \le x < 90$
F. $y = 35{,}000$	VI. $90 \le x \le 120$

3. Fill in the missing values for the piecewise function.

   A. $f(50) =$ ______ B. $f(20) =$ ______ C. $f(90) =$ ______

4. How can you determine the interval to which a function value belongs when it is at the end point of a segment on the graph?

Name ______________________________ Date ______________

Check for Understanding

Create and Analyze Piecewise Functions: Investigation 2

1. An injured climber is on a 5,245-foot mountain at an elevation of 3,875 feet. A rescue team is in the area. Fill in the missing values in the table.

Rescue Team's Elevation, x	2,925			4,430
Vertical Distance from Climber, y		585	305	

2. Which functions model the position for the rescue team's distance from the injured climber? *Select all that apply.*

A. $f(x) = \begin{cases} 3,875 - x, & \text{if } 0 \le x < 3,875 \\ x - 3,875, & \text{if } 3,875 \le x \le 5,245 \end{cases}$

B. $f(x) = |3,875 - x|$

C. $f(x) = \begin{cases} x - 3,875, & \text{if } 0 \le x < 3,875 \\ 3,875 - x, & \text{if } 3,875 \le x \le 5,245 \end{cases}$

D. $f(x) = |x - 3,875|$

3. Choose the graph that models the rescue team's distance to the injured climber.

A.
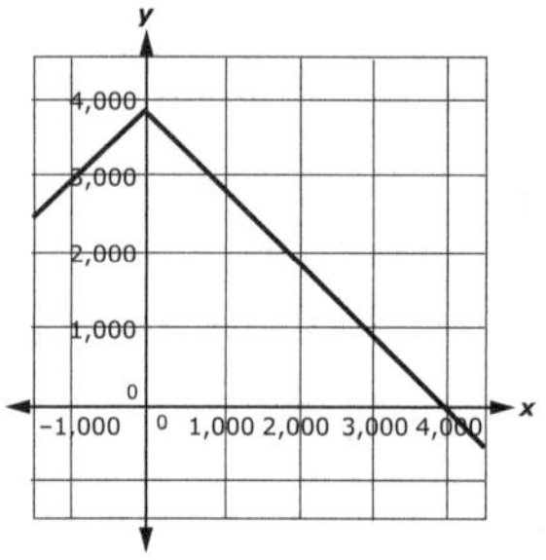

B.
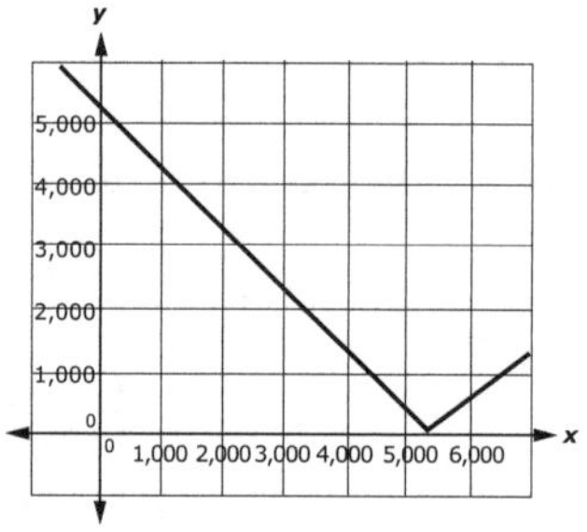

C.
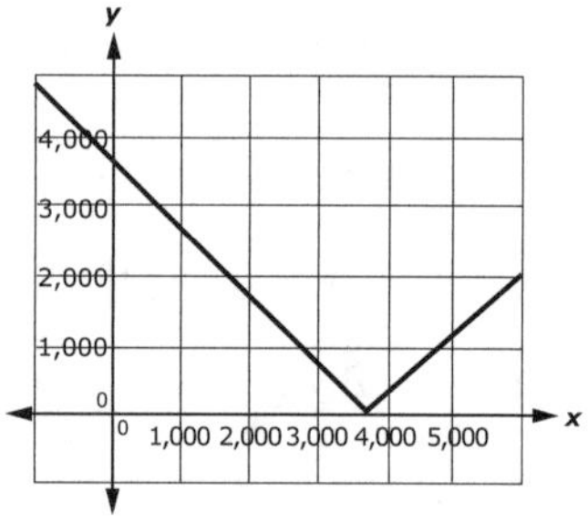

4. *Circle the terms to complete the following.* The graph of $f(x) = |x + 5| - 3$ is a shift of $f(x) = |x|$ to the [left 3 / left 5 / right 3 / right 5] units, and [up 3 / up 5 / down 3 / down 5] units.

5. Write $f(x) = |x - h| + k$ as a piecewise function on the domain $[0, c]$, and identify the minimum point on the graph.

Name ______________________________ Date ______________

Check for Understanding

Create and Analyze Piecewise Functions: Investigation 3

1. Define the intervals for the domain of the graph below.

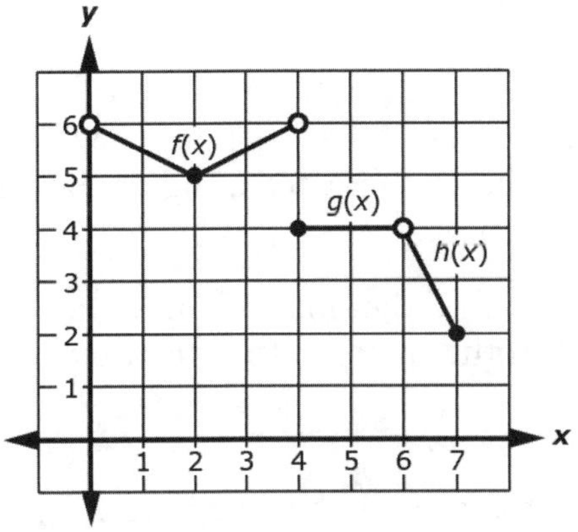

A. $f(x)$: ______________

B. $g(x)$: ______________

C. $h(x)$: ______________

2. Write the interval definitions for the following piecewise function to make it continuous at $x = 0$, discontinuous at $x = 3$, and continuous on the rest of its domain to $x = 8$.

$$f(x) = \begin{cases} 2x^2 & \text{if } \_\_\_\_\_\_ \\ -2x + 24 & \text{if } \_\_\_\_\_\_ \\ 2^x & \text{if } \_\_\_\_\_\_ \end{cases}$$

3. Choose the true statements from the following. *Select all that apply.*

   A. A piecewise function is continuous when it is defined for all values on its entire domain.

   B. A discontinuity can occur only at an end point of a defined interval.

   C. A shared interval endpoint should be included in both interval definitions.

   D. A break or gap in a piecewise graph is the result of adjacent function intervals that do not have the same y for the corresponding x.

4. Describe two types of discontinuities and how you can tell from a set of equations what type they are and whether the set describes a function.

Summary

Before you attempt the Practice Exercises, review what you've learned.

A **piecewise function**, also called a **piecewise-defined function**, is a function that behaves differently depending on the x value. A piecewise function is a combination of different equations applied to different non-overlapping parts of the function's domain.

EXAMPLE: Write Equations for Piecewise Functions

Write equations for the piecewise function shown. Identify the domain and range and where the graph is increasing and decreasing. Assume the graph extends to the left and right (not shown).

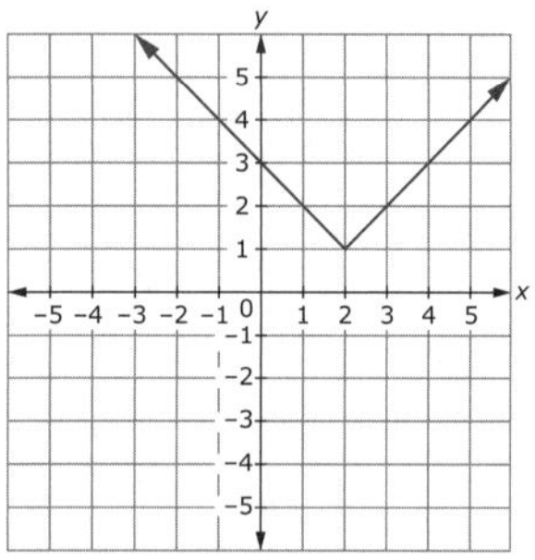

SOLUTION:

To find the equation for the function's piece where $x \le 2$, use two points to find the slope. Then, use one of the points to write the equation of the line.

$$y = -x + 3$$

The domain for this piece of the function is $x \le 2$ and the range for this piece of the function is $f(x) \ge 1$. This piece of the function is decreasing, as shown by its negative slope of −1.

Use two points to find the slope. Then write the equation for the other piece of this function.

$$y = x - 2$$

The domain for this piece of the function is $x > 2$, and the range for this piece of the function is $f(x) > 1$. This piece of the function is increasing, as shown by its positive slope of 1.

Use the two pieces to write the function shown on the graph and to define each piece's domain.

$$f(x) = \begin{cases} -x + 3 & \text{if } x \le 2 \\ x - 1 & \text{if } x > 2 \end{cases}$$

EXAMPLE: Graph a Piecewise Function

Graph the piecewise function below and find the values of $f(-1)$, $f(0)$, $f(2)$, $f(7)$, and $f(8)$.

$$f(x) = \begin{cases} 0.5x + 2 & \text{if } 0 \le x < 2 \\ 3 & \text{if } 2 \le x < 4 \\ -x + 7 & \text{if } 4 \le x \le 7 \end{cases}$$

For which of the equations is the function increasing? Decreasing?

SOLUTION:

To graph the function, graph each piece of the function and define the domain restrictions for each piece.

My Notes

Summary *(continued)*

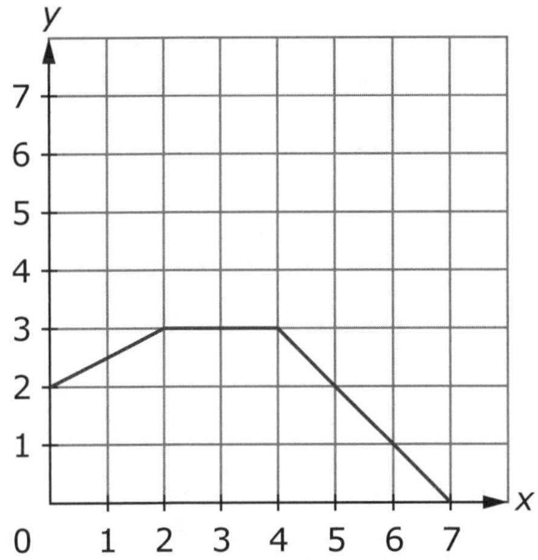

To find the values of $f(-1)$, $f(0)$, $f(2)$, $f(7)$, and $f(8)$, substitute the x values into the appropriate pieces of the piecewise function.

For $f(-1)$, the function is undefined.

For $f(0)$, the function is defined as $0.5x + 2$. Substitute 0 for x and solve.

$$f(0) = 0.5(0) + 2 = 2$$

For $f(2)$, the function is defined as 3. So, $f(2) = 3$.

For $f(7)$, the function is defined as $-x + 7$. Substitute 7 for x and solve.

$$f(7) = -7 + 7 = 0$$

For $f(8)$, the function is undefined.

The function is increasing when $0 \le x < 2$, and the function is decreasing when $4 \le x \le 7$. The function is neither increasing nor decreasing when $2 \le x \le 4$. This can be shown by the graph when you examine the domain for each piece of the function.

> **A** The **absolute value function** is a function defined by an expression with absolute value symbols. For example, $f(x) = |x|$ is an absolute value function that can also be written as the piecewise function defined below.
>
> $$f(x) = |x| = \begin{cases} x & \text{if } x \ge 0 \\ -x & \text{if } x < 0 \end{cases}$$

The absolute value function $f(x) = |x|$ is also called the **parent function** of all absolute value functions. Its vertex is at (0, 0).

The domain of $f(x) = |x|$ is all real numbers and the range is the set of non-negative real numbers. The vertex is a minimum located at (0, 0).

The function $f(x) = |x|$ can be transformed or translated by a, k, and h in the equation $g(x) = a \cdot |x - h| + k$.

- The graph of $g(x) = |x| + k$ is the graph of $f(x) = |x|$ shifted k units up if k is positive.
- The graph of $g(x) = |x| + k$ is the graph of $f(x) = |x|$ shifted k units down if k is negative.
- The graph of $g(x) = |x - h|$ is the graph of $f(x) = |x|$ shifted h units to the right if h is positive.
- The graph of $g(x) = |x - h|$ is the graph of $f(x) = |x|$ shifted h units to the left if h is negative.
- The graph of $g(x) = -|x|$ is the graph of $f(x) = |x|$ reflected across the x-axis.

EXAMPLE: The Absolute Value Function Translations

Translate the graph of $f(x) = |x|$ so that the coordinates of the vertex are $(-1, 3)$. What is an equation of the graph?

My Notes

Summary *(continued)*

To translate the graph of $f(x) = |x|$ 1 unit to the left of (0, 0), add 1 to x.

Then to translate the graph 3 units up, add 3 to $|x + 1|$. The result is shown in the graph of $f(x) = |x + 1| + 3$ below.

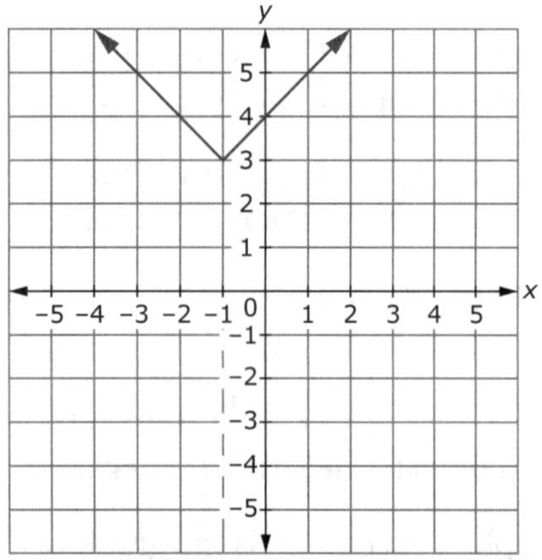

A **continuous function** is a function with a connected graph and no holes. An absolute value function is a continuous function defined below as a piecewise function.

$$f(x) = |x| = \begin{cases} x & \text{if } x \geq 0 \\ -x & \text{if } x < 0 \end{cases}$$

A **discontinuous function** is a function that is not defined at certain values for x or the function may have breaks or gaps.

The piecewise function shown is discontinuous at $x = 2$.

$$f(x) = \begin{cases} 1 & \text{if } 0 \leq x < 2 \\ x & \text{if } x \geq 2 \end{cases}$$

The two pieces have a different value at $x = 2$. The graph shows that the function seems to jump from one piece to the other.

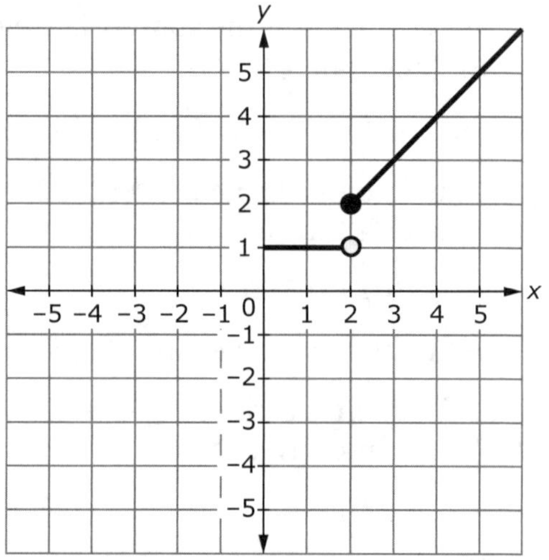

Many real-life situations are represented by a combination of equations or piecewise functions, including the absolute value function when calculating distance. For example, when calculating time compared to the speed of a roller coaster ride, the ride can start out slowly by going up an incline. The speed is greatly increased while the coaster goes down the incline, and then the speed changes yet again when the coaster approaches a sharp curve.

My Notes

Practice Exercises

Review what you've learned using these practice problems. For practice problems with feedback, try the Coach and Play items in the Practice section online.

1. The graph of $g(x)$ shown below is a translation of $f(x) = |x|$.

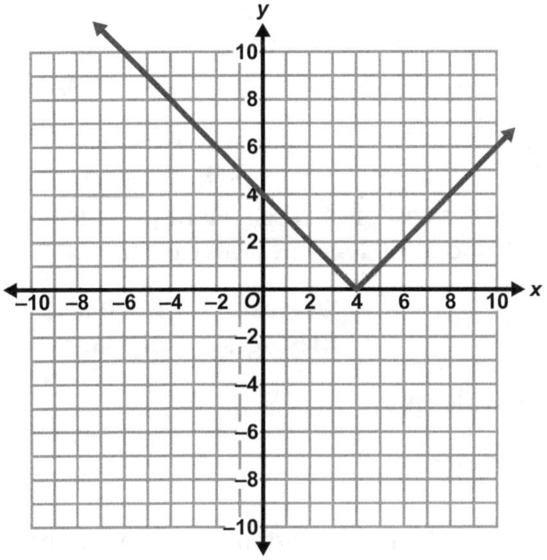

Write the equation for $g(x)$ in the format $g(x) = a\,|x - h| + k$.

Replace each of the variables a, h, and k in the equation with the correct number.

$g(x) =$ ____________________

2. The Sweet Factory makes candy and sells it by the pound to stores. The charge for candy in bulk is \$3.25 per pound for quantities of less than 40 pounds. For orders of 40 pounds or more, the Sweet Factory charges \$2.90 per pound for the entire order.

So that it does not have a drop in revenue where the price per pound changes, the Sweet Factory charges a processing fee of k dollars to make the cost continuous. If p is the number of pounds of candy ordered, then the piecewise function below gives the total cost C of the order in dollars.

$$C(p) = \begin{cases} 3.25p & \text{if } 0 < p < 40 \\ 2.90p + k & \text{if } p \geq 40 \end{cases}$$

Find the value of k so that the cost function $C(p)$ is continuous.

The processing fee is $k =$ \$__________.

3. Analyze the graph of a piecewise function $f(x)$.

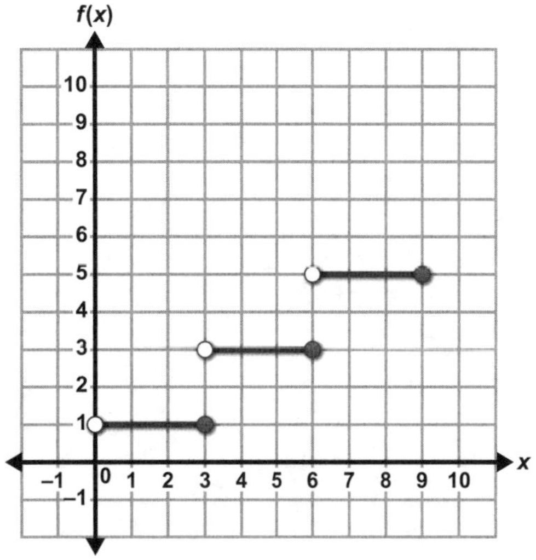

Define the intervals for this piecewise function.

$$f(x) = \begin{cases} 1 & \text{if } 0 \underline{\quad\quad} x \underline{\quad\quad} 3 \\ \underline{\quad\quad} & \text{if } 3 < x \underline{\quad\quad} 6 \\ 5 & \text{if } 6 \underline{\quad\quad} x \underline{\quad\quad} 9 \end{cases}$$

4. Rewrite the function $f(x)$ as an absolute value function $g(x)$.

$$f(x) = \begin{cases} x - 4 & \text{if } x \leq 3 \\ -x + 2 & \text{if } x > 3 \end{cases}.$$

$g(x) =$ ____________________

5. $f(x) = \begin{cases} 2x^2 - 1 & \text{if } x < 1 \\ x + 4 & \text{if } x \geq 1 \end{cases}$

What is the value of $f(-3)$?

A. −19

B. 3

C. 5

D. 17

Practice Exercises *(continued)*

6. A coordinate plane represents a rectangular pool table. A ball is on a pool table at the point (3, 4). The ball is rolled so that it hits the side of the pool table at the point (9, 10). Then it rolls toward the other side and stops at the point (13, 6) as shown in the diagram.

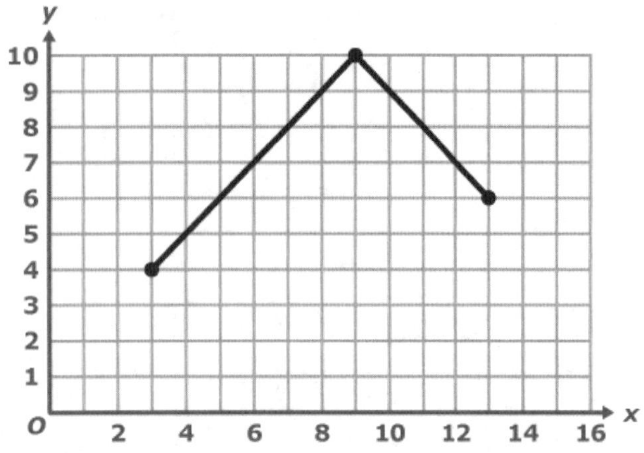

Complete the piecewise function so that it represents the path of the ball over its entire domain, x.

A. When $3 \leq x \leq 9$, $y =$ ______________.

B. When $9 < x \leq 13$, $y =$ ______________.

7. A single person filing taxes for 2013 owed 10% for all income up to and including \$8,925; 15% for income from \$8,926 up to and including \$36,250; and 25% for income from \$36,251 up to and including \$87,850.

Complete the equation form of the piecewise function $T(x)$ which describes the tax owed for income x dollars.

$$T(x) = \begin{cases} 0.10x & \text{if } \$0 < x < \$\_\_\_\_ \\ 0.15x & \text{if } \$\_\_\_\_ \leq x < \$36{,}251 \\ \_\_\_\_x & \text{if } \$\_\_\_\_ \leq x \leq \$87{,}850 \end{cases}$$

8. Cell phone provider A has a \$35/month plan that provides unlimited minutes and 1,000 free text messages per month. After that, each text message costs \$0.10.

Cell phone provider B has a monthly plan that also offers unlimited minutes and charges a monthly amount according to the piecewise function defined below, for x text messages.

$$f(x) = \begin{cases} \$25 & \text{if } 0 \leq x \leq 500 \\ \$25 + \$0.05(x - 500) & \text{if } x > 500 \end{cases}$$

Find the cost of each cell phone provider's plan for 900 and 1,500 text messages each month.

	900 text messages	**1,500 text messages**
Provider A	\$__________	\$__________
Provider B	\$__________	\$__________

9. The graph of $g(x)$ is a translation on the absolute value parent function, $f(x) = |x|$.

The graph of $g(x)$ is the graph of $f(x)$ reflected across the x-axis and shifted to the left 3 units and up 2 units.

Write the equation of $g(x)$.
$g(x) =$ ______________

10. Which of the following can be modeled with a piecewise function? *Select all that apply.*

A. the path of a baseball from the moment it is hit to when it is caught

B. the amount of a server's tip based on the total bill

C. the speed of a bike when a person rides it to and from school

D. the pay you earn at job, including time-and-a-half if more than 40 hours are worked

E. the relationship between Fahrenheit and Celsius

Apply

How Many T-Shirts Should You Purchase for a Fundraiser?

The student council is planning a fundraiser to help pay for field trips and school dances. They plan to purchase T-shirts online and sell them. They estimate that they can sell about 70 T-shirts. Below are price sheets from two different T-shirt companies.

How many T-shirts should they order to spend the least money?

T-Shirts D-Luxe

Quantity	Cost Per Shirt
Fewer than 24	\$5.64
24–35	\$4.88
36–71	\$4.64
72–142	\$4.08
143+	\$3.72

Best Tees

Quantity	Cost Per Shirt
Fewer than 24	\$4.50 (plus a one-time print fee of \$5.00)
24–35	\$4.25 (plus a one-time print fee of \$10.00)
36–71	\$3.50 (plus a one-time print fee of \$15.00)
72–142	\$3.00 (plus a one-time print fee of \$18.00)
143+	\$2.90 (plus a one-time print fee of \$20.00)

Write piecewise functions to model the relationship between the number of T-shirts ordered and the total cost of the shirts in dollars for each company.

Graph both piecewise functions. Use the graphs to determine how many T-shirts the student council representatives should purchase and from which company they should purchase the T-shirts.

Show what you've learned by completing the other performance tasks in the online Apply section.

Apply *(continued)*

Your answer to Apply will be assessed on the following criteria:

1. Writing a piecewise function to model the relationship between the number of T-shirts and total cost for T-Shirts D-Luxe
2. Writing a piecewise function to model the relationship between the number of T-shirts and total cost for Best Tees
3. Graphing both piecewise functions
4. Explaining and justifying how many T-shirts the student council representatives should buy

Scale / Criteria	4 Exceeds Criteria	3 Meets Criteria	2 Progressing to Criteria	1 Below Expectations	0 No Expectation
Function for T-Shirts D-Luxe	Writes a function that correctly models the relationship between the number of T-shirts and total cost for T-Shirts D-Luxe.	Writes a function that models the relationship between the number of T-shirts and total cost for T-Shirts D-Luxe with minor errors.	Writes a function that models the relationship between the number of T-shirts and total cost for T-Shirts D-Luxe with significant errors.	Writes a function that does not model the relationship between the number of T-shirts and total cost for T-Shirts D-Luxe.	Does not attempt.
Function for Best Tees	Writes a function that correctly models the relationship between the number of T-shirts and total cost for Best Tees.	Writes a function that models the relationship between the number of T-shirts and total cost for Best Tees with minor errors.	Writes a function that models the relationship between the number of T-shirts and total cost for Best Tees with significant errors.	Writes a function that does not model the relationship between the number of T-shirts and total cost for Best Tees.	Does not attempt.
Graphs	Correctly graphs both piecewise functions.	Graphs both piecewise functions with minor errors.	Correctly graphs one piecewise function.	Graphs both piecewise functions incorrectly.	Does not attempt.
Explanation and Justification	Clearly explains and justifies the answer.	Explains and justifies the answer but with lack of clarity.	Explains and justifies the answer but makes minor errors.	Explanation and justification are incomplete or incorrect.	Does not attempt.

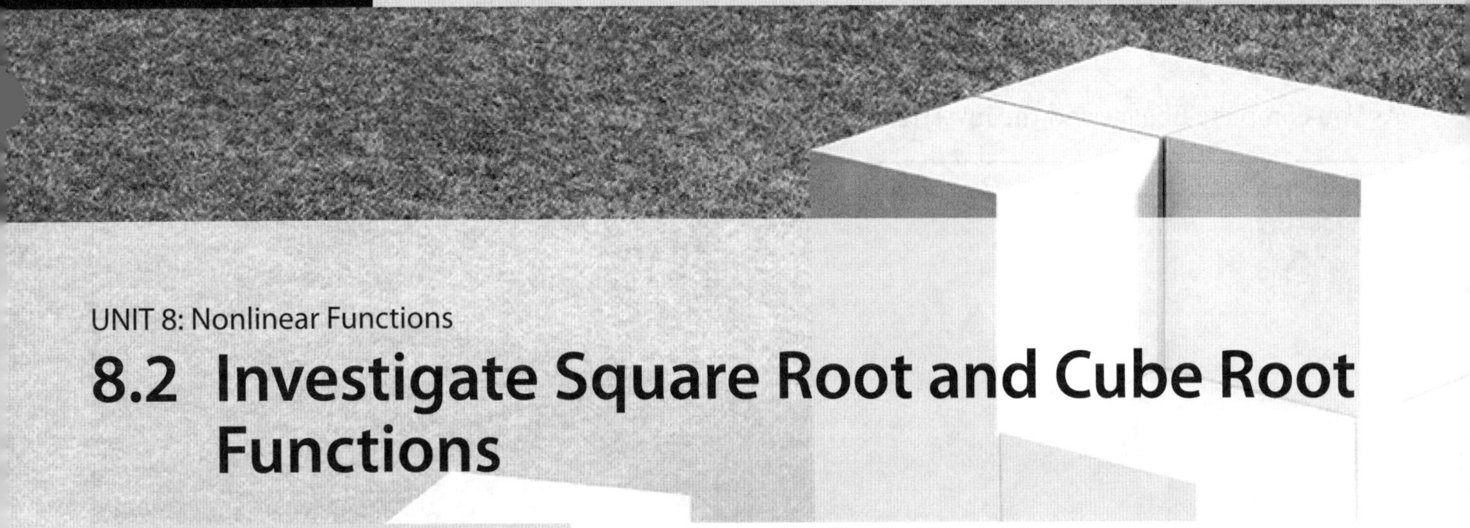

UNIT 8: Nonlinear Functions

8.2 Investigate Square Root and Cube Root Functions

photo: Getty Images

Lesson Objectives

- Graph square root and cube root functions.
- Analyze square root and cube root functions.

Essential Question

- How are algebraic, numeric, and graphic representations of radical functions related?

photo: Getty Images

Investigations

Building Squares and Cubes

How would an area equation look on a graph? How would it compare to a volume equation?

Getting to the Root of the Graphs!

Use road curvature and friction at an accident site to find the speed of a car.

Radical Transformers

Shift the parent function, and you may find a new radical.

Comparing Functions

So many functions, so many ways to represent them. Which model shows their best features?

Key Vocabulary

cube root, index, radical function, square root function

Discover

As you complete Engage and the investigations, record the most important ideas you've learned.

Engage	**Investigation 1**
Investigation 2	**Investigation 3**

Name ______________________________ Date ______________

Check for Understanding

Investigate Square Root and Cube Root Functions: Investigation 1

Compare the graphs of $y = x^2$ and $y = \sqrt{x}$, and the graphs of $y = x^3$ and $y = \sqrt[3]{x}$, and answer the questions that follow.

A.

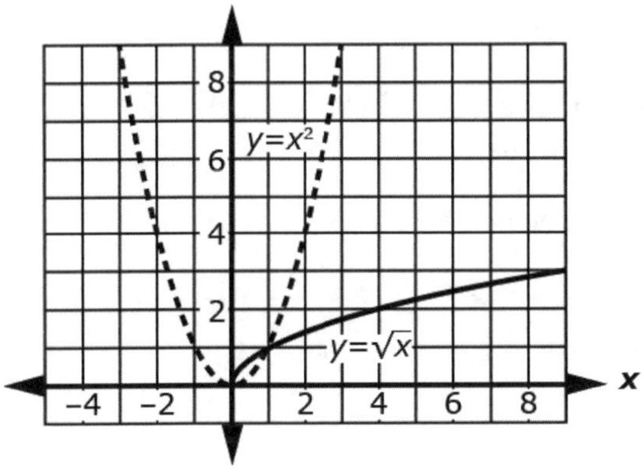

B.

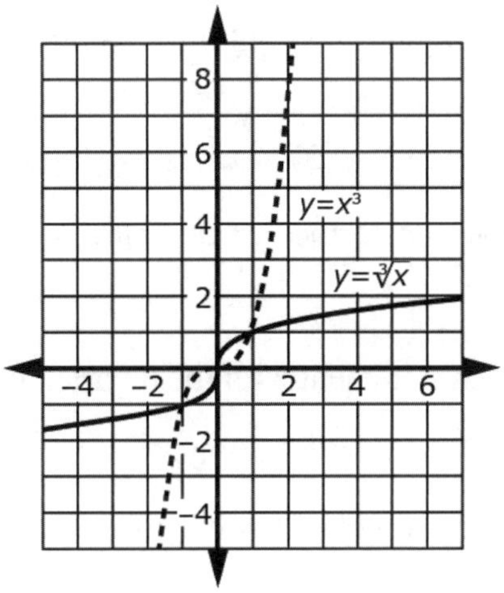

1. The functions in both graphs have their only x-intercept at ______________, and their y-intercept at ______________.

2. The domain of $y = \sqrt{x}$ is the same as the ______________ of $y = x^2$ on the interval ______________.
   This means that if a point (a, b) is on the graph of $y = \sqrt{x}$, the point (____, ____) is on the graph of $y = x^2$.

3. The domain of $y = \sqrt[3]{x}$ is the same as the ______________ of $y = x^3$ on the interval ______________.
   This means that if a point (c, d) is on the graph of $y = \sqrt[3]{x}$, the point (____, ____) is on the graph of $y = x^3$.

4. Explain why the graphs of functions that are even roots have restricted domains, and the graphs of functions having odd roots do not have any restrictions.

Name ______________________________ Date ______________

Check for Understanding

Investigate Square Root and Cube Root Functions: Investigation 2

1. Recall that a translation of the square root function has the form $y = \sqrt{x - h} + k$. *Circle the terms and values that make each statement true.*

   A. The graph of $g(x) = \sqrt{x - 6} - 3$ is a transformation on the parent function $f(x) = \sqrt{x}$ to the [left / right] by [3 / 6] units and [up / down] by [3 / 6] units.

   B. The graph of $h(x) = \sqrt[3]{x + 4} + 2$ is a transformation on the parent function $f(x) = \sqrt[3]{x}$ to the [left / right] by [2 / 4] units and [up / down] by [2 / 4] units.

   C. In the transformed graph of $f(x) = \sqrt{x + 7} - 5$, the value of h is [−5 / 5 / −7 / 7] and the value of k is [−5 / 5 / −7 / 7].

2. Complete each statement with the interval that makes it true.

   A. The domain of the parent square root function is ______________________.

   B. The domain of the parent cube root function is ______________________.

   C. The domain of a transformed function $y = \sqrt{x - h} + k$ is ______________________, and the range is ______________________.

   D. The domain of a transformed function $y = \sqrt[3]{x - h} + k$ is ______________________, and the range is ______________________.

3. Which of the following statements are true? *Select all that apply.*

   A. The graph of the square root function is increasing across its entire domain.

   B. The graph of the cube root function has a maximum value at $y = k$, where k is the value of the vertical transformation.

   C. The graph of the cube root function can be translated vertically so that it never has a negative value.

   D. The graphs of the parent square root and cube root functions intersect in two points.

4. Consider the functions $y = \sqrt{x}$ and $y = \sqrt[3]{x}$. On which intervals is $y = \sqrt{x}$ greater? On which intervals is $y = \sqrt[3]{x}$ greater? Explain why they differ on those intervals.

Name ______________________________ Date ____________

Check for Understanding

Investigate Square Root and Cube Root Functions: Investigation 3

Cheri is collecting plant samples on the ocean floor. A general guideline that some divers use for their ascent rate is 9 meters per minute. In addition, they must make decompression stops to adjust to the changes in pressure. Cheri plots a point for each of her stops and approximates her ascent using a square root function as a model.

Use the guideline and graph for questions 1–6.

Diver Ascent with Decompression Stops

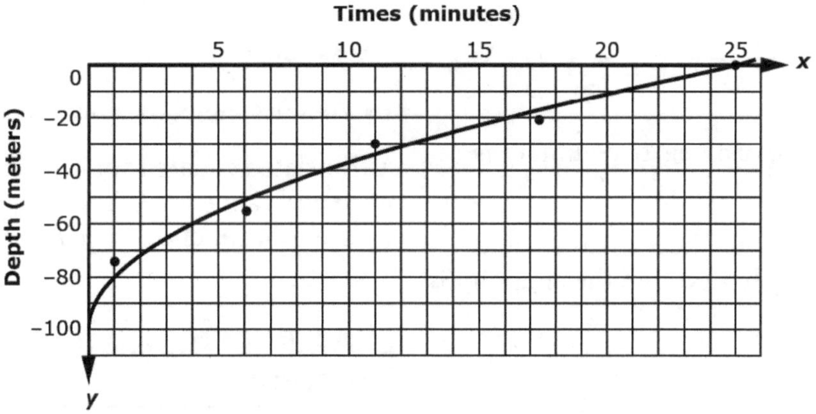

1. The y-intercept for the graph is ______________, and this represents Cheri's ______________.

2. The function model for Cheri's ascent without stops is ______________.

3. The function model with decompression stops is $f(x) = 20\sqrt{x}$ before translation. What is the model with the translation? ____________________

4. The average rate of change over Cheri's ascent without stops is ________ m/min and with the stops is __________ m/min.

5. Without decompression stops, Cheri would reach the surface after about _________ minutes, and with decompression stops, after _________ minutes.

6. The square root function is a model that approximates Cheri's ascent. Describe the general shape (no specific points) for the graph of a function that would model Cheri's actual location at each depth with a 9-meter ascent rate that includes 4 decompression stops.

Summary

Before you attempt the Practice Exercises, review what you've learned.

Square Root Functions

The square root parent function is $y = \sqrt{x}$.

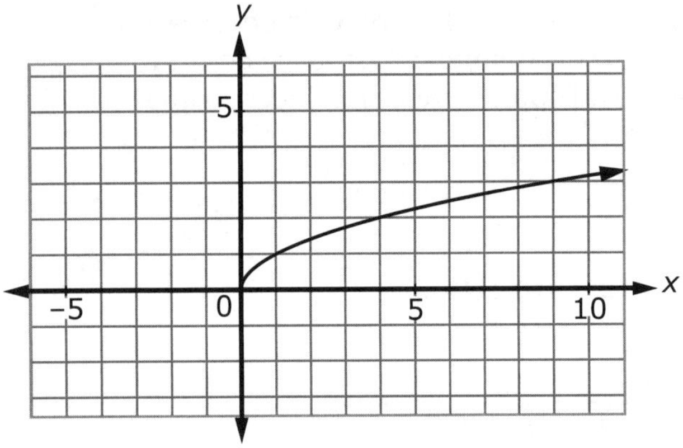

Because the square root of a negative number is not a real number, the domain is $x \geq 0$.

The range of $y = \sqrt{x}$ is $y \geq 0$. The x-intercept and y-intercept of $y = \sqrt{x}$ are both 0, and the endpoint is at (0, 0).

My Notes

Transforming Square Root Functions

The general equation for a square root function is $y = \sqrt{x-h} + k$.

The value of h tells how the graph is translated left or right. The value of k tells how the graph is translated up or down.

- As h increases, the graph shifts right.
- As h decreases, the graph shifts left.
- As k increases, the graph shifts up.
- As k decreases, the graph shifts down.

A translated square root function has the domain $x \geq h$ and range $y \geq k$. The endpoint of a translated square root function is (h, k).

EXAMPLE: Translating a Square Root Function

The graph shows a square root function that was translated. Find the end point of the graph and the equation, domain, and range for the function.

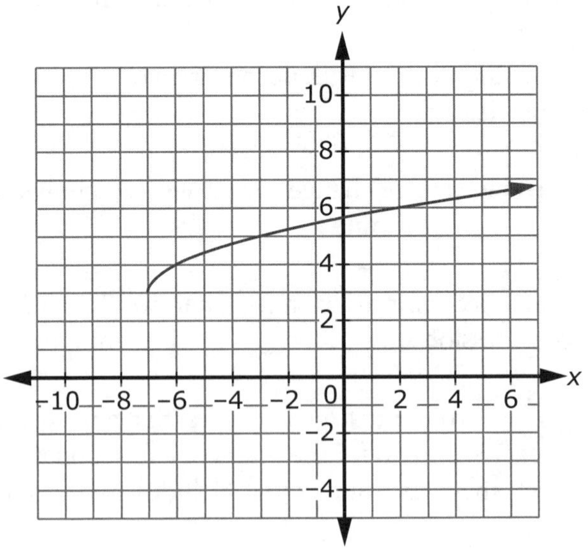

Summary *(continued)*

SOLUTION:

Read the coordinates of the end point from the graph: (−7, 3).

The equation is of the form $y = \sqrt{x - h} + k$, so substitute −7 for h and 3 for k: $y = \sqrt{x + 7} + 3$.

The domain is $x \geq h$, so it is $x \geq -7$.

The range is $y \geq k$, so it is $y \geq 3$.

Cube Root Functions

The cube root parent function is $y = \sqrt[3]{x}$.

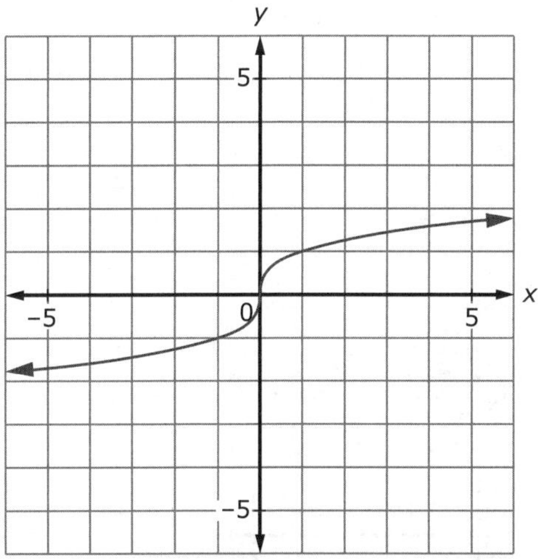

You can take the cube root of any number. The cube root of a positive number is a positive number, and the cube root of a negative number is a negative number.

This means there are no restrictions on the domain or the range. The domain and range of $y = \sqrt[3]{x}$ are all real numbers.

Transforming Cube Root Functions

The general equation for a cube root function is $y = \sqrt[3]{x-h} + k$. The value of h tells how the graph is translated left or right. The value of k tells how the graph is translated up or down.

- As h increases, the graph shifts right.
- As h decreases, the graph shifts left.
- As k increases, the graph shifts up.
- As k decreases, the graph shifts down.

For a translated cube root function, the domain and range are still all real numbers. The values of h and k translate the point at the origin of the parent function to the point (h, k).

My Notes

Summary *(continued)*

EXAMPLE: Translating a Cube Root Function

The point at the origin of the cube root function has been translated to (9, −2). Write an equation for the function. Then graph the function. Give the domain and range of the function.

SOLUTION:

The point (h, k) is given as (9, −2), so $h = 9$ and $k = -2$.

The general form of the equation is $y = \sqrt[3]{x - h} + k$.

Substitute 9 for h and −2 for k: $y = \sqrt[3]{x - 9} - 2$.

Use a graphing calculator to graph the function.

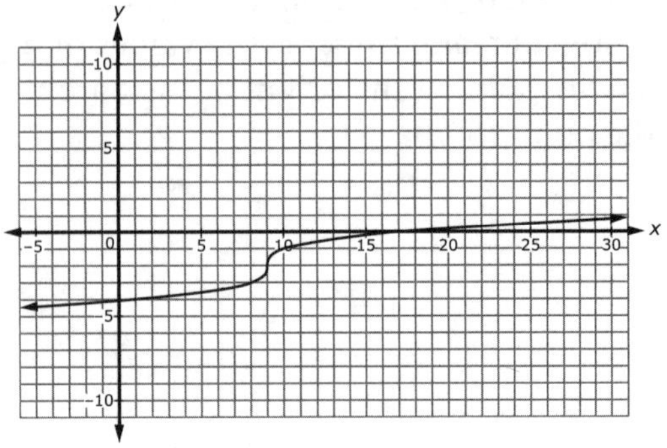

The domain is all real numbers.

The range is all real numbers.

My Notes

Comparing Functions in Various Representations

To compare different types of functions that are represented in different ways, you will compare features such as average rate of change over an interval, x- and y-intercepts, and domain and range.

EXAMPLE: Comparing Functions

Which of the following functions has a greater y-intercept?

Function A:

$y = \sqrt{x + 2} + 3$

Function B:

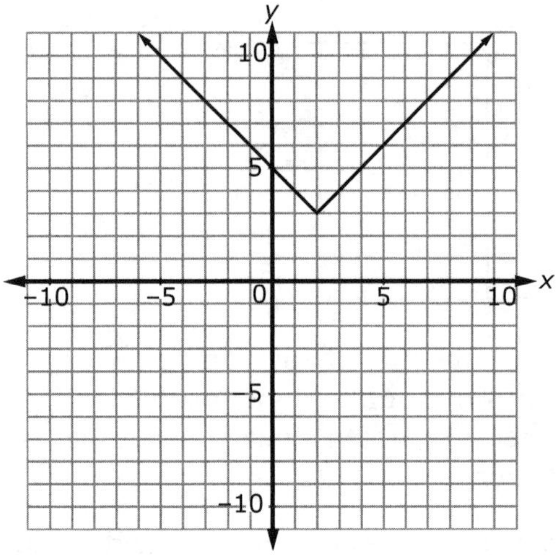

SOLUTION:

The y-intercept of function B is 5. For function A, evaluate the equation for $x = 0$.

$$y = \sqrt{0 + 2} + 3$$

$$y = \sqrt{2} + 3$$

$$y \approx 1.41 + 3$$

$$y \approx 4.41$$

The y-intercept of function A is approximately 4.41.

$$5 > 4.41$$

Function B has the greater y-intercept.

Practice Exercises

Review what you've learned using these practice problems. For practice problems with feedback, try the Coach and Play items in the Practice section online.

1. Which of these inequalities can be used to identify the domain of $f(x) = \sqrt{x+3} + 2$?

   A. $x \geq 0$

   B. $x \geq 3$

   C. $x + 2 \geq 0$

   D. $x + 3 \geq 0$

2. Mr. Perkins creates cubes for volume models. Each cube in a set for a volume model has a side length that is an integer multiple of one centimeter.

   The function $s = \sqrt[3]{V}$ represents the side length of a cube that can be created with V centimeter cubes.

   Which of these statements are correct about the appropriate domain and range for Mr. Perkins's cube sets? *Select all that apply.*

   A. The domain and the range of Mr. Perkins' cube sets are each the set of all real numbers because the cube root function is defined for all real numbers.

   B. The range of Mr. Perkins' cube sets will be the set of all positive integers because the models should all have volumes that are perfect cube numbers.

   C. The domain of Mr. Perkins' cube sets will be $V \geq 1$ because volume can have only nonnegative values.

   D. The range of Mr. Perkins' cube sets can be described as all nonnegative integers, s, where s is measured in centimeters.

   E. The domain of Mr. Perkins' volume models will be the set of positive perfect cube integers because the cubes in a set have integer side lengths.

3. Grant compares the graphs of $f(x) = \sqrt[3]{x}$, $g(x) = \sqrt[3]{x-2} + 2$, and $h(x) = \sqrt[3]{x+7} - 1$.

   Which of these features do all three functions have in common? *Select all that apply.*

   A. All of the functions decrease as x increases.

   B. All of the functions pass through the origin.

   C. All of the functions have a domain of the set of all real numbers.

   D. All of the functions have a range of $y \geq 0$.

   E. All of the functions pass through the point (1, 1).

4. The graph of $h(x)$ shown below is a translation of the square root function, $f(x) = \sqrt{x}$.

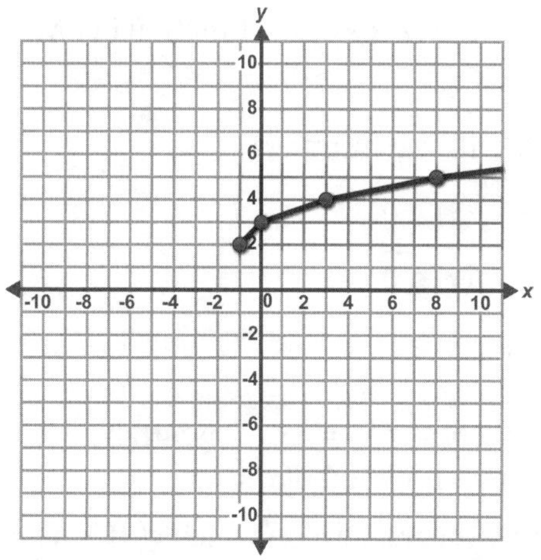

   Which of these points will also be on the graph of $h(x)$? *Select all that apply.*

   A. (–3, 1)

   B. (–4, –2)

   C. (15, 6)

   D. (48, 9)

   E. (99, 10)

Practice Exercises *(continued)*

5. The table represents some of the values for the function $f(x) = \sqrt{x-4} - 2$.

x	$f(x)$
4	−2
5	−1
8	0
13	1
20	2

Which of these statements are true about the graph of $f(x)$? *Select all that apply.*

A. $f(x)$ is increasing on its domain, $x \geq 4$.

B. $f(x)$ is negative on the interval $4 \leq x < 8$.

C. $f(x)$ has a minimum value of $f(x) = 4$.

D. $f(x)$ has a positive value when $x > 0$.

E. As x gets very large, $f(x)$ approaches infinity, and there is no horizontal asymptote.

6. What is the range of the function below?
$f(x) = \sqrt{x-5}$

A. all real numbers

B. $f(x) \geq 0$

C. $f(x) \leq 0$

D. $f(x) \geq 5$

7. The function $g(x) = \sqrt[3]{x} - 2$ is a transformation of the parent function, $f(x) = \sqrt[3]{x}$.

Identify the interval on which $g(x)$ is negative and the one on which $g(x)$ is positive.

A. $(-\infty, \infty)$

B. $(-\infty, 0)$

C. $(-\infty, -8)$

D. $(-\infty, -2)$

E. $(-\infty, 8)$

F. $(0, \infty)$

G. $(-2, \infty)$

H. $(8, \infty)$

The function $g(x)$ is negative on interval __________.

The function $g(x)$ is positive on interval __________.

8. The function $g(x)$ is a translation of $f(x) = \sqrt{x}$. The function $g(x)$ has an x-intercept of (18, 0).

Which two of the single translations below may have been applied to $f(x)$ to create the function $g(x)$? *Select two translations.*

A. a translation 9 units right

B. a translation 9 units left

C. a translation 9 units up

D. a translation 9 units down

E. a translation 3 units right

F. a translation 3 units left

G. a translation 3 units up

H. a translation 3 units down

9. Makayla wants to graph the function $f(x) = \sqrt{x+4} - 3$ by hand on a coordinate plane.

To make it as accurate as possible, she plans to graph the minimum and both intercepts, and she wants to be sure she labels appropriate scales for the axes.

Fill in the information that Makayla needs to graph the function.

The function's minimum is at the point ________.
The y-intercept is __________.
The x-intercept is __________.
The domain is $x \geq$ __________.
The range is $y \geq$ __________.

10. The function $f(x) = \sqrt{x}$ is transformed to create $g(x) = \sqrt{(x-h)} + k$.

Suppose that $h < 0$ and $k > 0$. Which of these statements is correct about the intercepts of $g(x)$?

A. $g(x)$ will have an x-intercept but will not have a y-intercept.

B. $g(x)$ will have a y-intercept but will not have an x-intercept.

C. $g(x)$ will have both an x-intercept and a y-intercept.

D. $g(x)$ will have neither an x-intercept nor a y-intercept.

Apply

How Far Can You See?

Have you ever noticed that the distance to the horizon seems to change when you look far off from different locations? Sometimes it looks close, but other times it looks far away. How can we determine the distance we can see to the horizon line? Watch the video to get started.

How far can you see?

The Horizon: View this video segment to see how far you can see.

This table shows the relationship between the height in feet of a person's eyes from the ground and the distance in miles they can see to the horizon. Find and graph an equation that models this relationship.

Height (ft)	0	10	20	30	40	50	60	70	80	90	100
Distance (mi)	0	3.9	5.5	6.7	7.8	8.7	9.5	10.3	11.0	11.6	12.3

Use the Internet to find a building you would like to visit and record its height. Using your model, predict how far you would be able to see from the top floor of that building.

Show what you've learned by completing the other performance tasks in the online Apply section.

Apply *(continued)*

Your answer to Apply will be assessed on the following criteria:

1. Graphing the data accurately and deciding which type of equation best models the data
2. Using the data to find an appropriate model and explaining the process
3. Researching a building and making a reasonable estimate for eye height on the top floor
4. Finding the distance you can see to the horizon from your chosen building

Scale / Criteria	4	3	2	1	0
	Exceeds Criteria	**Meets Criteria**	**Progressing to Criteria**	**Below Expectations**	**No Expectation**
Graph and Type of Function	Graphs the data accurately and decides which type of equation best models the data correctly.	Graphs the data accurately and decides which type of equation best models the data with a flaw in reasoning.	Graphs the data inaccurately or decides which type of equation best models the data with a flaw in reasoning.	Graphs the data inaccurately and decides which type of equation best models the data with a flaw in reasoning.	Does not attempt task.
Equation to Model Data	Uses the data to find a correct, appropriate model and explains the process.	Uses the data to find a nearly correct model and explains the process.	Incorrectly uses the data to find a model or does not explain the process.	Incorrectly uses the data to find a model and does not explain the process.	Does not attempt task.
Building Research and Height Estimation	Researches a building and makes a correct and reasonable estimate for eye height on the top floor.	Researches a building and makes an estimate for eye height on the top floor with minor errors.	Researches a building and makes an incorrect estimate for eye height on the top floor.	Shows no evidence of research, and makes an incorrect estimate for eye height on the top floor.	Does not attempt task.
Distance Calculation and Explanation	Correctly finds the distance to the horizon from the chosen building and supports work mathematically.	Finds the distance to the horizon from the chosen building with reasonable mathematical support.	Finds the distance to the horizon from the chosen building but uses faulty reasoning.	Finds the distance to the horizon from the chosen building incorrectly and uses faulty reasoning.	Does not attempt task.

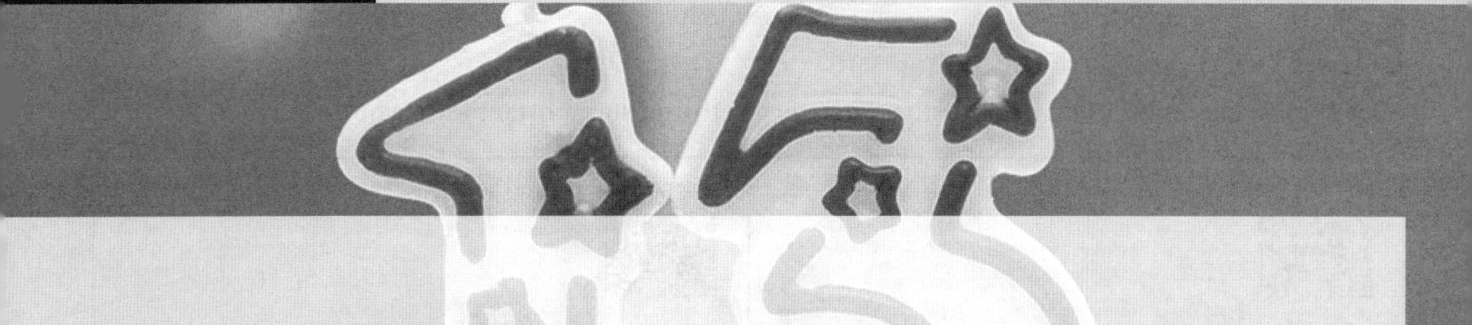

UNIT 8: Nonlinear Functions

8.3 Investigate Rational Exponents

photo: Getty Images

Lesson Objectives

- Rewrite expressions using the laws of exponents.
- Perform operations with rational exponents.
- Simplify radical expressions.

Essential Question

- How can you write different but equivalent forms of an expression that includes radicals or rational exponents?

photo: Getty Images

Investigations

Happy Numbers

Numbers can be expressed in many equivalent forms. Investigate the structure of numbers.

A Radical Change

How are exponents and radicals related? Is there a way to represent both with the same notation?

Irrational Cubes

Use rational exponents to design a house: a cubic house, to be precise.

Radical Properties

A square in a circle made simple with radicals. Just keep the properties in mind.

Key Vocabulary

index, radical, radicand, rational exponent

Discover

As you complete Engage and the investigations, record the most important ideas you've learned.

Engage	**Investigation 1**
Investigation 2	**Investigation 3**

Name ______________________________ Date ______________

Check for Understanding

Investigate Rational Exponents: Investigation 1

1. In the expression $(\sqrt[n]{a})^m$, a is the [exponent / index / radical / radicand], m is the [exponent / index / radical / radicand], and n is the [exponent / index / radical / radicand].

2. In the expression $x^{\frac{y}{z}}$, z is the [exponent / index / radical / radicand], x is the [exponent / index / radical / radicand], and y is the [exponent / index / radical / radicand].

3. Which of the following are equivalent to $8^{\frac{1}{3}}$? *Select all that apply.*

   A. $\sqrt{8}$ **B.** $\sqrt[3]{8}$ **C.** $\sqrt{2}$

   D. $\sqrt[8]{3}$ **E.** $\frac{8}{3}$ **F.** 2

4. Write an equivalent expression using a radical.

   A. $6^{\frac{1}{2}}$ = __________ **B.** $10^{\frac{1}{n}}$ = __________ **C.** $m^{\frac{3}{5}}$ = __________

5. Write an equivalent expression using a rational exponent.

   A. $\sqrt[4]{3}$ = __________ **B.** $(\sqrt{5})^3$ = __________ **C.** $(\sqrt[r]{p})^q$ = __________

6. Does $(\sqrt[n]{a})^m = \sqrt[n]{a^m}$ for all $a \geq 0$? Use rational exponents to explain why or why not.

Name ______________________________ Date ____________

Check for Understanding

Investigate Rational Exponents: Investigation 2

1. Match each product to its equivalent expression.

A. $4^{\frac{5}{3}} \cdot 4^{\frac{4}{3}}$ I. $3^{\frac{9}{5}}$

B. $\frac{4^{\frac{8}{3}}}{4}$ II. $6^{\frac{6}{5}}$

C. $\left(6^{\frac{2}{5}}\right)^4$ III. $4^{\frac{5}{3}}$

D. $\left(3^{\frac{2}{5}} \cdot 3^{\frac{1}{2}}\right)^2$ IV. 4^3

E. $3^{\frac{6}{5}} \cdot 2^{\frac{6}{5}}$ V. $6^{\frac{8}{5}}$

2. Which of the following expressions are equivalent to 4? *Select all that apply.*

A. $16^{\frac{1}{3}} \cdot 16^{\frac{1}{6}}$ B. $\sqrt{8}$ C. $2^{\frac{1}{2}} \cdot 8^{\frac{1}{2}}$ D. $64^{\frac{1}{3}}$

3. Write each of the following in simplest exponential form.

A. $2^{\frac{3}{4}} \cdot 2^2 =$ __________ B. $3^{\frac{1}{3}} \cdot 3^{\frac{1}{2}} =$ __________

4. Explain how you might use the properties of exponents to simplify $2^{\frac{4}{3}} \cdot 4^{\frac{1}{3}}$.

Name ______________________________ Date ______________

Check for Understanding

Investigate Rational Exponents: Investigation 3

1. Apply the product property of radicals to rewrite the expressions as indicated.

   A. Rewrite as two radicals: $\sqrt{2 \cdot 5}$ = ______________________

   B. Rewrite as a single radical: $\sqrt{3} \cdot \sqrt{7}$ = ______________________

2. Which of the following is the most efficient factorization for simplifying $\sqrt{72}$?

   A. $\sqrt{8 \cdot 9}$ B. $\sqrt{24 \cdot 3}$ C. $\sqrt{36 \cdot 2}$ D. $\sqrt{4 \cdot 18}$

3. Write each radical in simplified form.

   A. $\sqrt{72}$ = ______________________

   B. $\sqrt{96}$ = ______________________

   C. $\sqrt{120}$ = ______________________

4. Explain how you can use prime factorization to determine the most efficient factorization for simplifying $\sqrt{108}$. When might this method be useful?

Summary

Before you attempt the Practice Exercises, review what you've learned.

Rational Exponents and Radicals

A **rational exponent** is an exponent that can be written as a fraction.

You can write expressions that have exponents in radical notation.

The expression $\left(8^{\frac{1}{3}}\right)^3$ is equivalent to $8^{\frac{1}{3}\cdot 3}$, which equals 8^1, or 8.

The expression $\left(\sqrt[3]{8}\right)^3$ is equivalent to 2^3, or 8.

So, the two expressions we started with are equal.

$$\left(8^{\frac{1}{3}}\right)^3 = \left(\sqrt[3]{8}\right)^3$$

$$8^{\frac{1}{3}} = \sqrt[3]{8}$$

In the expression $8^{\frac{1}{3}}$, 8 is the base and $\frac{1}{3}$ is the rational exponent.In the equivalent form $\sqrt[3]{8}$, 8 is called the **radicand** and 3 is called the **index**.

This results in the following definition.

$a^{\frac{1}{n}} = \sqrt[n]{a}$ where n is a natural number.

Sometimes the rational exponent has a numerator that is not equal to 1.

The expression $\left(5^{\frac{2}{3}}\right)^3$ is equivalent to $5^{\frac{2}{3}\cdot 3}$, which equals 5^2.

The expression $\left(\sqrt[3]{5^2}\right)^3$ is equivalent to $\left((5^2)^{\frac{1}{3}}\right)^3$, which equals $(5^2)^{\frac{1}{3}\cdot 3}$, or 5^2.

So, the expressions we started with are equal.

$$\left(5^{\frac{2}{3}}\right)^3 = \left(\sqrt[3]{5^2}\right)^3$$

$$5^{\frac{2}{3}} = \sqrt[3]{5^2}$$

In general, the definition becomes:

$a^{\frac{m}{n}} = \sqrt[n]{a^m}$ where m and n are a natural numbers.

You can use this knowledge to show the following.

$$a^{\frac{m}{n}} = \left(\sqrt[n]{a}\right)^m$$

$$a^{\frac{m}{n}} = a^{\frac{1}{n}\cdot m} = \left(a^{\frac{1}{n}}\right)^m = \left(\sqrt[n]{a}\right)^m$$

EXAMPLE: Rewriting a Rational Exponent in Radical Form

Rewrite $13^{\frac{6}{5}}$ in radical form.

SOLUTION:

Think of $13^{\frac{6}{5}}$ as $(13^6)^{\frac{1}{5}}$.

$$\sqrt[5]{13^6}$$

My Notes

Summary *(continued)*

EXAMPLE: Rewriting a Radical Expression with a Rational Exponent

Rewrite $\sqrt[17]{126^3}$ with a rational exponent.

SOLUTION:

Write 126 with a fractional exponent. The number 126 is the base. For the fractional exponent, 17 is the index, so it will be the denominator. The 3 will be the numerator.

$$126^{\frac{3}{17}}$$

Properties of Rational Exponents

All the laws of exponents apply to rational exponents.

Property	Example
$(a^m)^n = a^{mn}$	$\left(2^{\frac{4}{3}}\right)^{\frac{5}{3}} = 2^{\frac{4}{3}\cdot\frac{5}{3}} = 2^{\frac{20}{9}}$
$a^m a^n = a^{m+n}$	$2^{\frac{4}{3}}2^{\frac{7}{3}} = 2^{\frac{4}{3}+\frac{7}{3}} = 2^{\frac{11}{3}}$
$\frac{a^m}{a^n} = a^{m-n}$	$\frac{2^{\frac{4}{3}}}{2^{\frac{5}{3}}} = 2^{\frac{4}{3}-\frac{5}{3}} = 2^{-\frac{1}{3}}$
$(ab)^m = a^m b^m$	$(2\cdot 5)^{\frac{4}{3}} = 2^{\frac{4}{3}}5^{\frac{4}{3}}$
$\left(\frac{a}{b}\right)^m = \frac{a^m}{b^m}$	$\left(\frac{2}{5}\right)^{\frac{4}{3}} = \frac{2^{\frac{4}{3}}}{5^{\frac{4}{3}}}$
$a^{-m} = \frac{1}{a^m}$	$2^{-\frac{4}{3}} = \frac{1}{2^{\frac{4}{3}}}$
$a^0 = 1, a \neq 0$	$2^{\frac{0}{3}} = 2^0 = 1$
$a^1 = a$	$2^{\frac{3}{3}} = 2^1 = 2$
$1^m = 1$	$1^{\frac{4}{3}} = 1$

My Notes

Summary *(continued)*

Simplifying Square Root Expressions

Product property of radicals: $\sqrt{a \cdot b} = \sqrt{a} \cdot \sqrt{b}$

You can use the following strategy to simplify square root expressions:

1. Rewrite the radicand as the product of its factors.
2. Look for a square among the factors.
3. Apply the product property of radicals.
4. Take the square root of the square number.

EXAMPLE: Simplifying Square Root Expressions

Simplify $\sqrt{125}$.

SOLUTION:

$\sqrt{125} = \sqrt{25 \cdot 5} = \sqrt{25}\sqrt{5} = 5\sqrt{5}$

My Notes

Practice Exercises

Review what you've learned using these practice problems. For practice problems with feedback, try the Coach and Play items in the Practice section online.

1. Order the following equations to explain how a term with a rational exponent, $5^{\frac{2}{3}}$, can be expressed as a radical.

   A. $\left(\sqrt[3]{5^2}\right)^3 = 5^2$

   B. $\frac{2}{3} \cdot 3 = 2$

   C. $5^{\frac{2}{3}} = \sqrt[3]{5^2}$

   D. $\left(5^{\frac{2}{3}}\right)^3 = 5^2$

   Multiply a fraction by its denominator to make a whole number. ____________

   Raise the rational exponent to make a whole number power. ____________

   Raise the root to a power to remove the radical.

   ____

   Conclusion from Steps 2 and 3. ____________

2. Rewrite the expression in simplified radical form.

   $\sqrt{32} =$ ____________

3. Which of the following expressions are equivalent to $\sqrt{192}$? *Select all that apply.*

   A. $\sqrt{2^6 \cdot 3}$

   B. 14

   C. $(2^2 \cdot 2^2 \cdot 2^2 \cdot 3)^{\frac{1}{2}}$

   D. 16

   E. $16\sqrt{3}$

   F. $\sqrt{2^8}\sqrt{3}$

   G. $(2^6 \cdot 3)^{\frac{1}{2}}$

   H. $8\sqrt{3}$

4. Evaluate the expression.

   $8^{\frac{2}{3}} + 9^{\frac{3}{2}} =$ ____________

5. Select the expressions that are equivalent to $8^{\frac{2}{3}}$. *Select all that apply.*

   A. $\sqrt[3]{8^2}$

   B. $\sqrt{8^3}$

   C. $(\sqrt{8})^3$

   D. 4

6. Which of the following expressions is not equivalent to 16?

   A. $4^{\frac{1}{2}}$

   B. 2^4

   C. $\left(16^{\frac{1}{2}}\right)^2$

   D. $\left(2^{\frac{1}{2}}\right)^8$

7. Use the properties of exponents to simplify the expression $\sqrt[5]{10^{12}}$.

   A. $10\sqrt[5]{10}$

   B. $100\sqrt[5]{100}$

   C. 20

   D. 2,000

8. Evaluate the expression.

   $\dfrac{25^{\frac{3}{2}}}{25^{\frac{1}{2}}} =$ ____________

Practice Exercises *(continued)*

9. Which of the following expressions are equivalent to $64^{-\frac{3}{2}}$? *Select all that apply.*

A. $64^{\frac{2}{3}}$

B. $\left(64^{\frac{3}{2}}\right)^{-1}$

C. -512

D. $(-64)^{\frac{3}{2}}$

E. $\frac{1}{512}$

F. $\left(\frac{1}{\sqrt{64}}\right)^{3}$

G. -8^{3}

H. $-\sqrt{64^{3}}$

10. Simplify the expression.

$\left(9^{\frac{1}{3}} \cdot 2^{\frac{2}{3}}\right)^{3} =$ ______________

11. Write the expression in its simplest form.

$\left(\frac{27}{64}\right)^{\frac{2}{3}} =$ ______________

12. Simplify the expression.

$\left(16^{\frac{1}{3}}\right)^{\frac{3}{2}} =$ ______________

13. Evaluate the expression.

$\left(\frac{24}{3}\right)^{\frac{2}{3}} =$ ______________

14. Rewrite the expression $\sqrt{96}$ in simplest radical form.

$\sqrt{96} =$ ______________

Apply

How Do Consumer Prices Change over Time?

People have been buying some of the same products for decades, even centuries, but they haven't always paid the same price. Choose two products, and research their average prices for two different years. For instance, you may choose to research the prices of two different products in 1950 and 2010.

What products can you think of that have been around for a long time?

Products at the Store

Find the average annual rate of change of the prices using the formula:

$$r = \left(\frac{\text{new price}}{\text{old price}}\right)^{\frac{1}{t}} - 1,$$

where r is rate as a decimal, and t is time in years. Express r as a percent.

Compare and contrast the rate of change in price for the products you chose.

Use the rates to predict the costs of those products 25 years after your second date (2010, in this example). Do you think your prediction is reliable? Why or why not?

Some employees receive an annual Cost of Living Adjustment (COLA). Suppose that a company gave its employees a 3% increase in pay each year. Would that make up for the change in price for the products you chose? Explain your answer.

Show what you've learned by completing the other performance tasks in the online Apply section.

Apply *(continued)*

Your answer to Apply will be assessed on the following criteria:

1. Researching and reporting the prices of two products for two years and explaining your choice
2. Using the formula to calculate the average rate of change for each product's price and comparing and contrasting the prices of the two products and their average rates of change
3. Using the rates to predict the future costs of those products and discussing the reliability of your prediction
4. Discussing the effect of COLAs on a consumer's buying power

Scale / Criteria	4 Exceeds Criteria	3 Meets Criteria	2 Progressing to Criteria	1 Below Expectations	0 No Expectation
Price Report	Accurately reports the prices of two products for the same two years; explains choices and cites sources.	Accurately reports the prices of two products for the same two years; explains choices or cites sources.	Reports the prices of two products for two different years; explains choices or cites sources.	Reports the prices of two products for two different years; does not explain choices or cite sources.	Does not attempt task.
Rates and Comparison	Correctly calculates both rates of change; compares and contrasts the rates.	Calculates both rates of change, with a minor error; compares and contrasts the rates.	Calculates both rates of change, with a major error; compares and contrasts the rates.	Calculates both rates of change, with a major error; does not compare and contrast the rates.	Does not attempt task.
Prediction	Uses the rates to accurately predict the costs of those products in 25 years; discusses the reliability of the prediction.	Uses the rates to predict the costs of those products in 25 years, with a minor error; discusses the reliability of the prediction.	Uses the rates to predict the costs of those products in 25 years, with a major error; discusses the reliability of the prediction.	Uses the rates to predict the costs of those products in 25 years, with a major error.	Does not attempt task.
Effect of COLA	Clearly explains the effect of the employees' COLA on their buying power.	Explains the effect of the employees' COLA on their buying power, with no more than one inaccuracy.	Explains the effect of the employees' COLA on their buying power, with more than one inaccuracy.	Attempts an explanation, but does not make a clear explanation of the effect of the COLA on buying power.	Does not attempt task.

UNIT 9: Exponential Functions

9.1 Represent Exponential Functions

photo: Getty Images

Lesson Objectives

- Describe how the growth rate and initial value influence an exponential function.
- Graph exponential functions, interpreting the impact of the value of a, b, and c in $f(x) = a b^x + c$.
- Identify the domain and range of exponential functions.
- Calculate the average rate of change of an exponential function from a graph and a table.
- Construct exponential functions to model real-world situations.

Essential Question

- How can you use an exponential function to interpret real-world and mathematical situations?

photo: Getty Images

Investigations

Filtered Out

Investigate filtration patterns, and you'll find an exponential function that describes your kidneys.

Interpreting Exponential Functions

What do graphs tell you about exponential functions? How do they change over intervals?

Exploring Transformations

Transform exponential functions, and you could speed up the process . . . or slow it down.

Comparing Linear, Exponential, and Quadratic Functions

Travel through space, but keep an eye on your fuel gauge! Warp drives run on different functions.

Key Vocabulary

exponential function, geometric sequence

Discover

As you complete Engage and the investigations, record the most important ideas you've learned.

Engage	Investigation 1
Investigation 2	Investigation 3

Name ______________________________ Date ______________

Check for Understanding

Represent Exponential Functions: Investigation 1

1. What is the range of the function $f(x) = 2e^{-x}$? *Select all that apply.*

   A. $y < 0$
   B. $(-\infty, \infty)$
   C. $y > 0$
   D. $(-\infty, 0)$
   E. $y \geq 0$
   F. $(0, \infty)$
   G. $y \leq 0$
   H. $[0, \infty)$

2. Match each function to the end behavior it demonstrates. *Select all that apply. Not all end behavior is included for all functions.*

   A. $y = 3e^{x}$
   B. $y = -\frac{1}{2}e^{-x}$
   C. $y = -5e^{x}$
   D. $y = 10e^{-2x}$

   I. $y \to -\infty$ as $x \to -\infty$
   II. $y \to \infty$ as $x \to \infty$
   III. $y \to 0$ as $x \to \infty$
   IV. $y \to 0$ as $x \to -\infty$

3. A student decides to give two people greeting cards with pleasant messages. He includes a note asking each recipient to give greeting cards to two other people. If everyone continues this chain of greeting cards, what is the function that would represent the number of people, y, who get greeting cards in the x round of giving?

   Hint: The initiating student giving out two cards is the first round, represented by $x = 1$.

   $y =$ ____________________

4. An exponential growth function, $g(t)$, describes the growth of a bacteria population in terms of time t. If the function has a y-intercept of 500, explain what you would expect this to represent in the function.

Name ______________________________ Date ____________

Check for Understanding

Represent Exponential Functions: Investigation 2

1. If a function $y = a \cdot b^x$ demonstrates exponential decay, which of the following must be true?

   A. $0 < a < 1$

   B. $a < 0$

   C. $0 < b < 1$

   D. $b < 0$

2. An e-mail chain is represented by the equation $y = (4)\,3^x$ to represent the number of people, y, who receive an e-mail in round x of the chain. Which statements are true about this e-mail chain? *Select all that apply.*

   A. The e-mail chain starts with 3 people.

   B. The e-mail chain starts with 4 people.

   C. In each round, each person forwards the e-mail to 3 people.

   D. In each round, each person forwards the e-mail to 4 people.

3. A general equation for exponential growth can take the form of $y = a^x + k$. How can this equation be modified to represent a y-intercept of 89?

   $y =$ ______________________

4. Match each equation with the effect it has on the graph of $y = 3^x$.

   A. $y = 3^x + 5$ I. Translated 5 units down

   B. $y = -5(3)^x$ II. Translated 5 units up

   C. $y = 3^x - 5$ III. Reflected, stretched by factor of 5

5. In an exponential growth function of the form $y = b^x$, where $b > 1$, how would the average rate of change vary from left to right on the graph? What does this look like on the graph of the equation?

Name ______________________________ Date ______________

Check for Understanding

Represent Exponential Functions: Investigation 3

1. Calculate the rate of change over the interval [2, 5] for the three functions.

$y = 3x + 5$ __________

$y = 3x^2 + 1$ __________

$y = 3^x - 10$ __________

2. Match each type of function to the term that describes its rate of change in y values.

A. Exponential Function	**I.** Common difference
B. Quadratic Function	**II.** Common ratio
C. Linear Function	**III.** No common difference or ratio

3. What is the range for each of the three functions?

$y = -\frac{2}{5}x + 1$ __________

$y = -4x^2$ __________

$y = 5^{-x} + 1$ __________

4. The average rate of change in y values for a set of data over the interval $(-\infty, 5)$ is a, and over the interval $(5, \infty)$ is $-a$. Consider linear, exponential, and quadratic models of this data. Explain which types of models cannot fit this data and why. Give an example of an equation or type of model that would fit this pattern.

Summary

Before you attempt the Practice Exercises, review what you've learned.

In an exponential relationship, every time the x value increases or decreases by 1, the y value changes by a constant ratio.

An exponential function can be written as $y = a(b)^x$.

a: initial amount when $x = 0$

b: rate of growth or decay

EXAMPLE: Write an Exponential Function

Hailey started a new website about pets, and 20 of her friends each visited the website. The next week, each of those friends told 5 more friends about the website, and each of them visited the website. The week after that, each of those friends told 5 more people, and so on. Everyone who was told about the website visited it and then told 5 more people. How many people will visit Hailey's website in week 5?

SOLUTION:

Write an exponential function to model the situation. The initial number of friends who visited the website was 20, so $a = 20$. Each week, the number of people who visit the website is multiplied by 5, so $b = 5$. Let x represent the week number and y represent the number of people who visit the website that week.

$$y = 20(5)^x$$

To find the number of people who visit the website in week 5, you can make a table of values using the pattern of multiplying the previous y value by 5. You can also trace the graph of the function or evaluate the function for $x = 5$.

To evaluate the function, substitute 5 for x in the function.

$$\begin{aligned} y &= 20(5)^x \\ &= 20(5)^5 \\ &= 20(3{,}125) \\ &= 62{,}500 \end{aligned}$$

Each method will show that 62,500 people will visit the website in week 5.

EXAMPLE: Identify Domain and Range

What are the reasonable domain and range for the website's context? Are the data continuous or discrete?

SOLUTION:

The domain includes each week that the pattern continues. So the reasonable domain is whole numbers.

The range includes the number of people who visit the website each week.

The reasonable range is represented by the geometric sequence 20, 100, 500, 2,500…. Because we do not include values between the numbers of the sequence, the data are discrete.

My Notes

Summary *(continued)*

EXAMPLE: Average Rate of Change

Describe how the average rate of change between the intervals below differs.

- week 0 to week 2
- week 2 to week 4
- week 4 to week 6

SOLUTION:

Find the average rate of change for each interval. This means finding the slope between the points at the boundaries of the interval.

week 0 to week 2:
boundary points: (0, 20) and (2, 500)
average rate of change $= \frac{500 - 20}{2 - 0} = \frac{480}{2} = 240$

week 2 to week 4:
boundary points: (2, 500) and (4, 12,500)
average rate of change $= \frac{12{,}500 - 500}{4 - 2} = \frac{12{,}000}{2}$
$= 6{,}000$

week 4 to week 6:
boundary points: (4, 12,500) and (6, 312,500)
average rate of change $= \frac{312{,}500 - 12{,}500}{6 - 4} = \frac{300{,}000}{2}$
$= 150{,}000$

You can show the average rates of change by drawing a line segment between the boundary points for each interval.

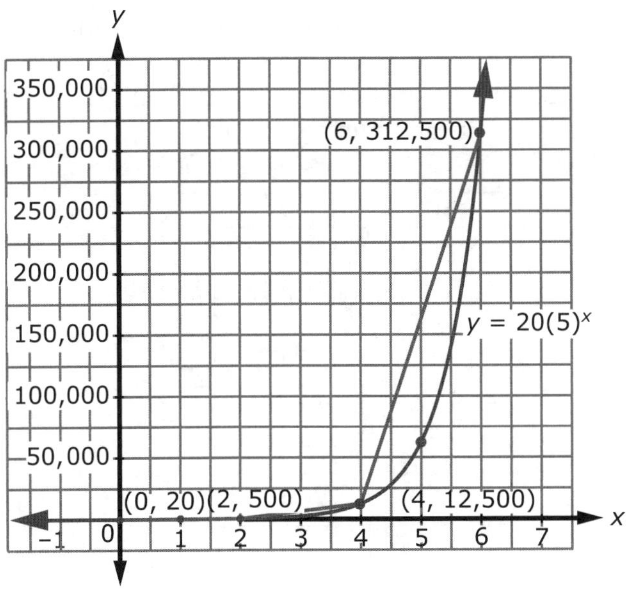

The average rate of change between each interval increases significantly as you move from left to right along the graph.

EXAMPLE: Transformations

Transform the graph $f(x) = 2^x$ as indicated. Describe how the graph changes.

A. $g(x) = 2^x + 3$

B. $h(x) = 3(2)^x$

SOLUTION:

A. Graph $f(x)$ and $g(x)$ on your graphing calculator.

Adding 3 to the function rule shifts the graph up 3 units. If you were to add −3 to the function, the graph would shift down 3 units on your graphing calculator.

B. Graph $f(x) = 2^x$ and $h(x) = 3(2)^x$.

Every output from $h(x) = 3(2^x)$ is 3 times greater than the corresponding output from $f(x) = 2^x$, resulting in a stretched curve. Changing the value of a to a number greater than one creates a vertical stretch.

My Notes

Summary *(continued)*

EXAMPLE: Explore Different Bases

Describe how the graph of $f(x) = 2^x$ differs from the graph of $k(x) = (4)^x$.

A. $k(x) = (4)^x$

SOLUTION:

A. Graph $f(x) = 2^x$ and $k(x) = (4)^x$ on your calculator.

Changing the base, b, of the exponential expression from 2 to 4 makes the function increase more quickly. The graph is steeper.

EXAMPLE: Compare Function Types

Explain whether each relationship shown in the graphs is linear, exponential, or neither.

SOLUTION:

A. Graph A shows a line. A table of values will show that as the x values increase by 1, the y values increase or decrease by –3, called the common difference. This function is linear.

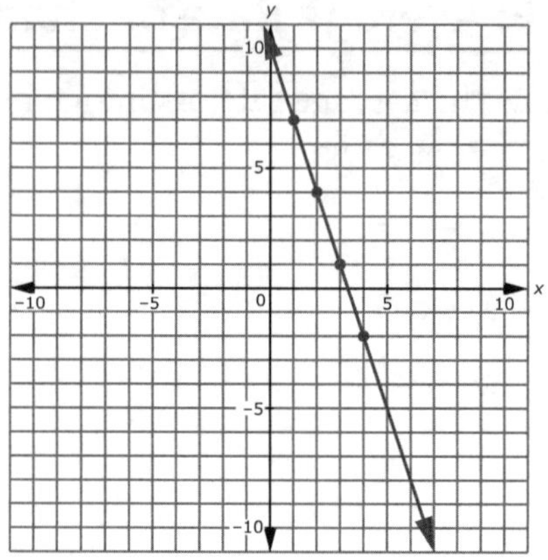

B. Graph B shows a curve that gets steeper as x increases. A table of values will show that as the x values increase by 1, the y values increase by a factor of 3, called the common ratio. This function is exponential.

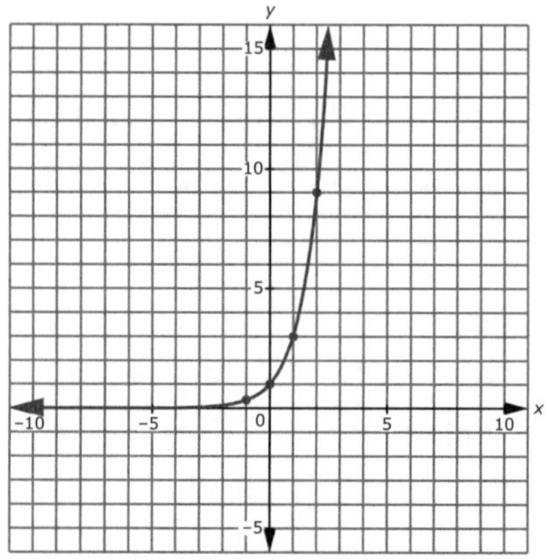

C. Graph C shows a U shape. A table of values will show that as the x values increase by 1, the y values do not increase or decrease by a common difference. It is not linear. The table also shows that the y values do not increase or decrease by a common ratio. It is not exponential. This function is neither linear nor exponential.

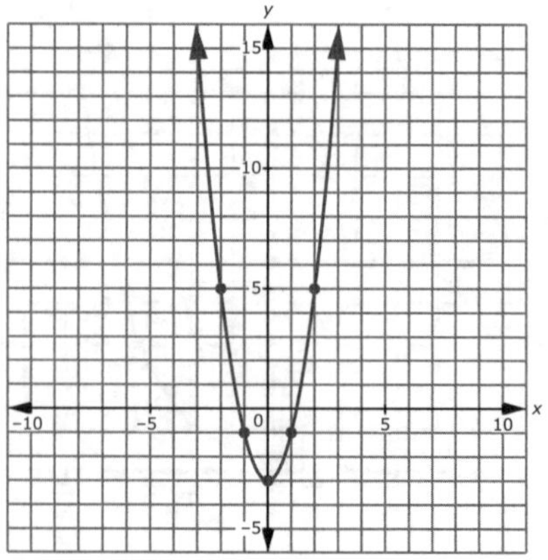

My Notes

Practice Exercises

Review what you've learned using these practice problems. For practice problems with feedback, try the Coach and Play items in the Practice section online.

1. Choose the exponential function with the smallest growth rate.

   A. $y = 55(10)^x$

   B. $y = 120(1.5)^x$

   C. $y = 95(3)^x$

   D. $y = 25(7)^x$

2. Describe the transformations that take the function $f(x) = 3^x$ to $g(x) = 5(3^x) - 7$ by giving the number of units $f(x)$ is stretched vertically and translated vertically.

   Vertical stretch: ______________

   Vertical translation: ______________

3. The valve on a water cooler is left open and the water is free to flow out. As the cooler drains, the water flows out at a slower and slower rate. However, due to the position of the valve, the water will not fully drain. The given function $w(t) = 9.8(0.6)^t + 0.2$ describes the amount of water in the cooler, in gallons, as a function of time, t, in minutes.

   What would be the expected range and domain for this scenario?

   A. Domain: all real numbers; Range: $y < 9.8$

   B. Domain: $t \geq 0$; Range: $0.2 < y \leq 10$

   C. Domain: all real numbers; Range: $0 < y < 9.8$

   D. Domain: $t \geq 0$; Range: $y \leq 10$

4. The graph of an exponential function is shown, with four points on the graph labeled A, B, C, and D.

   Determine the greatest average rate of change.

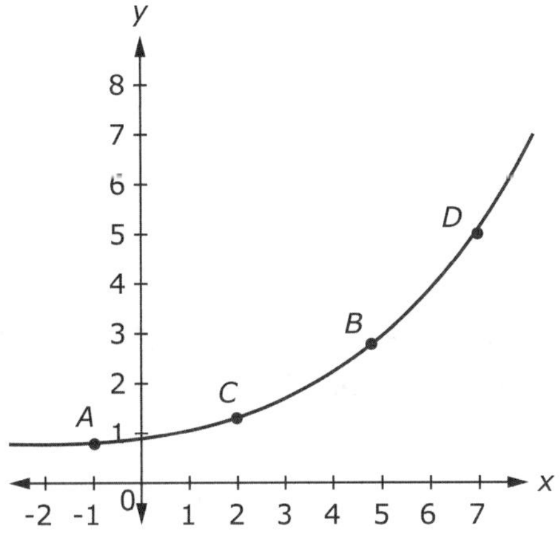

   The average rate of change is greatest between point ____________ and point ____________.

5. Three students at school catch a flu virus. One of the three students decides to stay home from school until he gets better. Each day, the number of infected students in school increases by 40%.

   Model the number of infected students with a function of the form $f(t) = ab^t + c$, where t is time in days. Write a value for a, b, and c.

   $a =$ ______ $b =$ ______ $c =$ ______

Practice Exercises *(continued)*

6. The graph shows the amount of money in a savings account over time.

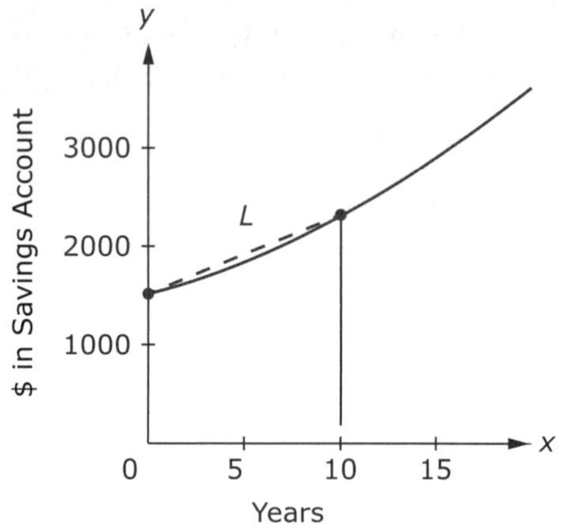

The line segment L has a slope of 100. Which option best describes the meaning of the slope of line segment L?

A. the amount of money earned by the savings account each year

B. the amount of money earned by the savings account over 10 years

C. the average amount of money that the savings account earns per year

D. the average amount of money that the savings account earns per year, over the first 10 years

7. Given an average rate of inflation of 4%, which functions, $V(t)$, could be used to model the value of a home worth \$150,000 today over a time, t, in years? *Select all that apply.*

A. $V(t) = 150{,}000(1.4)^t$

B. $V(t) = 150{,}000(1.04)^t$

C. $V(t) = 150{,}000^{1.04t}$

D. $V(t) = 150{,}000(0.04)^t$

E. $V(t) = 150{,}000(0.96)^t$

F. $V(t) = 150{,}000(1 + 0.04)^t$

G. $V(t) = 150{,}000(1 - 0.04)^t$

8. Barium-140 has a half-life of 13 days. The equation $A = 300\left(\frac{1}{2}\right)^{\frac{t}{13}}$ can be used to find the amount, A, of barium left after t days. Initially there are 300 milligrams. What is the average amount of barium lost per day between day 0 and day 26?

Express your answer to the nearest hundredth.

_______________ mg

9. Determine the domain and range for the function given.

$$f(x) = -3(2)^x - 4$$

Domain: __________

Range: __________

10. A set of nesting dolls has 12 layers that surround the innermost doll. The innermost doll has a height of 1.4 centimeters. The exponential function $f(n) = 1.4(1.2)^n$ describes the height of each layer of the nesting doll.

What is an appropriate domain for this function?

A. all integers

B. all real numbers

C. all integers from 0 to 12

D. all real numbers 0 to 12

Apply

How Much Will You Need to Save for Your Child's College Tuition?

The cost of a traditional four-year college education has been rising dramatically over the past several decades. It may seem crazy to be thinking about college for any children you may have in the future, when you haven't even gone to college yourself yet. But one of the most powerful factors in growing wealth is time.

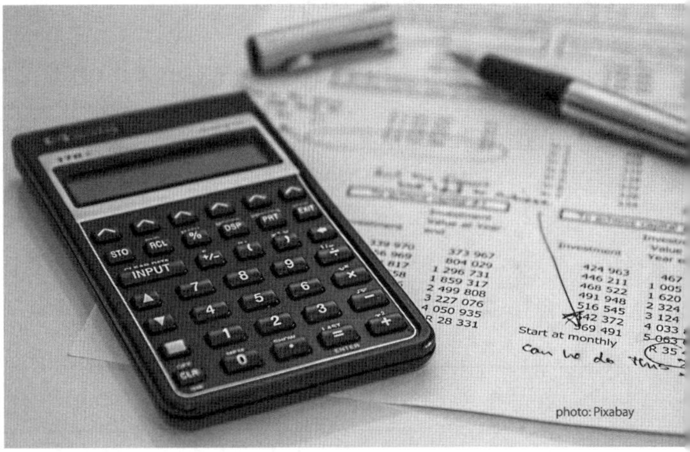

photo: Pixabay

Saving for College: View this video segment to learn how you can use trends to determine a plan to save for the future.

This problem focuses on college savings, but you can apply the same thinking to saving for any of your goals.

Consider a conservative estimate of \$50,000 for the projected yearly cost at a four-year public university, and \$85,000 for the projected yearly cost at a four-year private university. The following monthly payment function will help you determine how much you need to save each month so you can meet your goal, depending on the number of months you save:

$$f(t) = \frac{Cr}{(1+r)^t - 1}$$

where C is the cost of the first year of college, r is the annual percentage rate divided by 12 expressed as a decimal, and t is the number of months you will be saving.

- Assume that the annual percentage rate during the time that you are saving for college will remain steady at 3%. Choose three different times before your child starts college to start the savings plan.
- Use the monthly payment function to find the amount you would need to deposit each month to cover the cost of the first year at a public college for each starting time. Graph the function, and label the three points. Calculate the total amount of money you will have deposited with each of your three starting times.
- Use the same function and values of t to find the amount you would need to deposit each month to cover the cost of the first year at a private college for each starting time. Graph the function, and label the three points. Calculate the total amount of money you will have deposited with each of your three starting times.
- What is the moral of the story? Research alternatives to a traditional four-year college education, and write about one that you think is a good model.

Show what you've learned by completing the other performance tasks in the online Apply section.

Apply *(continued)*

Your answer to Apply will be assessed on the following criteria:

1. Choosing three times to start saving, and evaluating the monthly payment function for private and public schools for all three times
2. Graphing the two monthly payment functions; plotting the three points for each function
3. Discussing the total amounts deposited into savings and the "moral of the story"
4. Describing a reasonable alternative to a traditional four-year college education

Scale / Criteria	4 Exceeds Criteria	3 Meets Criteria	2 Progressing to Criteria	1 Below Expectations	0 No Expectation
Times and Functions	Correctly evaluates the monthly payment function for private and public schools for three chosen times.	Evaluates the monthly payment function for private and public schools for three chosen times, with minor errors.	Evaluates the monthly payment function incorrectly for private and public schools for three chosen times.	Evaluates the monthly payment function for private schools only or public schools only.	Does not attempt.
Graphs and Comparisons	Correctly graphs the two monthly payment functions and plots the three points for each function.	Graphs the two monthly payment functions and plots the three points for each function, with minor errors.	Graphs the two monthly payment functions incorrectly, and plots the three points for each function.	Graphs the monthly payment function for private schools only or public schools only.	Does not attempt.
Savings Totals and Moral of the Story	Provides a strong and accurate comparison of the amounts deposited, and draws a sound conclusion.	Provides a reasonable comparison of the amounts deposited, and draws a sound conclusion.	Provides a comparison of the amounts deposited, but uses faulty logic in conclusions made.	Includes a weak comparison of the amounts deposited, and uses faulty logic in conclusions made.	Does not attempt.
Alternative to a Four-Year University	Clearly describes a reasonable alternative to a traditional four-year college education.	Describes a reasonable alternative to a traditional four-year college education, with some lack of clarity.	Describes a somewhat reasonable alternative to a traditional four-year college education.	Describes an unreasonable alternative to a traditional four-year college education.	Does not attempt.

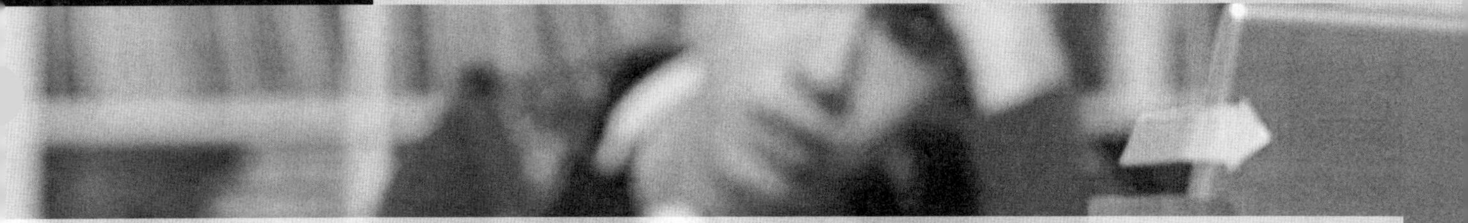

UNIT 9: Exponential Functions

9.2 Analyze Exponential Growth and Decay Models

photo: Getty Images

Lesson Objectives

- Describe exponential growth and decay in the context of real-world scenarios.
- Identify exponential growth and decay from equations and graphs.
- Rewrite exponential functions to interpret the function in context.
- Solve exponential equations graphically.

Essential Question

- How does the structure of an exponential expression reveal the growth or decay behavior of an exponential function?

photo: Getty Images

Investigations

Breaking the Bank

Maximize your earnings in the Investment Challenge! Watch out for the effects of the variables.

Exponential Behavior

Compare investment options by their equations. What features help you make the best choice?

Explore Compound Interest

Look for patterns to discover the formula for compound interest.

Using an Exponential Model

What's similar about populations? Whether people or bacteria, the growth can be exponential.

Comparing Exponential Models

Growth or decay? Model fish populations to make predictions.

Rewriting Exponential Functions

Form follows function. Write exponential functions in different forms to identify properties.

Key Vocabulary

exponential decay, exponential growth

Discover

As you complete Engage and the investigations, record the most important ideas you've learned.

Engage	**Investigation 1**
Investigation 2	**Investigation 3**
Investigation 4	**Investigation 5**

Name ______________________________ Date ____________

Check for Understanding

Analyze Exponential Growth and Decay Models: Investigation 1

1. Which equation represents \$4,000 invested at a rate of 5% interest per year?

   A. $y = 5x + 4{,}000$

   B. $y = 4{,}000\,(1.05)^x$

   C. $y = 4{,}000x + 5$

   D. $y = 5\,(4{,}000)^x$

The table shows the amount in an investment for a period of four years. The amount invested in year 0 in each option is \$2,500.

Use the options shown to answer questions 2 and 3.

Time (years)	Option 1 Amount	Option 2 Amount
1	\$2,600	\$2,650
2	\$2,704	\$2,809
3	\$2,812	\$2,978
4	\$2,925	\$3,156

2. Which statement is true about options 1 and 2?

   A. Option 1 returns 104% interest. Option 2 returns 150% interest.

   B. Option 1 returns 4% interest. Option 2 returns 6% interest.

   C. Option 1 returns 6% interest. Option 2 returns 4% interest.

   D. Option 1 returns 150% interest. Option 2 returns 104% interest.

3. How much will each option be worth after year 6, rounded to the nearest dollar?

   Option 1: ________ Option 2: ________

4. An investment grows according to an equation in the form $y = a(b)^x$. Explain the meaning of the values of a and b in this equation.

Name ______________________________ Date ______________

Check for Understanding

Analyze Exponential Growth and Decay Models: Investigation 2

1. Which expression represents the total value of a $2,000 investment after earning one year of 3% interest?

   A. 2,000 + 3

   B. 2,000 + 1.03

   C. 0.03(2,000)

   D. 1.03(2,000)

2. Which equations are increasing and which are decreasing? *Circle* Increasing *or* Decreasing *for each equation.*

   A. $y = 5{,}000(0.84)^t$ Increasing   Decreasing

   B. $y = 0.84(1.07)^t$ Increasing   Decreasing

   C. $y = 5{,}000(1.07)^t$ Increasing   Decreasing

   D. $y = 1.07(0.84)^t$ Increasing   Decreasing

3. Complete the following description of a situation that can be modeled by the equation $4{,}437 = 10{,}000(0.85)^5$.

   Talia buys a car for $ ________. It depreciates at a rate of ________ % every ________ year(s). After ________ year(s), the value of the car is $ ________.

4. A student rewrites a compound interest equation of $y = a(b)^x$ in the form $y = a(1 + r)^x$.

   What does having the compound interest equation in this form tell you about the interest rate? How can you use this equation to tell if the balance in the account is increasing or decreasing?

Name ______________________________ Date ______________

Check for Understanding

Analyze Exponential Growth and Decay Models: Investigation 3

1. Which unit can represent a population density?

   A. square miles

   B. people per hour

   C. millions of people

   D. people per square mile

2. For the growth factor $(1 + r)$ that represents population density that grows at a rate of 0.05% each year, what is the value of r?

   $r =$ ______________

3. Look at the table below.

Time since Antibiotic Applied (hours)	Bacteria Colony Population (millions)
0	900
2	576
4	369
6	236

   Write an equation to represent this decay.

   ____

4. A population changes over time t according to the equation $y = p(1 + r)^t$. If $0 < r < 1$, what do you know about the range of the function for domain $t \geq 0$? How does this relate to whether the function is increasing or decreasing?

Name ______________________________ Date ______________

Check for Understanding

Analyze Exponential Growth and Decay Models: Investigation 4

1. An invasive plant species is driving a native plant species out of an area. The population density of the invasive plant species is modeled by the equation $y = ab^x$. The population density of the native plant species in the area is modeled by $y = cd^x$. In both equations, x represents the amount of time since the invasive species was introduced. Which relationship is most likely true?

A. $a > c$ **B.** $d > b$

C. $b > d$ **D.** $d > c$

2. What is the approximate decreasing percent rate of change for a species whose population is modeled by the equation $y = 6{,}579(0.943)^x$?

A. 0.943% **B.** 5.7%

C. 94.3% **D.** 0.057%

3. Which characteristics represent exponential growth and exponential decay models? *Circle either* Exponential Growth *or* Exponential Decay *for each equation with the given characteristics.*

A. $y = 0.95b^x$ with $b > 1$	Exponential Growth	Exponential Decay
B. $y = 5{,}000(1 - r)^x$ with $r > 0$	Exponential Growth	Exponential Decay
C. $y = 254.8b^t$ with $0 < b < 1$	Exponential Growth	Exponential Decay
D. $y = 0.001(1 + r)^t$ with $r > 0$	Exponential Growth	Exponential Decay

4. A graph shows two curves of population density in a region versus time since an invasive species was introduced. Curve A represents the exponential growth of an invasive species. Curve B represents the exponential decay of a native species. Explain how you would find the first year when the invasive species exceeds the native species.

Name ______________________________ Date ____________

Check for Understanding

Analyze Exponential Growth and Decay Models: Investigation 5

1. Which function grows at the fastest rate?

   A. $V = 200(1 + 0.1)^t$

   B. $V = 200(1 - 0.1)^t$

   C. $V = 200(1 + 0.1)^{12t}$

   D. $V = 200(1 - 0.1)^{12t}$

2. Which equations are equivalent to $y = 25(0.2)^{4t}$? *Select all that apply.*

   A. $y = 5^{4t}$

   B. $y = 25(1 - 0.8)^{4t}$

   C. $y = (25 \cdot 0.2)^{4t}$

   D. $y = 25(1 + 0.2)^{4t}$

   E. $y = 25(0.2^4)^t$

   F. $y = 25(0.0016)^t$

3. Sales at a clothing website are modeled by the equation $y = 10(1 + 0.02)^{12t}$, where t represents the number of years since the website began selling clothes. What is the annual percentage increase in sales, to the nearest whole percent?

   ____

4. A blogger models the growth in traffic to his blog using the equation $y = a(1.00111)^{365t}$, where t represents the number of years since starting the blog. What does the structure of this equation suggest about the rate of growth of the traffic to the blog? What is the significance of the exponent in this equation? Explain your answer.

Summary

Before you attempt the Practice Exercises, review what you've learned.

Exponential Growth and Exponential Decay

An exponential function that can be modeled by an equation in the form $y = ab^t$, where $a > 0$ is either an **exponential growth** function or an **exponential decay** function.

- y is the value of the quantity at time t.
- a is the initial value of the quantity at time $t = 0$.

Exponential Growth

$y = ab^t$ where $a > 0$ and $b > 1$

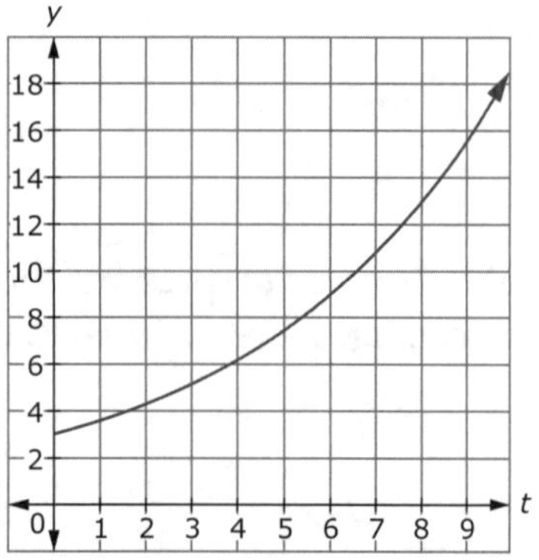

b is the growth factor—the factor by which the quantity is increasing per time period.

Exponential Decay

$y = ab^t$ where $a > 0$ and $0 < b < 1$

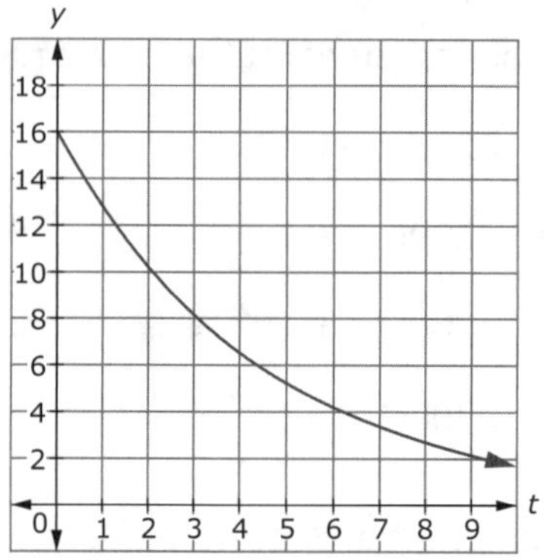

b is the decay factor—the factor by which the quantity is decreasing per time period.

When an exponential function is written in the form $y = a(1 + r)^t$, r is the percent rate of change in the quantity per time period t.

When $a > 0$ and $r > 0$, the function represents growth. When $a > 0$ and $r < 0$, the function represents decay.

Compound Interest

Compound interest is interest that is earned on both the principal and any previously earned interest. When interest on an account is compounded annually, interest is added to the account once per year.

You can use the formula $A = P(1 + r)^t$, to determine the amount A in an account after t years, where P is the principal, or initial amount in the account, and r is the annual compound interest rate, expressed as a decimal. Notice that this formula is an exponential growth function.

EXAMPLE: Applying Compound Interest

Mae deposits \$150 in an account that earns 2% interest compounded annually. At the same time, Soren deposits \$100 in an account that earns 4% interest compounded annually. After how many years will the amount in Soren's account first exceed the amount in Mae's account?

My Notes

Summary *(continued)*

SOLUTION:

Write a function to model the amount in each account.

Mae: The principal is $150, so $P = 150$. The annual compound interest rate is 2%, so $r = 0.02$. The amount A in Mae's account after t years is given by $A = 150\ (1 + 0.02)^t$.

Soren: The principal is $100, so $P = 100$. The annual compound interest rate is 4%, so $r = 0.04$. The amount A in Soren's account after t years is given by $A = 100\ (1 + 0.04)^t$.

Next, graph the functions using the Graphing Calculator. You will need to substitute y for A and x for t to enter the equations.

The graphs of the functions intersect at approximately (20.88, 226.81). For values of x less than 20.88, the amount in Soren's account is less than the amount in Mae's account. For values of x greater than 20.88, the amount in Soren's account is greater than the amount in Mae's account.

Interest is deposited once per year, so the amount in Soren's account will first exceed the amount in Mae's account after 21 years.

Modeling Exponential Functions

You can model some sets of data using exponential growth or exponential decay functions.

EXAMPLE: Developing an Exponential Model

The table shows how the value of a car is predicted to change over time.

Value of Car Over Time						
Time since purchase (years)	0	1	2	3	4	5
Value ($)	19,135	17,600	16,200	14,900	13,700	12,600

Write the equation of a function that models the data in the table.

SOLUTION:

Method 1: Use Mathematical Reasoning

First, check whether the value of the car is predicted to decrease by a constant factor each year.

Value of Car Over Time

Change in time	Time Since Purchase (years)	Value ($)	Change in value
	0	19,135	
+1	1	17,600	× 0.9198
+1	2	16,200	× 0.9205
+1	3	14,900	× 0.9198
+1	4	13,700	× 0.9195
+1	5	12,600	× 0.9197

Each year, the value of the car decreases by a factor of about 0.92, which shows that an exponential model is appropriate for the data.

Let V represent the value of the car in year t. Then, write an exponential decay function in the form $V = ab^t$.

The initial value of the car is $19,135, so substitute 19,135 for a. The decay factor for the car's value is about 0.92, so substitute 0.92 for b.

An exponential model for the car's value is $V = 19{,}135(0.92)^t$. You can rewrite this exponential decay function in the form $V = 19{,}135(1 - 0.08)^t$ to show that the value of the car is decreasing by 8% each year.

My Notes

Summary *(continued)*

Method 2: Use the Graphing Calculator

Enter the data as a table in the Graphing Calculator. Then, view the best-fit exponential curve for the data set.

The equation of the best-fit exponential curve is $y = 19{,}138.2765\ (0.9198)^x$. If we let V represent the value of the car after t years, we can write the equation as $V = 19{,}138.2765\ (0.9198)^t$. Notice that the parameters of this equation are very close to those we found using Method 1.

The equation $V = 19{,}138.2765\ (0.9198)^t$ can be rewritten as $V = 19{,}138.2765\ (1 - 0.0802)^t$, which confirms that the value of the car is decreasing by about 8% per year.

EXAMPLE: Using an Exponential Model to Solve Problems

Predict what the value of the car will be 8 years after it is purchased.

SOLUTION:

Substitute 8 for t in the exponential model.

$V = 19{,}135(0.92)^t$	Use the model from Method 1.
$V = 19{,}135(0.92)^8$	Substitute.
$V \approx 9{,}820$	Evaluate with a calculator.

After 8 years, the car will be worth about $9,820.

My Notes

Rewriting Exponential Functions

Sometimes you may need to rewrite an exponential growth or exponential decay function in the form $y = a(1 + r)^x$, or $y = a\ (1 - r)^x$, in order to determine the percent increase or decrease in the value of the function per time period.

EXAMPLE: Rewrite an Exponential Function

Gemma deposited $3,000 in a savings account that has an annual interest rate of 4.8%. The function $A = 3{,}000(1 + 0.048)^{\frac{t}{12}}$ gives the amount A in an account after t months. What is the percent increase in the amount in the account per month?

SOLUTION:

Notice that the equation of the function is not written in the form $A = P\ (1 + r)^t$, because the exponent is equal to $\frac{t}{12}$ rather than t. To find the percent increase per month, we need to rewrite the function in the form $A = P\ (1 + r)^t$.

$A = 3{,}000(1 + 0.048)^{\frac{t}{12}}$	
$A = 3{,}000(1.048)^{\frac{t}{12}}$	Add within the parentheses.
$A = 3,\ 000(1.048)^{\left(\frac{1}{12} \cdot t\right)}$	Write the exponent as a product.
$A = 3{,}000(1.048^{\frac{1}{12}})^t$	Apply the power of a power property.
$A = 3{,}000(1.0039)^t$	Use a calculator to approximate the expression in parentheses.
$A = 3{,}000(1 + 1.0039)^t$	Write 1.0039 as the sum $1 + 0.0039$.

The equation of the function is equivalent to $A = 3{,}000\ (1 + 0.0039)^t$, which has the form $A = 1\ (1 + r)^t$, with $r \approx 0.0039$. So, the percent increase in the value of the account per month is about 0.39%.

Practice Exercises

Review what you've learned using these practice problems. For practice problems with feedback, try the Coach and Play items in the Practice section online.

1. Two little towns, Franklin and Pierce, both had a population of 1,000 people each at the beginning of 2010. Since then, the populations of both towns have decreased as people move to the bigger cities.

Franklin loses 20 people per year. Pierce loses 3% of its population each year.

If this trend continues, what will be the approximate difference in population between Franklin and Pierce at the beginning of 2020? *Round your answer to the nearest person.*

The difference will be __________ people.

2. Brianna learns in science class that a typical cup of coffee has 130 milligrams of caffeine. She also discovers that her body eliminates 15% of the caffeine each hour.

Brianna quickly drinks a full cup of coffee on the way to school.

Complete the statements below about how much caffeine Brianna has in her system hours after drinking the coffee. Write your answers to the nearest whole number.

To the nearest milligram, 8 hours after drinking the coffee, Brianna still has __________ milligrams of caffeine from that coffee in her system.

Brianna has 49 milligrams of caffeine in her system after __________ hours.

3. Martin buys a new SUV for $28,000. It depreciates at the rate of 12% each year. He plans to sell it after 10 years. How much does Martin expect it to be worth then?

To the nearest 10 dollars, Martin expects the vehicle to be worth $__________.

4. The value of Jayme's car (in dollars) as a function of time (in years) is given by $f(x) = 24{,}000(0.83)^x$. What is the rate of depreciation?

The rate of depreciation is ______________%.

5. When Mitch invested $1,000, he believed that it would increase at a rate of 2%. The actual function that describes the growth of this investment over time is $f(x) = 1{,}000(1.04)^x$. What is the actual rate of growth?

The actual rate of growth is ______________%.

6. Velma takes 100 milligrams of a drug, which dissipates at a constant rate of 10% per hour.

Which function gives the amount of the drug in her body as a function of the number of hours?

A. $f(x) = 100(10)^x$

B. $f(x) = 10(1 + 100)^x$

C. $f(x) = 100(0.9)^x$

D. $f(x) = 100(1.1)^x$

E. $f(x) = 10(100)^x$

7. Margaret sells 1,500 tickets for her school's first baseball game, but then she notices that attendance changes by a certain rate at each game.

If the function giving the attendance as a function of the game number is $y = 1{,}500(0.95)^x$, then what is that change?

A. 95% increase

B. 95% decline

C. 95% increase

D. 5% decline

E. 15% increase

Practice Exercises *(continued)*

8. A Petri dish of bacteria begins with 100 bacteria, increasing in population at the rate of 10% per hour.

In another dish, 10,000 bacteria are exposed to a chemical that reduces their population by 25% per hour.

Graph the functions to determine at what time the two dishes briefly have the same number of live bacteria.

The populations are identical after __________ hours, to the nearest tenth.

9. Maria has a savings of $3,186.00. The rate on her account is 2.6% compounded yearly. Without depositing any other funds, when will Maria have at least $5,000 in her savings account?

- **A.** after 6 years
- **B.** after 12 years
- **C.** after 18 years
- **D.** after 24 years

10. Derek places 1,000 bacteria in a vial. Their change in population per hour is modeled by the function $f(x) = 1{,}000(1.03)^x$. Does the function show growth or decay, and at what rate?

The function shows __________ at a rate of __________%.

11. Match the following descriptions of growth or decay with the equations below. *Select all that apply. Descriptions may be used more than once.*

- **A.** Growth of 5%
- **B.** Decay of 5%
- **C.** Growth of 50%
- **D.** Decay of 50%
- **E.** Growth factor 0.5
- **F.** Growth factor 1.5

$y = a(1 - 0.05)^x$ __________

$y = a(1 - 0.5)^x$ __________

$y = a(1 + 0.5)^x$ __________

$y = a(1 + 0.05)^x$ __________

12. One country has a population of 42 million and a growth rate of 6% per year. Another has a population of 80 million and a growth rate of 2% per year. Graph each function to find at what time the two countries have the same population. *Write your answer to the nearest tenth.*

They have the same population at approximately __________ years.

Apply

What Is Happening to Some Animal Populations?

Animal populations can change for a variety of reasons. Watch the video to learn more.

Learn about endangered species.

Endangered Species: View this video segment about tracking the population levels of endangered animals.

- Conduct research to find a graph showing the population of an animal species whose population is increasing or decreasing exponentially. The graph must include time and population values.
- Using the graph, create a data table and write a function that models the data.
- Use your model to predict the population of that species 20 years from now. Do you feel that this is a reliable prediction? Why or why not?
- If an animal population is increasing or decreasing exponentially, do you think it will continue to do so forever? Write a short paragraph answering both parts of this question and explaining your answer.

Show what you've learned by completing the other performance tasks in the online Apply section.

Apply *(continued)*

Your answer to Apply will be assessed on the following criteria:

1. Researching and submitting an appropriate graph and converting the data to a table
2. Creating a mathematical function to model the data and explaining your method
3. Estimating the population 20 years from now and commenting on your prediction's reliability
4. Discussing the potential for continuing exponential growth and decays

Scale / Criteria	4	3	2	1	0
	Exceeds Criteria	**Meets Criteria**	**Progressing to Criteria**	**Below Expectations**	**No Expectation**
Data Presentation	Chooses an appropriate graph; clearly and correctly presents the data in a table.	Chooses a somewhat appropriate graph; correctly presents the data in a table.	Chooses an appropriate graph; does not clearly and correctly present the data in a table.	Chooses an inappropriate graph; does not clearly and correctly present the data in a table.	Does not attempt.
Correct Model	Correctly finds a model that represents the population; includes clear work and explanation.	Correctly finds a model that represents the population; includes mostly clear work or explanation.	Finds a model that represents the population with minor errors; includes mostly clear work and explanation.	Finds a model that represents the population with major errors; includes unclear work and explanation.	Does not attempt.
Correct Prediction	Uses data and calculations to make a correct prediction; discusses reliability.	Uses data and calculations to make a prediction with a minor error; discusses reliability.	Uses data and calculations to make a prediction with a major error; discusses reliability.	Uses data and calculations to make a prediction with a major error; does not discuss reliability.	Does not attempt.
Continued Growth or Decay	Discusses the idea of continuing exponential growth and decay in a well-developed paragraph that is strengthened by relevant arguments.	Discusses the idea of continuing exponential growth and decay in a well-developed paragraph with a weak argument.	Discusses continuing exponential growth and decay without any reasons to back up the discussion.	Does not discuss the idea of continuing exponential growth and decay in any meaningful way.	Does not attempt.

UNIT 10: Polynomials

10.1 Perform Operations on Polynomials

photo: Getty Images

Lesson Objectives

- Classify polynomials, identify key features, and write polynomials in a variety of forms.
- Add, subtract, and multiply polynomials.

Essential Question

- How do properties of integer operations apply to operations on polynomials?

photo: Getty Images

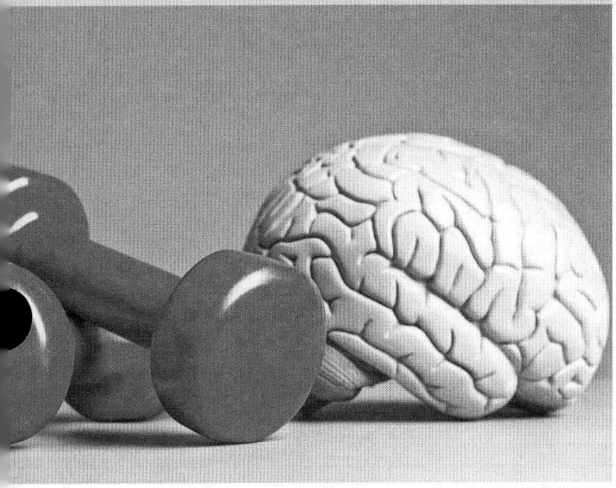

Investigations

Metabolic Equations

What are polynomials? What are their characteristics? How can you describe them?

Vital Capacity

Use properties of exponents to identify equivalent exponential expressions.

Tug of War

Add and subtract polynomials visually using polynomial tiles. Verify the sum or difference with graphs.

Oxygen Consumption

Discover a more efficient way to add and subtract polynomials.

Brainercise?

Visualize polynomial products with polynomial tiles.

Products of Linear Binomials

What does the graph of a polynomial function look like? Predict x- and y-intercepts.

Target Heart Rate

Remember the distributive property? It works for polynomials too.

Key Vocabulary

conjugates, degree of a polynomial, trinomial

Discover

As you complete Engage and the investigations, record the most important ideas you've learned.

Engage	
Investigation 1	**Investigation 2**
Investigation 3	**Investigation 4**
Investigation 5	**Investigation 6**

Name ______________________________ Date ______________

Check for Understanding

Perform Operations on Polynomials: Investigation 1

1. Which of the following expressions are equivalent to x^8? *Select all that apply.*

A. $x^2 \cdot x^6$ **B.** $(x^2)^6$

C. $x^2 \cdot x^4$ **D.** $(x^2)^4$

E. $\frac{x^{10}}{x^2}$ **F.** $x^2 \cdot x^2 \cdot x^2 \cdot x^2$

G. $\frac{x^{16}}{x^2}$ **H.** $x^{\frac{1}{8}} \cdot x^{\frac{1}{8}} \cdot x^{\frac{1}{8}} \cdot x^{\frac{1}{8}}$

2. Which properties are demonstrated in the following equations? *Draw a line from the equation to the property it demonstrates.*

A. $(c^4)^3 = c^{12}$ **I.** product of powers property

B. $\frac{x^5}{x^2} = x^3$ **II.** quotient of powers property

C. $m^4 \cdot m^3 = m^7$ **III.** power of a power property

3. Write the expression $\frac{12x^5y^{-2}}{3x^2y}$ in simplified form.

____

4. What is the connection between rational exponents and radical expressions? Explain how this connection allows you to rewrite the expression $\left(3^{\frac{1}{3}}x^{\frac{2}{3}}\right)^5$ as a radical expression. What is the equivalent expression in radical form?

Name ______________________________ Date ____________

Check for Understanding

Perform Operations on Polynomials: Investigation 2

1. What does it mean to say that a pair of numbers is a zero pair?

   A. One of the numbers is zero.

   B. The product of the pair of numbers is zero.

   C. The sum of the pair of numbers is zero.

2. What is the other member of the zero pair including $5x$?

   ____

3. Look at the geometric model of a polynomial expression.

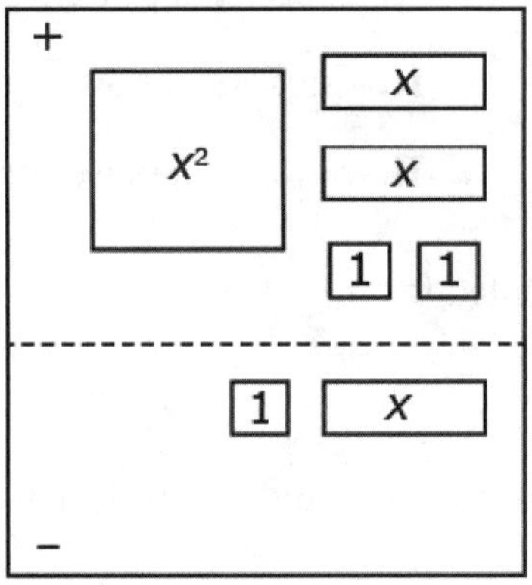

Which expression is equivalent to the above geometric model? *Select only one.*

   A. $(x^2 + 2x + 2) + (x + 1) = x^2 + 3x + 3$

   B. $(x^2 + 2x + 2) - (x + 1) = x^2 + x + 1$

   C. $(x^2 + x + 1) - (x + 1) = x^2$

   D. $(x^2 + x + 1) + (x + 1) = x^2 + 2x + 2$

4. Explain how you can use your understanding of the sum and difference of polynomials to identify the y-intercept of $y = (x^3 + 2x) + (3x + 4) - (x^2 - 1)$. What is the y-intercept of $y = (x^3 + 2x) + (3x + 4) - (x^2 - 1)$?

Name ________________________________ Date ______________

Check for Understanding

Perform Operations on Polynomials: Investigation 3

1. In the space provided, create geometric models to show that $(2x^2 + x + 4) - (x^2 + x + 1)$ is equivalent to $(2x^2 + x + 4) + (-x^2 - x - 1)$. *Write your simplified equivalent expression on the line.*

+	+
–	–

____

2. Look at the work a student writes to add two polynomials horizontally.

$$(x^2 + 2x + 3) + (2x^2 + x + 1) = x^2 + 2x^2 + 2x + x + 3 + 1$$

What property allows the student to rearrange terms so that like terms are adjacent?

A. distributive property

B. commutative property

C. reflexive property

3. What is the simplified form of $(2x^2 + x + 1.5) - (5x^2 - x + 1.25)$?

____

4. Explain why the difference of two polynomials is also a polynomial.

Name ______________________________ Date ____________

Check for Understanding

Perform Operations on Polynomials: Investigation 4

Consider the 4-Region board shown for questions 1 and 2.

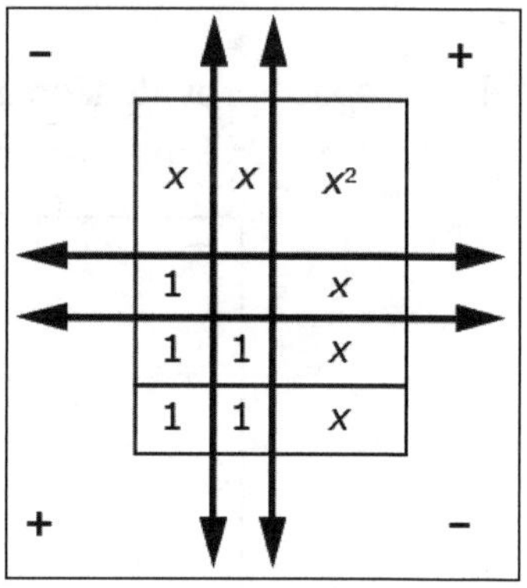

1. What expression is represented in the 4-Region board?

   A. $(x + 2)(x + 1)$ **B.** $(x + 2)(x - 1)$

   C. $(x - 2)(x - 1)$ **D.** $(x - 2)(x + 1)$

2. What is the final solution from the 4-Region board?

   ____

3. Complete the 4-Region board below to determine the product of $(x - 2)(x + 3)$. *Write your final answer on the line.*

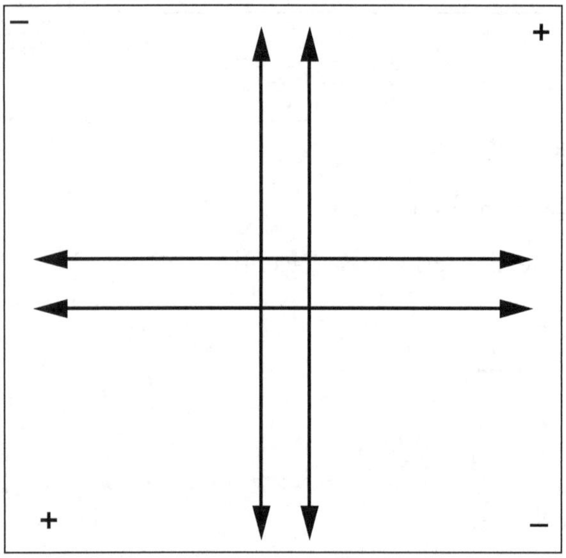

____

4. Consider the expression $(x + a)(x + b)$, where x is a variable and a and b are integers. If the product of these two binomials contains a negative term, what must be true about a and b?

Name ______________________________ Date ______________

Check for Understanding

Perform Operations on Polynomials: Investigation 5

The graph of the linear equation $f(x)$ has a x-intercept of (2, 0) and a y-intercept of (0, 4). The graph of the linear equation $g(x)$ has a x-intercept of (–1, 0) and a y-intercept of (0, –3).

For questions 1 and 2, consider the equation $y = f(x) \cdot g(x)$.

1. What are the x- and y-intercepts of $y = f(x) \cdot g(x)$?

x-intercept(s) ______________ y-intercept(s) ______________

2. What type of equation is $y = f(x) \cdot g(x)$?

A. linear equation **B.** quadratic equation **C.** exponential equation

3. Use the 4-Region Board below to find the product of the binomials $2x - 2$ and $2 - x$. *Write your final answer on the line.*

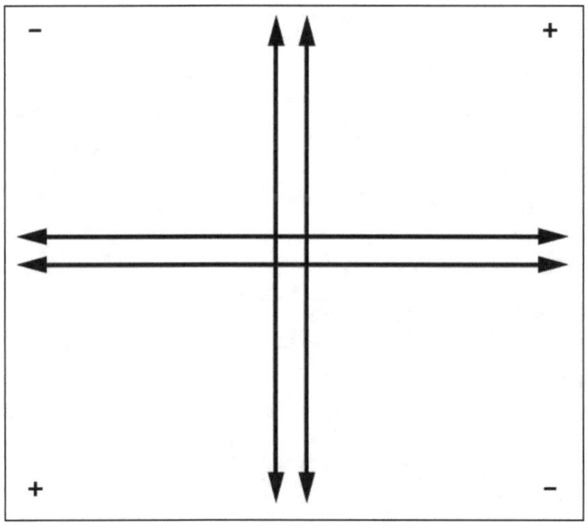

____

4. If a polynomial is a product of four first-degree binomials, explain how the x- and y-intercepts are related to those binomials. What degree is the polynomial?

Name ______________________________ Date ____________

Check for Understanding

Perform Operations on Polynomials: Investigation 6

1. Which of the following expressions are equivalent to $(2x - 5)(x + 5)$? *Select all that apply.*

A. $2(x + 5) - 5(x + 5)$ **B.** $2x(x + 5) - 5(x + 5)$

C. $x(2x - 5) + 5(2x - 5)$ **D.** $x(2x - 5) - 5(x + 5)$

2. Which of the following expressions are equivalent to $3x(4x - 6) - 2(4x - 6)$? *Select all that apply.*

A. $(3x - 2)(4x - 6)$ **B.** $12x^2 - 18x - 8x + 12$

C. $(12x - 18)(8x - 12)$ **D.** $12x^2 - 18x - 8x - 12$

E. $12x^2 - 26x + 12$ **F.** $12x^2 - 26x - 12$

3. Write an equivalent expression for each of the products of powers below, in simplified form.

$(x + 8)(x - 8) =$ ________________

$(x + 5)^2 =$ ________________

4. Expand and simplify the product of $(x - a)^2$.

Summary

Before you attempt the Practice Exercises, review what you've learned.

A **polynomial** is an algebraic expression that consists of the sum or difference of two or more terms. The terms in a polynomial can have constants, coefficients, variables, and whole number exponents. A polynomial cannot have division by a variable.

The set of polynomials is closed under the operations of addition, subtraction, and multiplication. This means that the sum of polynomials is a polynomial, the difference of polynomials is a polynomial, and the product of polynomials is a polynomial.

A **monomial** is a polynomial with only 1 term.

A **binomial** is a polynomial with 2 terms.

A **trinomial** is a polynomial with 3 terms.

Geometric Models

You can model polynomials using polynomial tiles. A large square models x^2, a rectangle represents x, and a small square represents the constant 1.

EXAMPLE: Modeling Polynomials

Use polynomial tiles to model the expression $2x^2 + 7x + 3$.

SOLUTION:

Use the appropriate number of tiles to create a picture of the polynomial.

Use 2 x^2-tiles to model $2x^2$. Use 7 x-tiles to model $7x$. Use 3 unit tiles (1-tiles) to model the constant 3.

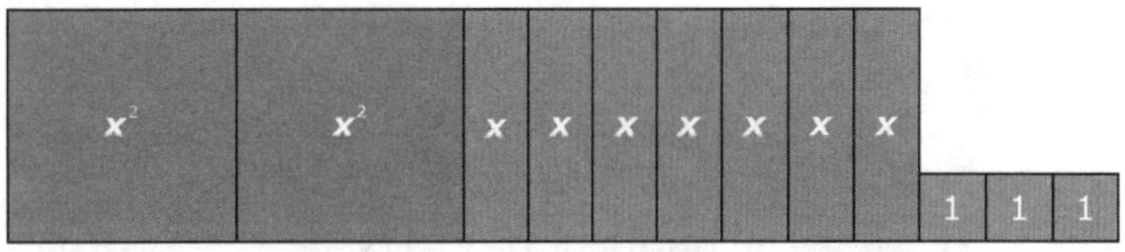

This is an area model. The area of the model is $2x^2 + 7x + 3$.

Rewriting Exponential Expressions

The properties of exponents that you have used with numerical bases also apply to expressions with variable bases.

My Notes

Summary *(continued)*

EXAMPLE: Simplifying Variable Expressions

Rewrite each expression.

A. Simplify: $x^4 x^7$

B. Simplify: $\frac{x^8}{x^3}$

C. Simplify: $(x^3)^4$

D. Write without a radical: $\sqrt{3x^7y^4}$

SOLUTION:

A. Use the product of powers property.

$x^4x^7 = x^{4+7} = x^{11}$

B. Use the quotient property.

$\frac{x^8}{x^3} = x^{8-3} = x^5$

C. Use the power of a power property.

$(x^3)^4 = x^{3 \cdot 4} = x^{12}$

D. A radical with an index of 2 is equivalent to raising the radicand to the $\frac{1}{2}$ power.

$\sqrt{3x^7y^4} = \sqrt[2]{3x^7y^4} = (3x^7y^4)^{\frac{1}{2}} = 3^{\frac{1}{2}}x^{\frac{7}{2}}y^2$

My Notes

Terms in a polynomial expression are called like terms if they have the same variables raised to the same powers. You can add or subtract polynomials by modeling with polynomial tiles or solve algebraically by using the commutative and associative properties.

Adding and Subtracting Polynomials

EXAMPLE: Adding Polynomials Algebraically

Find the sum of $x^2 + 4x + 5$ and $2x^2 + x$.

SOLUTION:

$(x^2 + 4x + 5) + (2x^2 + x)$

To add vertically, align the like terms, and then add them. To add horizontally, use the commutative property to reorder the terms so that like terms are together. Then, add the like terms.

Vertical Method:

$$\begin{array}{r} x^2 + 4x + 5 \\ +\ 2x^2 + x \\ \hline 3x^2 + 5x + 5 \end{array}$$

Horizontal Method:

$$(x^2 + 4x + 5) + (2x^2 + x)$$

$$(x^2 + 2x^2) + (4x + x) + 5$$

$$3x^2 + 5x + 5$$

The sum is $3x^2 + 5x + 5$.

Summary *(continued)*

EXAMPLE: Subtracting Polynomials Algebraically

Find the difference of $2x^2 + 6x - 3$ and $x^2 + 5x - 1$.

SOLUTION:

Write the subtraction problem, and then rewrite it as addition of the opposite.

$$(2x^2 + 6x - 3) - (x^2 + 5x - 1)$$
$$(2x^2 + 6x + (-3)) + (-x^2 - 5x + 1)$$

Then collect like terms and add.

$$2x^2 + 6x + (-3) + (-x^2) + (-5x) + 1$$
$$2x^2 + (-x^2) + 6x + (-5x) + (-3) + 1$$
$$x^2 + x + (-2)$$
$$x^2 + x - 2$$

The difference is $x^2 + x - 2$.

Multiplying Polynomials

To multiply polynomials, you will use the commutative, associate, and distributive properties.

EXAMPLE: Multiplying Binomials Algebraically

Find the product $(x - 2)(3x + 1)$ by using the distributive property.

SOLUTION:

$(x - 2)(3x + 1)$

$x(3x + 1) + (-2)(3x + 1)$	Distribute $(3x + 1)$ to both terms of the other binomial.
$3x^2 + x + (-6x) + (-2)$	Distribute x and -2 to $(3x + 1)$.
$3x^2 - 5x - 2$	Combine like terms.

My Notes

Summary *(continued)*

The products of some binomial factors follow predictable patterns. We call these **special products**.

Special Products	
Perfect Square Trinomial The product of a squared binomial.	$(x + b)^2$ $(x + b)(x + b)$ $x^2 + 2bx + b^2$ $(x - b)^2$ $(x - b)(x - b)$ $x^2 - 2bx + b^2$
Difference of Squares The product of conjugate pair binomials.	$(x + b)(x - b)$ $x^2 - b^2$

EXAMPLE: Using Special Products

Tell what type of special product each problem has.

Then, find the product without multiplying the binomials.

A. $(x + 8)^2$

B. $(x - 11)(x + 11)$

C. $(x - 5)(x - 5)$

SOLUTION:

A. The expression is a square of a binomial, so the product is a perfect square trinomial.

Substitute 8 for b in the pattern for a perfect square trinomial.

$x^2 + 2 \cdot 8x + 8^2$

$x^2 + 16x + 64$

B. The expression is the product of binomial conjugates, so the product is a difference of squares. Substitute 11 for b in the pattern in the pattern for a difference of squares.

$x^2 - 11^2$
$x^2 - 121$

C. The expression is a square of a binomial, so the product is a perfect square trinomial. Substitute 5 for b in the pattern for a perfect square trinomial.

$x^2 - 2 \cdot 5x + 5^2$
$x^2 - 10x + 25$

My Notes

Practice Exercises

Review what you've learned using these practice problems. For practice problems with feedback, try the Coach and Play items in the Practice section online.

1. Simplify the expression.

   $(4x^3)(3x^2 - 2x) =$ ___________

2. What is the area of the trapezoid whose bases are x and $3x + 4$ and whose height is $x + 2$?

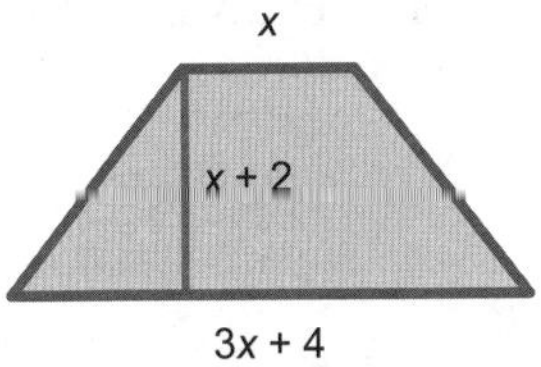

   The formula for the area of a trapezoid is

   $A = \frac{1}{2}h(b_1 + b_2)$.

   A. $4x + 6$ B. $3x^2 + 4$

   C. $2x^2 + 6x + 4$ D. $3x^2 + 6x + 8$

   E. $3x^3 + 6x^2 + 8x$

3. Examine each expression and decide whether or not it is a polynomial. *Select all that apply.*

   A. $2x^2$ B. $5 - x\sqrt{3}$

   C. $5\sqrt{2x} - 14$ D. $2x^3 - 2x^2 - x + 2$

   E. $\frac{x^2 + 5}{x^3 + 5}$ F. $\frac{6}{x} + 3$

   G. $\frac{x^3}{5} + 4x^2 - 9$ H. $(x^2 + 9)^{\frac{1}{2}}$

   Not a polynomial: __________

   Polynomial: __________

4. Larry's custom T-shirt business makes a revenue of \$15 for every T-shirt sold. His expenses for materials and advertising are represented by the expression $4t + 10$, where t is the total number of T-shirts made.

   Which expression represents his profit for making and selling t T-shirts?

   A. $19t + 10$ B. $11t - 10$ C. $29t$

   D. $11t + 10$ E. $10 - 19t$

5. What is $\left(\frac{2m^{10}}{4}\right)\left(\frac{m^{-2}}{3}\right)\left(\frac{6}{m^2}\right)^2$ reduced to simplest form? ___________

6. Paula subtracts the polynomial $4x^2 - 3x + 2$ from the polynomial $9x^2 - 3x + 2$.

   Use the choices below each answer line to complete the statement that describes her final answer.

   The difference between the two polynomials is ___________.

   A. a polynomial B. not a polynomial

   This is because ___________.

   A. $5x^2$ is not a polynomial

   B. $5x^2 - 6x + 4$ is a polynomial

   C. subtracting polynomials always results in a polynomial

   D. subtracting polynomials results in nonpolynomials

Practice Exercises *(continued)*

7. Complete the statements below. Write **A** for *always*, **S** for *sometimes*, and **N** for *never*.

   The terms of a polynomial are ____________ monomials.

   The sum of two trinomials is ____________ a trinomial.

   Like terms ____________ have the same coefficient and the same variable factor.

   Subtraction of a term can ____________ be rewritten as addition of the opposite term.

   A binomial is ____________ a polynomial of degree 3.

8. Four different students tried to multiply the binomials $(2x - 3)(x - 2)$. Which student's work is correct?

   A. $(2x - 3)(x - 2)$
   $2x(x - 2) - 3(x - 2)$
   $2x^2 - 4x - 3x - 6$
   $2x^2 - 7x - 6$

   B. $(2x - 3)(x - 2)$
   $2x(x - 2) - 3(x - 2)$
   $2x^2 - 4 - 3x + 6$
   $2x^2 - 3x + 2$

   C. $(2x - 3)(x - 2)$
   $2x(x - 2) - 3(x - 2)$
   $2x^2 - 4x - 3x + 6$
   $2x^2 - 7x + 6$

   D. $(2x - 3)(x - 2)$
   $2x(x - 2) - 3(x - 2)$
   $2x^2 - 4 - 3x + 2$
   $2x^2 - 3x - 2$

9. Subtract the polynomial $5y - 7y^2 - 5$ from $9y + 7 - 3y^2$.

   A. $-10y^2 - 14y - 2$

   B. $-10y^2 + 14y + 2$

   C. $-4y^2 - 4y + 12$

   D. $4y^2 + 4y + 12$

10. Match the equivalent expressions.

$-3(x + 1)^2$ ____	**A.** $9x^2 - 9$
$(3 - 3x)^2$ ____	**B.** $9x^2 + 6x - 3$
$(3x + 3)(3x - 3)$ ____	**C.** $9x^2 - 18x + 9$
$(3x + 3)(3x - 1)$ ____	**D.** $-3x^2 - 6x - 3$

11. Which expression is equivalent to $2t(t - 3) - (t + 1)^2$?

    A. $t^2 - 4$

    B. $t^2 - 8t - 1$

    C. $t^2 + 2t - 2$

    D. $t^2 - 7t + 1$

12. Photos with a height to width ratio of 4 to 5 are framed by wooden frames that add 6 inches to the height and 4 inches to the width. This makes the total height $4x + 6$ inches and the total width $5x + 4$ inches.

    Which expression represents the total area in square inches that is bounded by the wooden frame?

    A. $18x + 20$

    B. $20x^2 + 24$

    C. $20x^2 + 16x + 24$

    D. $20x^2 + 46x + 24$

Apply

How Can Polynomials Help Predict Genetics?

Punnett squares are used in biology to determine the possible genetic combinations of physical traits in the offspring of two parents.

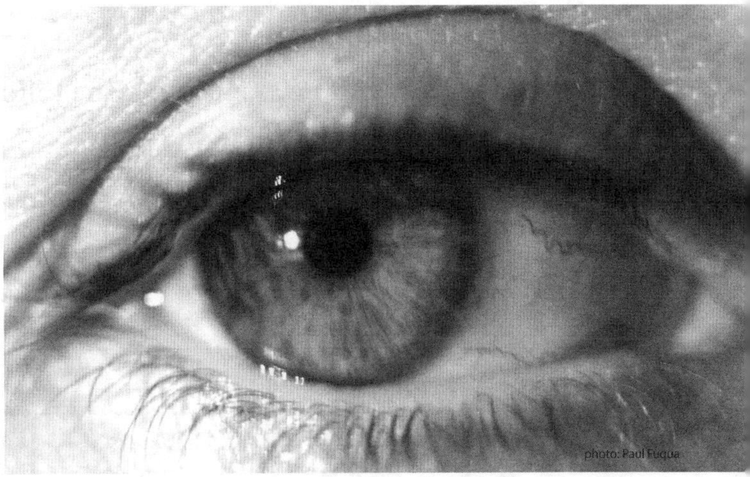
photo: Paul Fuqua

Punnett Squares: View this video segment about using Punnett squares to model operations with polynomials.

Research Punnett squares and how they are used. How can you use a square similar to a Punnett square to represent the product of two binomials?

- Write a short paragraph about Punnett squares and how they work.
- Explain how a square similar to a Punnett square could be used to multiply two binomials.
- Create a model with algebra tiles that illustrates your explanation.
- Use your model to demonstrate how a square similar to a Punnett square could be used to multiply trinomials.

Show what you've learned by completing the other performance tasks in the online Apply section.

Apply *(continued)*

Your answer to Apply will be assessed on the following criteria:

1. Writing a short paragraph about Punnett squares and how they work
2. Showing how to multiply two binomials with a square similar to a Punnett square
3. Making an algebra tile model showing the multiplication of two binomials
4. Showing how a square similar to a Punnett square can be used to multiply trinomials

Criteria \ Scale	4	3	2	1	0
	Exceeds Criteria	**Meets Criteria**	**Progressing to Criteria**	**Below Expectations**	**No Expectation**
Punnett Squares	Writes a paragraph correctly describing Punnett squares and how they work.	Writes a paragraph about Punnett squares, but does not completely describe how they work.	Writes a paragraph about Punnett squares, but does not describe how they work.	Writes a paragraph that mentions Punnett squares, but incorrectly describes how they work.	Does not attempt task.
Multiplication of Binomials Using Squares	Sets up the square properly, multiplies all four squares correctly, and adds all squares together to write the correct polynomial.	Sets up the square properly, multiplies the squares, and adds them together, with a minor error.	Sets up the square properly, multiplies the squares, and adds them together, with a major error.	Sets up the square properly, but does not perform the correct calculations.	Does not attempt task.
Model	Creates an algebra tile model that accurately multiplies all four terms together.	Creates an algebra tile model that multiplies all four terms together, with a minor error.	Creates an algebra tile model that multiplies all four terms together, with a major error.	Creates an algebra tile model that does not multiply all four terms together.	Does not attempt task.
Multiplication of Trinomials Using Squares	Sets up the square properly, multiplies all nine squares correctly, and adds all squares together to write the correct polynomial.	Sets up the square properly, multiplies the squares, and adds them together, with a minor error.	Sets up the square properly, multiplies the squares, and adds them together, with a major error.	Sets up the square properly, but does not perform the correct calculations.	Does not attempt task.

UNIT 10: Polynomials

10.2 Factor Polynomials

photo: Getty Images

Lesson Objectives

- Factor polynomials.
- Rewrite polynomials to reveal the contextual interpretation.

Essential Questions

- What is the relationship between multiplying polynomials and factoring polynomials?
- What is the relationship between the factors of a polynomial expression, the zeros of a function, and the x-intercepts of a graph?
- How can the characteristics of a polynomial be used to factor the polynomial?

photo: Getty Images

Investigations

Step It Up

x-intercepts play a special role in polynomial functions. What makes them so special?

Factors Remix

Second verse, now in reverse. Use polynomial tiles to find factors of an expression.

Uncover Hidden Structures

You recognize perfect square numbers, but what about square polynomial expressions?

Number Puzzles

Number puzzles are based on patterns. Look for factors, products, addends, and sums to solve puzzles.

Puzzling Trinomials – Part 1

Use what you learned solving number puzzles to factor quadratic expressions.

Puzzling Trinomials – Part 2

Puzzle me this! Factor more complex trinomials.

Key Vocabulary

degree of a polynomial, term (in an expression), trinomial

Discovery

As you complete Engage and the investigations, record the most important ideas you've learned.

Engage	**Investigation 1**
Investigation 2	**Investigation 3**
Investigation 4	**Investigation 5**

Name ______________________________ Date ______________

Check for Understanding

Factor Polynomials: Investigation 1

1. Based on your knowledge of the 4-Region board, identify the product that represents the area of the rectangle shown to the right.

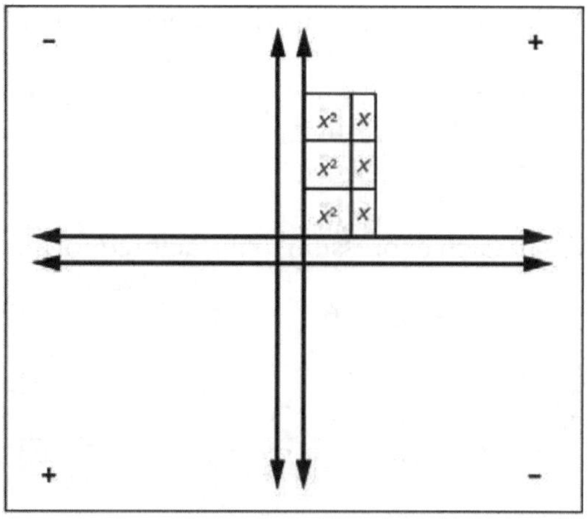

   A. $x(3x + 3)$

   B. $3x(x + 1)$

   C. $3x(x + 3)$

   D. $3x(3x + 3)$

2. Complete the table below by factoring the expressions with the given lengths.

Expression	Length	Rectangle or Square?	Width
$3x^2 + 5x + 2$	$3x + 2$		
$9x^2 + 12x + 4$	$3x + 2$		
$9x^2 - 4$	$3x + 2$		

3. Complete the table below by factoring the expressions.

Expression	Rectangle or Square?	Product of Factors (Length)(Width)
$x^2 + 10x + 25$		
$x^2 - 12x + 36$		
$2x^2 + 5x + 3$		

4. How could you write a quadratic trinomial that you know creates a square? Include specific examples to explain your reasoning.

Name ____________________ Date ____________

Check for Understanding

Factor Polynomials: Investigation 2

1. Which polynomials are differences of squares? *Select all that apply.*

A. $x^2 - 12x - 36$ **B.** $x^2 - 36$

C. $w^2 - 100$ **D.** $w^2 - 20w + 100$

2. Match each expression to its factored form.

A. $x^2 - 8x + 16$ **I.** $(x + 4)(x - 4)$

B. $x^2 + 8x + 16$ **II.** $(x - 4)^2$

C. $x^2 - 16$ **III.** $(x + 4)^2$

3. Factor the following expressions. Identify whether each is a square trinomial, difference of squares, or neither.

Expression	Factored Form	Square Trinomial, Difference of Squares, or Neither
$x^2 + 16x + 64$		
$x^2 + 12x - 64$		
$x^2 - 16x + 64$		
$x^2 - 64$		

4. For a function $f(x) = x^2 + px + q$, which is a square trinomial, explain how the factored form would relate to the graph of the function.

Name ______________________________ Date ______________

Check for Understanding

Factor Polynomials: Investigation 3

1. Complete the following number puzzles.

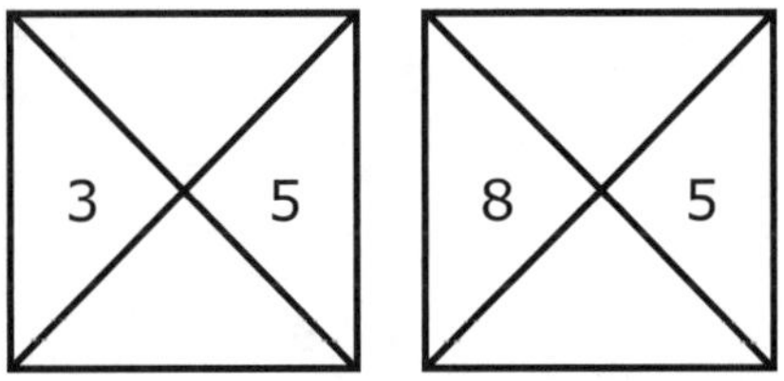

2. Complete the following number puzzles.

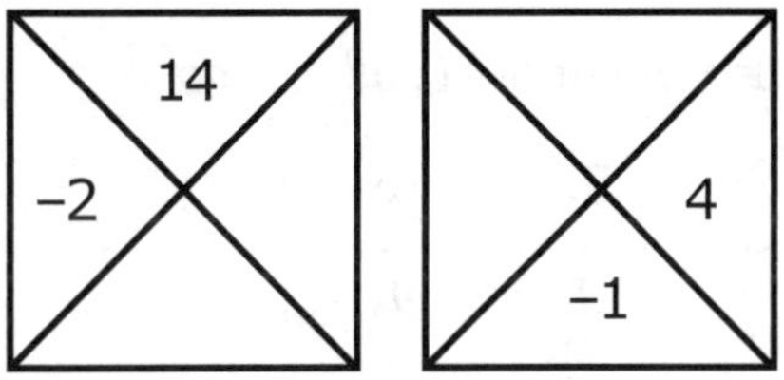

3. Which of the following are prime puzzles? *Select all that apply.*

A.

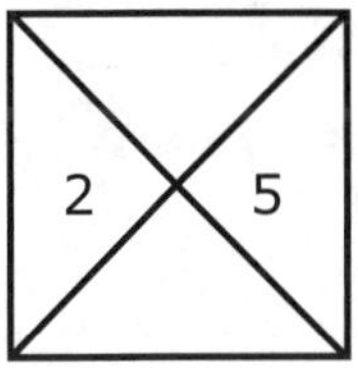

B.

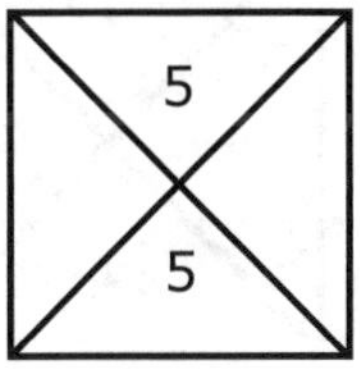

C.

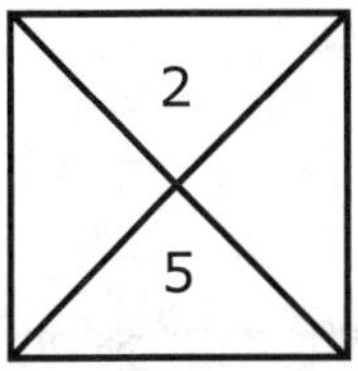

D.

4. Explain the relationships among the numbers in the number puzzle.

Name ________________________ Date ____________

Check for Understanding

Factor Polynomials: Investigation 4

1. Complete the number puzzle below to factor the trinomial $x^2 + 5x + 6$. Write the factored form of the trinomial on the line.

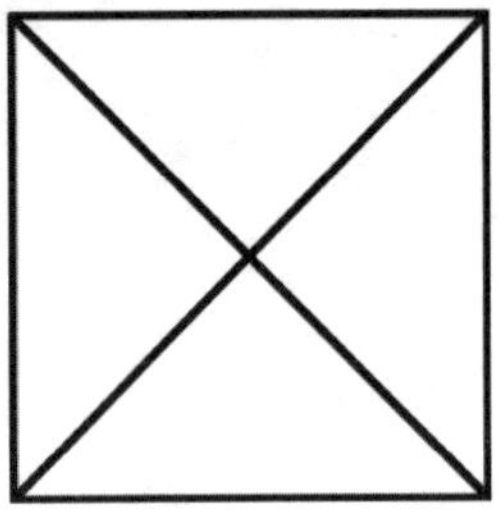

____

2. Which of the following is a prime trinomial? *Select all that apply.*

   A. $x^2 + 9x - 6$ **B.** $x^2 + 6x + 9$

   C. $x^2 - 6x + 9$ **D.** $x^2 + 9x - 3$

3. Consider the number puzzle shown.

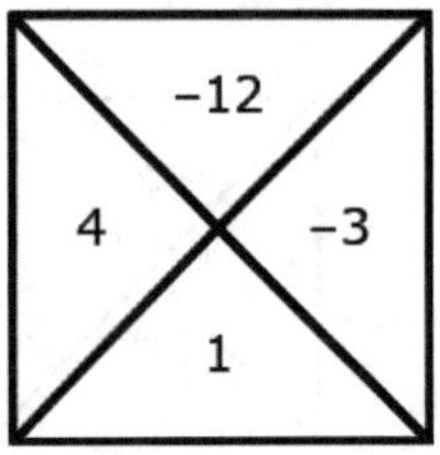

   What is the expanded form and factored form of the trinomial shown in this number puzzle?

   Expanded form: ______________

   Factored form: ______________

4. How does the factored form of a trinomial, $(x + a)(x + b)$, relate to the zeros of the expression? Explain why these values are called the “zeros” of the expression.

Name ____________________ Date __________

Check for Understanding

Factor Polynomials: Investigation 5

1. Which of the following is equivalent to $5x^2 + 30x + 40$?

   A. $5(x^2 + 25x + 35)$
   B. $5(x^2 + 35x + 45)$
   C. $5(x^2 + 6x + 8)$
   D. $5(x^2 + 30x + 40)$

2. Which of the following is equivalent to $x^2 + 6x - 40$? *Select all that apply.*

   A. $x^2 - 4x + 10x - 40$
   B. $x(x + 4) - 10(x + 4)$
   C. $x^2 + 4x - 10x - 40$
   D. $x(x - 4) + 10(x - 4)$

3. Follow the steps described to factor the trinomial expression $4x^2 - 20x + 24$.

   ____________________ Factor out the GCF.

   ____________________ Rewrite the x-term.

   ____________________ Apply the distributive property.

   ____________________ Write the trinomial in factored form.

4. How can you use the graph $y = ax^2 + bx + c$ to determine the x values that make the trinomial equal 0?

Summary

Before you attempt the Practice Exercises, review what you've learned.

Using Graphs to Find Zeros and Factors

EXAMPLE: Find Factors by Graphing

Use a graph to find the factors of $x^2 + 2x - 8$.

SOLUTION:

Graph $x^2 + 2x - 8$ in the coordinate plane.

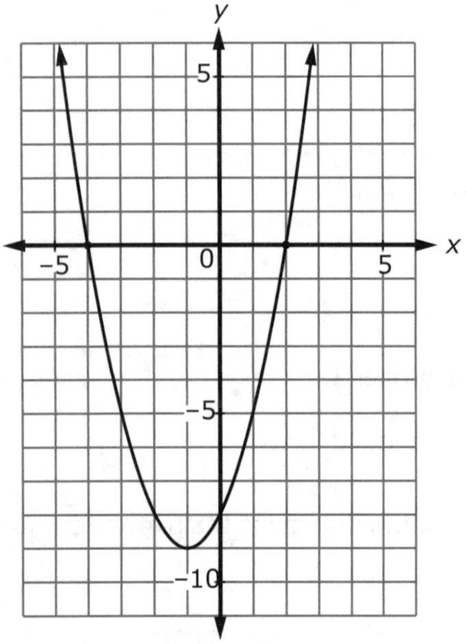

The zeros of the function are the x-intercepts, so locate where the graph crosses the x-axis. The zeros are $x = -4$ and $x = 2$. This means that $x + 4 = 0$ and $x - 2 = 0$, so the factors of $x^2 + 2x - 8$ are $(x + 4)(x - 2)$.

Factoring by Inspection

Some polynomials exhibit special patterns. Once you recognize the pattern, you can factor the polynomial.

Special Products	
Perfect Square Trinomial The product of a squared binomial.	$x^2 + 2bx + b^2$ $(x + b)(x + b)$ $(x + b)^2$ $x^2 - 2bx + b^2$ $(x - b)(x - b)$ $(x - b)^2$
Difference of Squares The product of conjugate pair binomials.	$x^2 - b^2$ $(x + b)(x - b)$

EXAMPLE: Factor a Perfect Square Trinomial

Factor $x^2 - 12x + 36$.

SOLUTION:

The trinomial fits the pattern of a perfect square trinomial $x^2 - 2bx + b^2$, since it can be written as $x^2 - 2(6)x + 6^2$.

Since $b = 6$, $(x - b)(x - b)$ can be written as $(x - 6)(x - 6)$.

So, $x^2 - 12x + 36 = (x - 6)^2$.

My Notes

Summary *(continued)*

EXAMPLE: Factor the Difference of Squares

Factor $16x^2 - 49$.

SOLUTION:

This polynomial fits the pattern of the difference of two perfect squares $x^2 - b^2$, since it can be written as $(4x)^2 - 7^2$.

And $(4x)^2 - 7^2$ can be written as $(4x + 7)(4x - 7)$.

So, $16x^2 - 49 = (4x + 7)(4x - 7)$.

Most polynomials, however, do not follow these special patterns. For these polynomials, you can use a different method for factoring.

Using Number Puzzles to Factor Trinomials

A number puzzle is useful when you need to identify the relationships among the coefficients of a trinomial.

EXAMPLE: Factor a Trinomial Where $a = 1$

Factor $x^2 + 8x - 48$.

SOLUTION:

Begin by placing the constant term in the top section of the number puzzle. Place the coefficient of the x term in the bottom section. Then complete the number puzzle by finding two numbers that multiply to give the value in the top section and add to give the value in the bottom section.

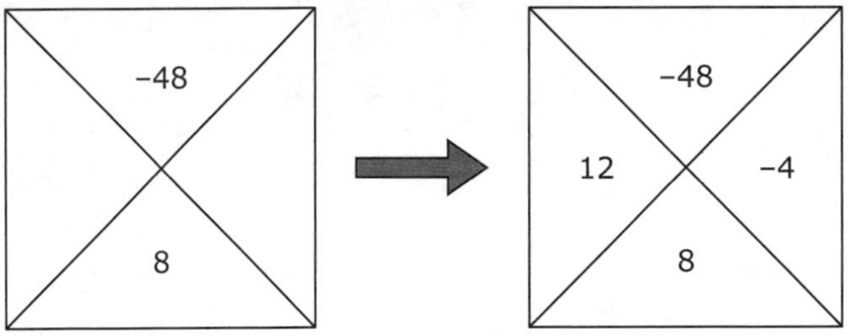

Use the side values in the number puzzle to rewrite the x term in the original trinomial.

$$x^2 + 8x - 48 = x^2 + 12x - 4x - 48$$

Next, identify the greatest common factor in each pair of terms.

Use the distributive property to factor out each GCF.

$$x^2 + 12x - 4x - 48 = x(x + 12) + (-4)(x + 12)$$

Now use the distributive property once more.

$$x(x + 12) + (-4)(x + 12) = (x - 4)(x + 12)$$

You can verify the solution by multiplying the binomials.

$$x^2 + 8x - 48 = (x - 4)(x + 12)$$

To factor a trinomial with a leading coefficient other than 1, use the AC method.

My Notes

Summary *(continued)*

EXAMPLE: Factor a Trinomial Where $a \neq 1$

Factor $6x^2 - x - 15$.

SOLUTION:

Since the coefficient of x^2 is not 1, multiply the coefficient 6 by the constant term –15. Place this number in the top section of the puzzle box. Place the coefficient of the x-term in the bottom section of the puzzle box.

Multiply the leading coefficient and the constant term: $6(-15) = -90$

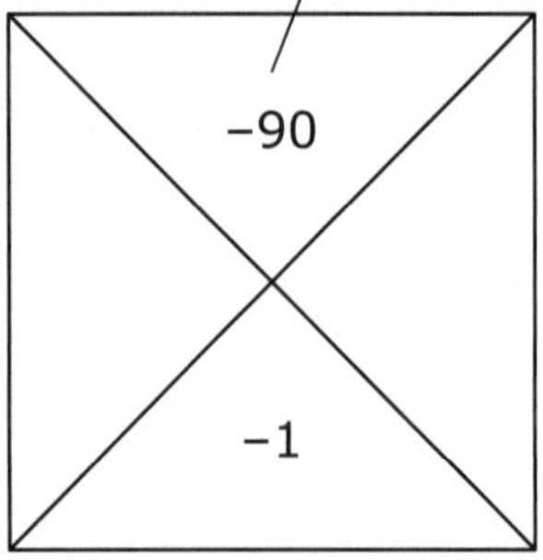

Now complete the puzzle box by finding two numbers whose product is –90 and whose sum is –1. Place these numbers in the left and right sections of the puzzle box.

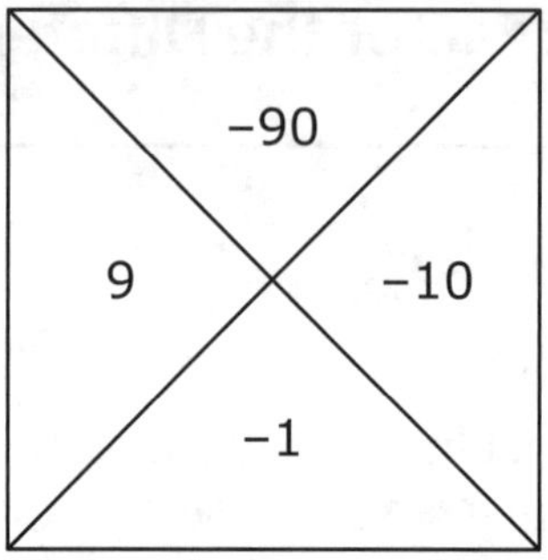

You can use 9 and –10 to form the x terms of the expression.

$6x^2 - x - 15$

$= 6x^2 + 9x - 10x - 15$

$= 3x(2x + 3) - 5(2x + 3)$ Factor out $3x$ and –5.

$= (3x - 5)(2x + 3)$ Use the distributive property.

My Notes

Practice Exercises

Review what you've learned using these practice problems. For practice problems with feedback, try the Coach and Play items in the Practice section online.

1. The graphs of the functions related to the expressions $6x^2 + 12x - 48$ and $x^2 + 2x - 8$ both cross the x-axis at -4 and 2.

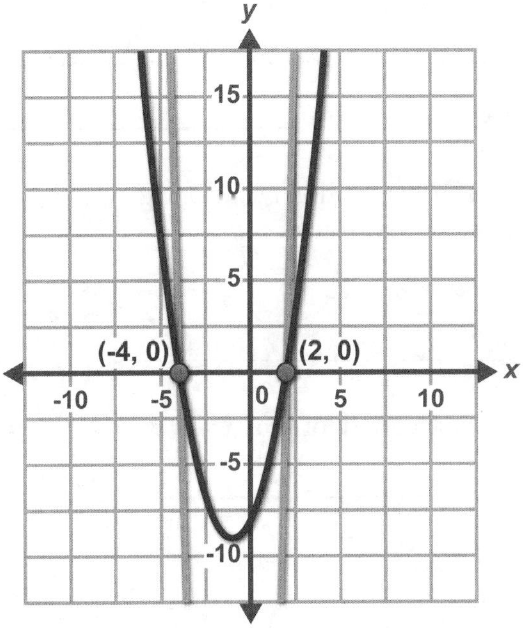

Write the two factors that the expressions have in common.

____________ and ____________

2. Which of the following are factors of the expression $2x^2 - 18x + 40$? *Select all that apply.*

A. x
B. 2
C. -18
D. 40
E. $(x - 2)$
F. $(x - 4)$
G. $(x - 5)$
H. $(x + 2)$
I. $(x + 4)$
J. $(x + 5)$

3. The area of a square is given by the expression $4x^2 + 28x + 49$.

Write an expression that represents the length of a side of the square.

Side length = ____________

4. The graph of the equation $y = x^2 - kx + 12$ passes through the points (3, 0) and (4, 0).

What is the value of k?

k = ____________

5. Which expression shows $6x^2 + 29xy - 5y^2$ in factored form?

A. $(6x + y)(x - 5y)$
B. $(6x^2 + y)(x - 5y^2)$
C. $(6x^2 - y)(x + 5y^2)$
D. $(6x - y)(x + 5y)$
E. $(6x - y)(x - 5y)$

6. Which of these expressions are equivalent to $3t(t + 1) - 4(t + 1)$? *Select all that apply.*

A. $3t^2 - t - 4$
B. $3t^2 + t - 1$
C. $(3t + 4)(t + 1)$
D. $(3t - 4)(t + 1)$
E. $3t - 4(t + 1)$

Practice Exercises *(continued)*

7. What value of c would make $x^2 - 14x + c$ a perfect square?

$c =$ ____________

What is the factored form of this expression for this value of c?

$(\_\_\_\_\_\_\_\_\_\_\_\_)^2$

8. What are the factors of $3x^2 + 24x + 48$?

A. 3, $(x + 2)$, and $(x + 2)$

B. $(3x + 6)$ and $(x + 8)$

C. 3, $(x + 4)$, and $(x + 4)$

D. $(3x + 4)$ and $(x + 4)$

9. Adriana factored the expression $x^2 + 6x + 9$ and concluded that the function has one zero at $x = 3$.

Is Adriana correct, and why?

A. Yes, because $3 \cdot 3 = 9$ and $3 + 3 = 6$.

B. Yes, because the two factors are $(x - 3)(x - 3)$.

C. No, because there is more than one zero.

D. No, because the zero is at $x = -3$.

10. Factor the polynomial by choosing the letters of the appropriate factors, in either order.

A. $x + 1$	**B.** $x + 3$	**C.** $x + 5$
D. $2x + 1$	**E.** $2x + 3$	**F.** $2x + 5$
G. $3x + 1$	**H.** $3x + 3$	**I.** $3x + 5$
J. $6x + 1$	**K.** $6x + 3$	**L.** $6x + 5$

$6x^2 + 19x + 15 = (\_\_\_\_\_\_\_\_)(\_\_\_\_\_\_\_\_)$

11. Rewrite the expression by factoring out the greatest common factor.

$-6x^2 + 18x - 72$

____

12. Write the factored form of $x^2 - 9x + 18$.

____

Apply

How Can You Predict How Many Diagonals Are in a Polygon?

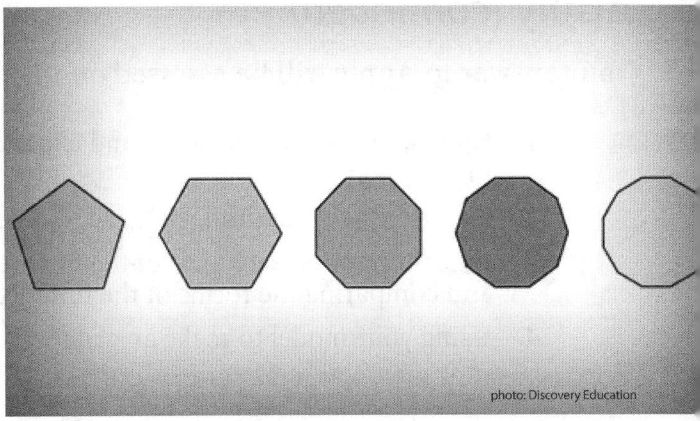
photo: Discovery Education

How can you figure out how many diagonals are in a polygon? One way is by drawing and counting them. But what if the polygon has many sides, or you don't even know how many sides it has?

An n-gon is a polygon with n sides.

- Write a function that will give you the number of diagonals in an n-gon. Then write the function in a different form, either factored or expanded, depending on the form you chose for your original function.
- Describe the method you used to discover your model.
- Explain what each part of the factored function means. Then compare both forms of the function. What type of function is it?
- Use your model to determine the number sides of a polygon with 135 diagonals.

Show what you've learned by completing the other performance tasks in the online Apply section.

Apply *(continued)*

Your answer to Apply will be assessed on the following criteria:

1. Writing two forms (factored and expanded) of a function for finding the number of diagonals in an *n*-gon
2. Describing the method you used to determine the model
3. Explaining what each part of the factored function means; identifying the type of function and comparing the forms of the functions
4. Using your model to make an accurate calculation

Criteria \ Scale	4	3	2	1	0
	Exceeds Criteria	**Meets Criteria**	**Progressing to Criteria**	**Below Expectations**	**No Expectation**
Two Forms of Function	Correctly writes two forms (factored and expanded) of a function for finding the number of diagonals in an *n*-gon.	Writes two forms (factored and expanded) of a function for finding the number of diagonals in an *n*-gon, with minor errors.	Correctly writes one form of a function for finding the number of diagonals in an *n*-gon.	Writes one or two functions that do not find the number of diagonals in an *n*-gon.	Does not attempt.
Method	Describes the method used to determine the model.	Describes the method used to determine the model, but with a lack of clarity.	Describes the method used to determine the model, but includes errors.	Description is incomplete or irrelevant.	Does not attempt.
Explain Function	Correctly explains what each part of the factored function means; identifies the type of function and compares the forms of the functions.	Explains what each part of the factored function means; identifies the type of function and compares the forms of the functions, with minor errors.	Correctly explains what each part of the factored function means or correctly identifies the type of function and compares the forms of the functions.	Explanation and identification are incorrect or incomplete.	Does not attempt.
Calculation	Uses the model to make an accurate calculation.	Uses the model to make a calculation, with minor errors.	Uses the model to make a partially accurate calculation.	Makes a calculation without the model.	Does not attempt.

UNIT 11: Quadratic Expressions and Equations

11.1 Solve Quadratics

photo: Discovery Education

Lesson Objectives

- Use a variety of methods to solve quadratic equations: by graphing, factoring, and completing the square.
- Determine if a solution to a quadratic equation is rational or irrational, and explain why.

Essential Questions

- How can you model and solve real-world problems using quadratic equations?
- How does the zero product property empower you to analyze a quadratic equation?
- How does completing the square help you analyze a quadratic equation?

photo: Getty Images

Investigations

What Goes Up Must Come Down

Sketch graphs of sporty quadratic relationships. What do they mean in real time?

Factors and Roots

Use algebra to find the roots. It's more precise than digging.

Solving Quadratics by Factoring

Add factors to your solution tool kit to solve real-world problems.

Estimating Roots

Get around irrational roots with estimation.

Solve Quadratic Equations by Completing the Square

Complete the polynomial tile square. It's in completion that you will find a solution.

Getting to the Root of It All

Complete the square algebraically to solve quadratic equations with rational and irrational roots.

Key Vocabulary

trinomial, x-intercept, y-intercept, zero of a function

Discover

As you complete Engage and the investigations, record the most important ideas you've learned.

Engage	**Investigation 1**
Investigation 2	**Investigation 3**
Investigation 4	**Investigation 5**

Name ______________________________ Date ______________

Check for Understanding

Solve Quadratics: Investigation 1

1. What value makes the following equations true?

$9x = 0$ $x =$ ________

$4(y - 5) = 0$ $y =$ ________

$2(8 + z) = 0$ $z =$ ________

2. What values of x make the equation $(x + 7)(2 - x) = 0$ true? *Select all that apply.*

A. 0 **B.** 7 **C.** -2

D. 2 **E.** 14 **F.** -7

3. Which equation's related function graph is concave down and has two x-intercepts?

A. $x^2 + 2x + 1 = 0$ **B.** $x^2 - 2x - 1 = 0$

C. $-x^2 + 2x - 1 = 0$ **D.** $-x^2 + 2x + 1 = 0$

4. Which of the following statements describes the graph of the related function to $-x^2 = 0$? *Select all that apply.*

A. The graph is concave up.

B. The graph is concave down.

C. The graph has one x-intercept.

D. The graph has two x-intercepts.

E. The graph has no x-intercepts.

5. Write the factored form of the equation $\frac{2}{5}x^2 + 4x = 0$. Explain how the zero product property relates the linear factors to the x-intercepts of the graph.

Name ______________________________ Date ______________

Check for Understanding

Solve Quadratics: Investigation 2

1. What are the solutions to $x^2 - 8x = -15$? *Select all that apply.*

A. $x = -3$

B. $x = 3$

C. $x = -5$

D. $x = 5$

E. $x = -8$

F. $x = 8$

2. Sketch the graph of $y = x^2 + x - 12$ on the coordinate plane. The sketch should include points to show the roots of the related equation, the axis of symmetry, and the minimum or maximum point on the graph.

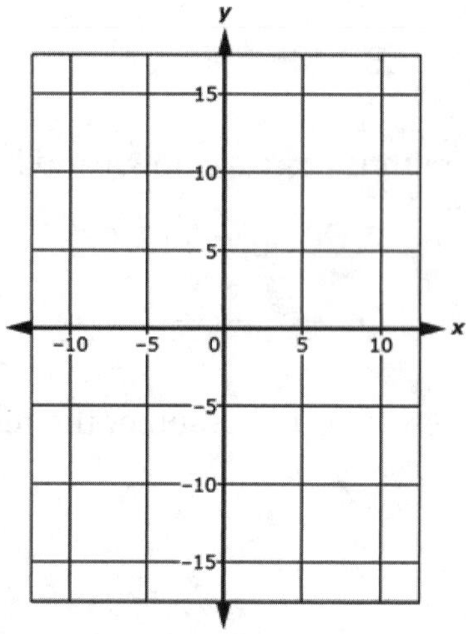

3. Which property tells you that if $2x(x - 50) = 0$, then the solutions are $x = 0$ and $x = 50$?

A. associative property

B. distributive property

C. reflective property

D. zero product property

4. A student launches a rocket two times. Each launch has the height, in feet, modeled by a different function, $f(t)$ and $g(t)$, in terms of time t, in seconds, from the moment of launch. The height of the first launch is modeled by $f(t) = -16t^2 + 96t$ and the second launch is modeled by $g(t) = -16t^2 + 96t + 432$. Which rocket stays in the air longer? Show your work and justify your answer.

Name ______________________________ Date ______________

Check for Understanding

Solve Quadratics: Investigation 3

1. What are the solutions to $x^2 - 64 = 0$? *Select all that apply.*

   A. $x = 32$
   B. $x = 8$
   C. $x = -32$
   D. $x = -8$

2. What are the solutions to $x^2 - 50 = 0$? *Select all that apply.*

   A. $x = \pm 5$
   B. $x = \pm 5\sqrt{10}$
   C. $x \approx \pm 15.81$
   D. $x = \pm 2\sqrt{5}$
   E. $x \approx \pm 4.47$
   F. $x = \pm 5\sqrt{2}$
   G. $x \approx \pm 7.07$
   H. $x = \pm 10\sqrt{5}$

3. A quadratic function $f(x)$ has an axis of symmetry of $x = 4$. If one of the roots is at $x = 4 + \sqrt{7}$, what is the value of the other root?

   $x =$ ______________________

4. The two functions $y = x^2 + 5$ and $y = 4x - 7$ are graphed on the same coordinate plane. This can be used to solve $x^2 - 4x + 12 = 0$. What features of the graph provide the solutions for x? Justify your answer.

Name ____________________ Date ____________

Check for Understanding

Solve Quadratics: Investigation 4

1. Which of the following expressions is equivalent to $x^2 - 10x + 25 = 0$?

   A. $x^2 + 10x = 25$ B. $x^2 - 10x = -25$

   C. $x^2 - 10x = 25$ D. $x^2 + 10x = -25$

2. Complete the steps in completing the square for $x^2 + 8x - 3 = 0$ by filling in the blank lines below.

   A. $x^2 + 8x =$ ______

   B. $x^2 + 8x +$ ______ $= 3 +$ ______

   C. $(x +$ ______$)^2 =$ ______

   D. x ______ $=$ ______

   E. $x =$ ____________

3. What are the solutions to $x^2 - 14x + 23 = 0$? *Select all that apply.*

   A. $x = 7 + \sqrt{26}$ B. $x = 7 - \sqrt{26}$ C. $x = 7 - \sqrt{23}$

   D. $x = 7 + \sqrt{23}$ E. $x = 9 - \sqrt{14}$ F. $x = 9 + \sqrt{14}$

4. In what cases do you not need to use completing the square to solve a quadratic equation? Consider a quadratic equation of the form $x^2 + bx + c = 0$. What relationship between the values of b and c indicate that completing the square would be a better method for solving a quadratic than factoring?

Name ______________________________ Date ______________

Check for Understanding

Solve Quadratics: Investigation 5

1. Gideon wants to use the method of completing the square to solve $5x^2 + 100x - 80 = 0$. Which of the following is the equivalent equation that can be used to begin completing the square?

   A. $10x^2 + 200x - 160 = 0$

   B. $x^2 + 95x - 75 = 0$

   C. $10x^2 + 105x - 85 = 0$

   D. $x^2 + 20x - 16 = 0$

Consider the function shown on the graph for questions 2 and 3.

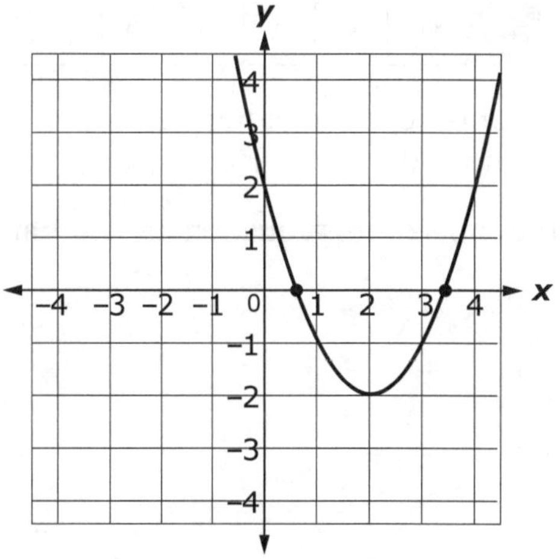

2. What is the axis of symmetry for the function?

   ____

3. Which equation represents this function?

   A. $x^2 + 4x + 2 = 0$

   B. $x^2 - 4x + 2 = 0$

   C. $x^2 + 4x - 2 = 0$

   D. $x^2 - 4x - 2 = 0$

4. A quadratic equation has a double root at $x = a$. Explain what this means and how you can use that information to find the form of the quadratic equation.

Summary

Before you attempt the Practice Exercises, review what you've learned.

Solve Quadratics

You can solve a quadratic equation algebraically by using a table or by analyzing a graph.

EXAMPLE: Using a Table

Make a table to find the roots of $x^2 - 4x + 3 = 0$.

SOLUTION:

Write the related function $y = x^2 - 4x + 3$ and make a table of values for x and y.

x	−2	−1	0	1	2	3	4	5	6
y	15	8	3	0	−1	0	3	8	15

The roots of the quadratic equation are the x values when $y = 0$.

The roots of this equation are $x = 1$ and $x = 3$.

EXAMPLE: Using the Zero Product Property

Solve the equation $x^2 - 4x = -3$ algebraically.

SOLUTION:

$x^2 - 4x + 3 = 0$		Set the equation equal to 0.
$(x - 1)(x - 3) = 0$		Factor the quadratic expression.
$(x - 1) = 0$	$(x - 3) = 0$	Apply the zero product property.
$x = 1$	$x = 3$	Solve each equation.

The roots are $x = 1$ and $x = 3$.

EXAMPLE: Using a Graph

Solve the equation $x^2 - 4x + 3 = 0$ by graphing.

SOLUTION:

Graph the related function $y = x^2 - 4x + 3$.

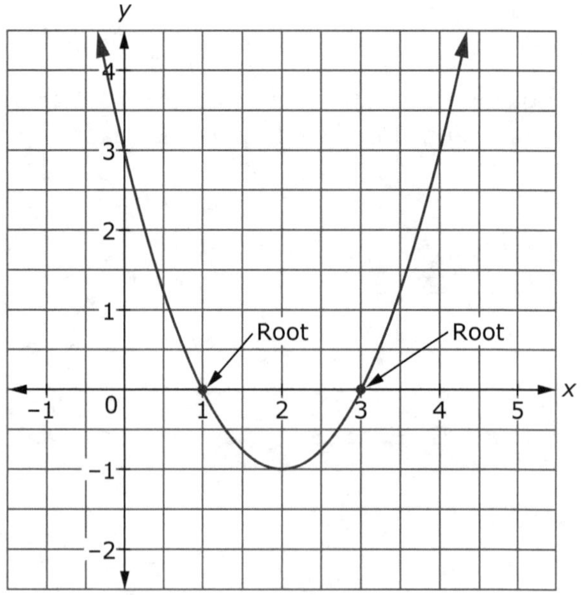

The roots of $x^2 - 4x + 3 = 0$ are $x = 1$ and $x = 3$.

My Notes

Summary *(continued)*

EXAMPLE: Solving by Factoring

Solve the quadratic equation by factoring.

$$9x^2 - 16 = 0$$

SOLUTION:

$9x^2 - 16 = 0$	Identify the expression as a difference of two squares.
$(3x + 4)(3x - 4) = 0$	Factor the quadratic expression.
$3x + 4 = 0 \quad 3x - 4 = 0$	Apply the zero product property.
$x = -\frac{4}{3} \quad x = \frac{4}{3}$	Solve each equation.

The roots are $x = -\frac{4}{3}$ and $x = \frac{4}{3}$.

EXAMPLE: Solve by Taking Square Roots

Find all possible solutions for x in each equation by taking the square root of both sides.

$$(x - 2)^2 = 7$$

SOLUTION:

$(x - 2)^2 = 7$	The variable expression is already isolated.
$\sqrt{(x - 2)^2} = \pm\sqrt{7}$	Take the square root of both sides.
$x - 2 = \pm\sqrt{7}$	Simplify.
$x = 2 \pm\sqrt{7}$	Solve for x.

The solutions are $x = 2 + \sqrt{7}$ and $x = 2 - \sqrt{7}$.

To complete the square, you will add a value to both sides of the equation so that the quadratic expression is a perfect square trinomial. The value you add will be $\left(\frac{b}{2}\right)^2$, where b is the coefficient of the x term of the quadratic expression.

EXAMPLE: Completing the Square

Solve $x^2 + 10x + 19 = 0$ by completing the square.

SOLUTION:

$x^2 + 10x + 19 = 0$	The quadratic expression is not a perfect square trinomial.
$x^2 + 10x = -19$	Subtract 19 from both sides.
$x^2 + 10x + \left(\frac{10}{2}\right)^2 = -19 + \left(\frac{10}{2}\right)^2$	Add $\left(\frac{10}{2}\right)^2$, or 25, to both sides.
$(x + 5)^2 = 6$	Simplify.
$x + 5 = \pm\sqrt{6}$	Take the square root of both sides.
$x = -5 \pm\sqrt{6}$	Solve for x.

The solutions are $x = -5 + \sqrt{6}$ and $x = -5 - \sqrt{6}$.

My Notes

Summary *(continued)*

Approximating Roots
EXAMPLE: Approximating from a Graph or Table

Solve $x^2 + x - 5 = 0$. Give approximate values for the roots.

SOLUTION:

One way to approximate the roots is by graphing.

Graph the related function $y = x^2 + x - 5$.

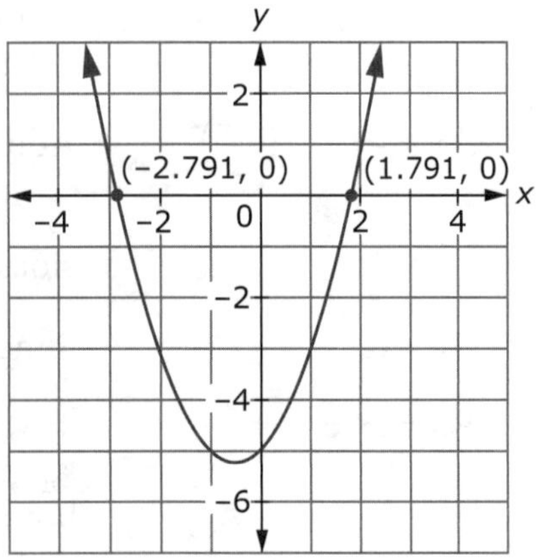

The traced values from the graph are $x = -2.791$ and $x = 1.791$.

Another way to approximate roots is to use a table.

Generate a table of values for the related function $y = x^2 + x - 5$.

x	$y = x^2 + x - 5$
−3	1
−2	−3
−1	−5
0	−5
1	−3
2	1
3	7

Note that the y value is positive when $x = -3$ and negative when $x = -2$. One root lies between −3 and −2.

The y value stays negative through $x = 1$ and then becomes positive when $x = 2$. A second root lies between 1 and 2.

To refine your approximation, generate additional values for your table. Choose values of x that lie between the values you have identified. Continue choosing values that result in y getting closer to zero.

Reasonable approximations of the roots are $x = -2.79$ and $x = 1.79$.

My Notes

Practice Exercises

Review what you've learned using these practice problems. For practice problems with feedback, try the Coach and Play items in the Practice section online.

1. Arthur is comparing a quadratic equation to its graph. However, the right side of the equation is smudged as shown.

$$x^2 - 8 = ?$$

Arthur has the graph of the equation that shows that the solutions are $x = \pm 6$.

What is the smudged value?

A. −6
B. 4
C. 6
D. 28
E. 36
F. 44
G. 24
H. 12

2. Barbara sets off a water-powered rocket. The height of the rocket at time t seconds is given by $h(t) = -6t^2 + 96t$.

Use completing the square to estimate when into the trip the rocket is 64 feet high.

A. at about 0.6 sec and about 5.4 sec
B. at about 0.7 sec and about 15.3 sec
C. at about 1.2 sec and about 4.8 sec
D. at about 2.2 sec and about 3.8 sec
E. at about 3 sec

3. What are the solutions to $2x^2 - 19 = 29$?

Write the exact solutions. Reduce completely so your answer is in the form of $a\sqrt{b}$, where a and b are both integers.

The exact solutions are $x =$ ________ and $x =$ ________.

4. What are the exact solutions to $(x - 2)^2 - 4 = 12$?

$x =$ ________ and $x =$ ________

5. Determine the roots for the equation $3x^2 - 20x + 16 = 4$. *Write the exact solutions.*

The solutions are $x =$ ________ and $x =$ ________.

6. Mellie plans to make a patio 12 feet by 12 feet. She decides to add a border of equal width around the patio to increase the total area to 180 square feet.

What is the width of this border, to the nearest tenth of a foot?

A. $x = 0.4$
B. $x = 0.7$
C. $x = 6.7$
D. $x = 7.4$
E. $x = 12.7$

7. Martin has the quadratic equation $x^2 = 64$. He knows he can solve it by graphing, but he can also recognize the solutions just by looking at the equation.

What are all the possible solutions for the equation?

A. $x = 8$
B. $x = 32$
C. $x = \pm\sqrt{8}$
D. $x = \pm 8$
E. $x = \pm 32$
F. $x = \pm\sqrt{32}$

Practice Exercises *(continued)*

8. Natasha uses the method of completing the square to find the solution of $2x^2 + 10x + 10 = 0$. She makes an error in the steps shown below.

Identify the step that contains her initial error.

A. $2x^2 + 10x + 10 = 0$

B. $x^2 + 5x + 5 = 0$

C. $x^2 + 5x = -5$

D. $x^2 + 5x + \frac{25}{4} = -5 + \frac{25}{4}$

E. $\left(x + \frac{5}{4}\right)^2 = \frac{5}{4}$

F. $x + \frac{5}{4} = \pm\frac{\sqrt{5}}{2}$

G. $x = -\frac{5}{4} \pm \frac{\sqrt{5}}{2}$

9. The time and height of a high jumper's jump is shown on the graph.

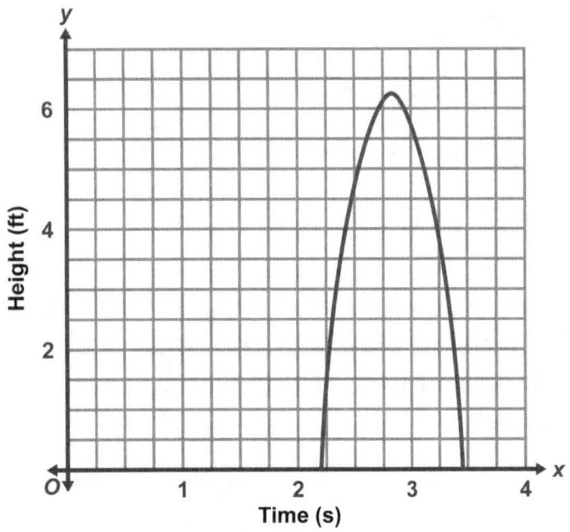

What is the approximate time and height of the jump?

Write values to complete the statements below about this high jump.

The high jumper was in the air for approximately __________ seconds and reached a maximum height of __________ feet.

10. Jess's age is six years less than three times Ethan's age. The product of their ages is 45. What is each of their ages?

Hint: Write an equation to represent the product of their ages, using x to represent Ethan's age, and then solve this quadratic equation.

A. 1 **B.** 3

C. 5 **D.** 9

E. 15 **F.** 45

Jess's age: ____________

Ethan's age: ____________

11. What are the solutions to $36x^2 = 18x$? *Write both solutions.*

$x =$ ____________ and $x =$ ____________

12. What are the exact and approximate solutions to $(x + 3)^2 - 10 = 8$? *Select all that apply.*

A. $x = -3$ and $x = \pm 3\sqrt{2}$

B. $x = -1 + \sqrt{2}$ and $x = -1 - \sqrt{2}$

C. $x = -3 + 3\sqrt{2}$ and $x = -3 - 3\sqrt{2}$

D. $x = -3$ and $x \approx \pm 4.24$

E. $x \approx 1.24$ and $x \approx -7.24$

F. $x \approx 0.41$ and $x \approx -2.41$

13. Solve $3x - 6x^2 = -9$.

A. $x = -1;\ x = \frac{2}{3}$

B. $x = 1;\ x = -3$

C. $x = -3;\ x = -\frac{1}{2}$

D. $x = 1;\ x = -\frac{3}{2}$

E. $x = -1;\ x = \frac{3}{2}$

F. $x = -3;\ x = -5$

Apply

A Very Special Number

What does this symbol represent?

Some numbers in mathematics are so important that they have their own names and symbols. This image shows one of them. Can you think of any others?

Complete this activity to learn about this special number in mathematics.

Segment AB is divided by point C so that $\frac{AC}{AB} = \frac{CB}{AC}$. Find the exact value of AC.

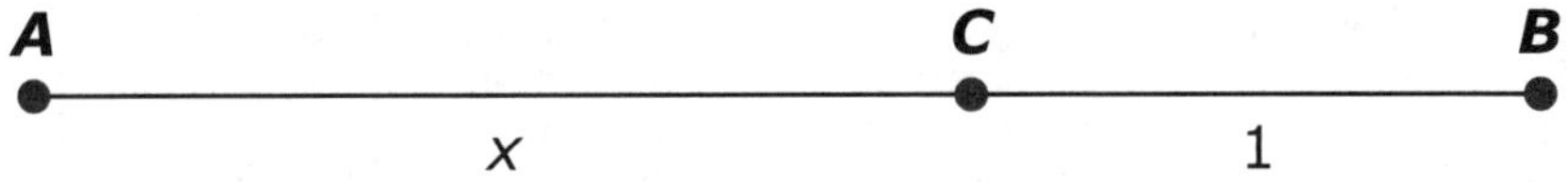

Segment EF is divided by point G so that $\frac{EG}{EF} = \frac{GF}{EG}$. Find the exact value of GF.

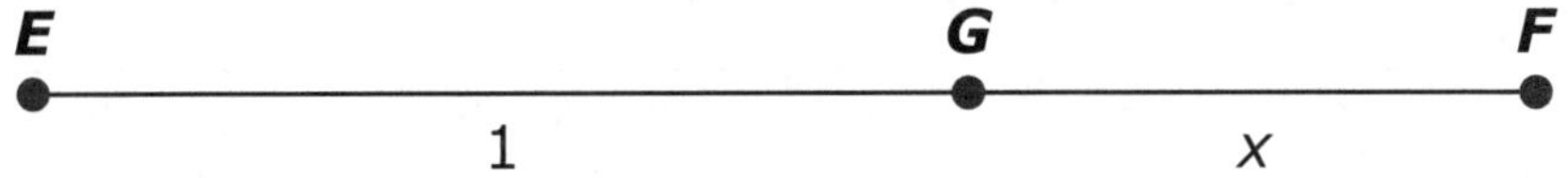

- Explain how AC and GF are related.
- Research the values you found, and share your findings.
- Find the name of the symbol in the image above.

Show what you've learned by completing the other performance tasks in the online Apply section.

Apply *(continued)*

Your answer to Apply will be assessed on the following criteria:

1. Finding the exact value of *AC* in radical form
2. Finding the exact value of *GF* in radical form
3. Explaining how *AC* and *GF* are related
4. Sharing research on the numbers you found, where they occur, and the name of the symbol

Criteria \ Scale	4 Exceeds Criteria	3 Meets Criteria	2 Progressing to Criteria	1 Below Expectations	0 No Expectation
Value of *AC*	Correctly sets up the proportion using the model, and solves for the exact value of *AC* in radical form.	Correctly sets up the proportion using the model, and solves for the value of *AC* as a decimal approximation.	Correctly sets up the proportion using the model, but solves for *AC* incorrectly.	Correctly sets up the proportion using the model, but does not solve for *AC*.	Does not set up the proportion and solve for *AC*.
Value of *GF*	Correctly sets up the proportion using the model, and solves for the exact value of *GF* in radical form.	Correctly sets up the proportion using the model, and solves for the value of *GF* as a decimal approximation.	Correctly sets up the proportion using the model, but solves for *GF* incorrectly.	Correctly sets up the proportion using the model, but does not solve for *GF*.	Does not set up the proportion and solve for *GF*.
Relationship between *AC* and *GF*	Correctly determines the relationship between *AC* and *GF*, and proves it mathematically.	Correctly determines the relationship between *AC* and *GF*, but is not able to completely justify it.	Correctly determines the relationship between *AC* and *GF*, with no justification.	Determines an incorrect relationship between *AC* and *GF*, with no justification.	Does not determine a relationship or a justification.
Research	Correctly finds the name of the number, at least two uses of the number, and the name of the symbol.	Correctly finds three of the following: the name of the number, two uses of the number, and the name of the symbol.	Correctly finds two of the following: the name of the number, two uses of the number, and the name of the symbol.	Correctly finds one of the following: the name of the number, a use of the number, or the name of the symbol.	Does not do research.

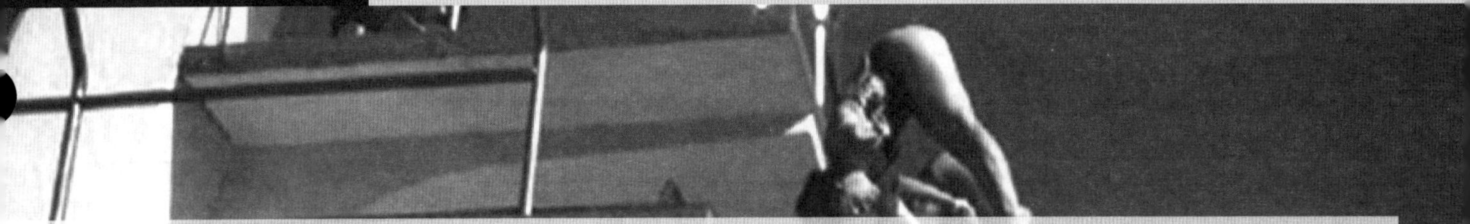

UNIT 11: Quadratic Expressions and Equations

11.2 Analyze Quadratic Equations

photo: Discovery Education

Lesson Objectives

- Derive the quadratic formula.
- Use the quadratic formula to solve quadratic equations.
- Determine if a solution to a quadratic equation is rational or irrational, and explain why.

Essential Question

- How do you use the quadratic formula and its discriminant to analyze problem situations and identify possible types of solutions?

photo: Getty Images

Investigations

So … What If It Will Not Factor?

Some quadratics simply will not factor, and the square will not complete. Now what? Take a dive!

Developing the Quadratic Formula

Develop the quadratic formula—the Swiss army knife of solving quadratic equations.

Using the Discriminant

Discriminate the roots. How can you tell without factoring or graphing?

Using the Quadratic Formula

Solve real-world problems with your new-found skills.

Key Vocabulary

irrational number, quadratic equation, radicand, symmetry, trinomial, vertex, x-intercept, y-intercept, zero of a function

Discover

As you complete Engage and the investigations, record the most important ideas you've learned.

Engage	Investigation 1
Investigation 2	Investigation 3

Name ______________________________ Date ____________

Check for Understanding

Analyze Quadratic Equations: Investigation 1

1. What are the justifications for each step in determining the solution to $ax^2 + bx + c = 0$? *Draw a line between the justifications on the left with the steps they justify on the right.*

Divide each term by the lead coefficient.
Take the square root of both sides.
Subtract the constant from both sides.
Move all terms to the right and simplify the square root.
Add the square of half of the linear term to both sides.
Write the squared binomial and simplify the right side.

Step 1:	$ax^2 + bx = -c$
Step 2:	$x^2 + \frac{b}{a}x = -\frac{c}{a}$
Step 3:	$x^2 + \frac{b}{a}x + \left(\frac{b}{2a}\right)^2 = -\frac{c}{a} + \left(\frac{b}{2a}\right)^2$
Step 4:	$\left(x + \frac{b}{2a}\right)^2 = \frac{b^2 - 4ac}{4a^2}$
Step 5:	$x + \frac{b}{2a} = \pm\sqrt{\frac{b^2 - 4ac}{4a^2}}$
Step 6:	$x = -\frac{b}{2a} \pm \frac{\sqrt{b^2 - 4ac}}{2a}$

2. What does the term $-\frac{b}{2a}$ in the quadratic formula represent for the related quadratic function?

A. It represents the axis of symmetry.

B. It represents the distance from the axis of symmetry.

C. It represents the maximum or minimum value.

D. It represents the end behavior.

3. Describe what the quadratic formula tells you about the cases for which a quadratic equation will not have real roots and why. Include an example of a quadratic equation with no real roots.

Name ____ Date ______________

Check for Understanding

Analyze Quadratic Equations: Investigation 2

1. What is the discriminant in the quadratic formula?

   A. $-\frac{b}{2a}$

   B. $b^2 - 4ac$

   C. $2a$

   D. $4ac$

2. If the graph of a quadratic function does not touch the x-axis, what is known about the related quadratic equation?

   A. It has no real roots.

   B. It has only irrational roots.

   C. It has exactly one real root.

   D. It has exactly two negative real roots.

3. Which discriminants, D, will result in rational roots for the quadratic equation they represent? *Select all that apply.*

   A. $D = -1$ B. $D = 50$ C. $D = 1$

   D. $D = 25$ E. $D = 0$ F. $D = -25$

4. What is the discriminant of $3x^2 - 12x + 2 = 0$?

   A. 2 B. 6 C. 24 D. 120

5. Explain how the value of the discriminant, D, of the quadratic formula can be used to determine the number and types of roots to the related quadratic equation. Explain for the cases of $D > 0$, $D = 0$, and $D < 0$.

Name ______________________________ Date ______________

Check for Understanding

Analyze Quadratic Equations: Investigation 3

A rocket is launched from an elevated platform. Its height, in feet, is determined by the equation $h(t) = -16t^2 + 56t + 15$, where t represents the time, in seconds, after launch.

Consider this situation to answer questions 1–3.

1. What is the height, in feet, of the elevated platform?

A. 4

B. 15

C. 16

D. 56

2. What is the maximum height of the rocket's path?

__________ feet

3. How long did it take for the rocket to return to the ground?

__________ seconds

4. The path of an object is modeled with a quadratic function $f(x) = ax^2 + bx + c$. Explain how you could use the quadratic formula to determine the maximum or minimum of the object's path.

Summary

Before you attempt the Practice Exercises, review what you've learned.

Analyzing Roots of Quadratic Equations

The **quadratic formula** can be used to solve any quadratic equation that is written in standard form. Given $ax^2 + bx + c = 0$, $x = \frac{-b}{2a} \pm \frac{\sqrt{b^2 - 4ac}}{2a}$. There may be one, two, or no real roots.

Since the roots can be of different types and numbers, we can evaluate the discriminant to anticipate the nature of the roots we will find.

The expression $b^2 - 4ac$, which is the expression under the radical in the quadratic formula, is called the **discriminant**. The discriminant is used to determine the number and types of roots of a quadratic equation. You can represent the discriminant with the variable D.

Discriminant	Number and Type of Roots
$b^2 - 4ac > 0$, and D is a perfect square.	Two rational roots
$b^2 - 4ac > 0$, and D is not a perfect square.	Two irrational roots
$b^2 - 4ac = 0$	One root
$b^2 - 4ac < 0$	No real roots

EXAMPLE: Determine Number and Type of Roots

Tell how many and what type of roots each quadratic equation has.

A. $9x^2 - 12x + 4 = 0$

B. $3x^2 + 2x - 8 = 0$

C. $x^2 - 10x + 22 = 0$

D. $3x^2 - x + 2 = 0$

SOLUTION:

A. $9x^2 - 12x + 4 = 0$
Find the value of $b^2 - 4ac$.
$a = 9$, $b = -12$, and $c = 4$.
$b^2 - 4ac = (-12)^2 - 4(9)(4) = 144 - 144 = 0$
The discriminant is 0, so there is one root.

B. $3x^2 + 2x - 8 = 0$
Find the value of $b^2 - 4ac$.
$a = 3$, $b = 2$, and $c = -8$.
$b^2 - 4ac = (2)^2 - 4(3)(-8) = 4 + 96 = 100$
The discriminant is greater than 0, and it is a perfect square. So, there are two rational roots.

C. $x^2 - 10x + 22 = 0$
Find the value of $b^2 - 4ac$.
$a = 1$, $b = -10$, and $c = 22$.
$b^2 - 4ac = (-10)^2 - 4(1)(22) = 100 - 88 = 12$
The discriminant is greater than 0, but it is not a perfect square. So, there are two irrational roots.

My Notes

Summary *(continued)*

D. $3x^2 - x + 2 = 0$
Find the value of $b^2 - 4ac$.
$a = 3$, $b = -1$, and $c = 2$.
$b^2 - 4ac = (-1)^2 - 4(3)(2) = 1 - 24 = -23$
The discriminant is less than 0, so there are no real roots.

You can visualize the roots of a quadratic equation by looking at the graph of the related quadratic function.

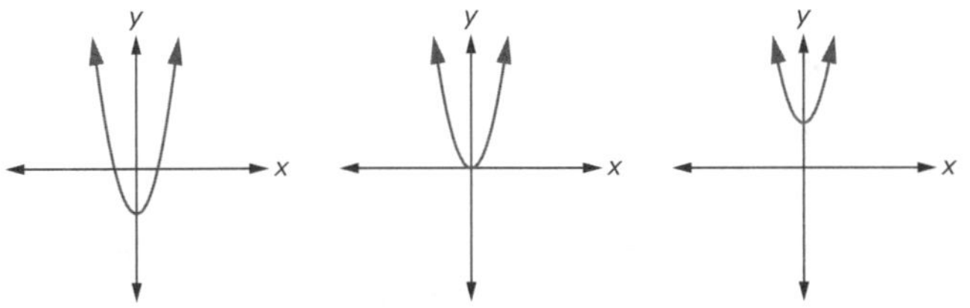

Two Real Solutions | **One Real Solution** | **No Real Roots**

Solving Quadratic Equations

By substituting the variable D for the discriminant in the quadratic formula, you can write the formula as $x = \frac{-b}{2a} \pm \frac{\sqrt{D}}{2a}$. When written in this form, the expression $\frac{-b}{2a}$ gives the x-coordinate of the axis of symmetry of the related parabola, and the expression $\frac{\sqrt{D}}{2a}$ gives the horizontal distance to the root(s).

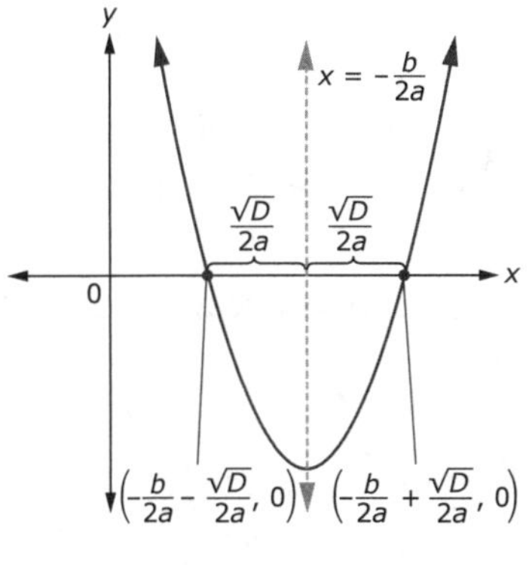

EXAMPLE: Solve Quadratic Equations

Solve each equation by using the quadratic formula. Give the axis of symmetry of the graph of the related quadratic function.

A. $9x^2 - 12x + 4 = 0$

B. $3x^2 + 2x - 8 = 0$

C. $x^2 - 10x + 22 = 0$

D. $3x^2 - x + 2 = 0$

SOLUTION:

A. $9x^2 - 12x + 4 = 0$

$a = 9$, $b = -12$, and $c = 4$.

$x = \frac{-b}{2a} \pm \frac{\sqrt{D}}{2a}$

$x = \frac{12}{18} \pm \frac{0}{18}$ The discriminant is 0.

$x = \frac{2}{3} \pm 0$ Simplify.

$x = \frac{2}{3}$ There is one root.

There is one root at $x = \frac{2}{3}$.

The axis of symmetry is $x = \frac{2}{3}$.

My Notes

Summary *(continued)*

B. $3x^2 + 2x - 8 = 0$

$a = 3, b = 2,$ and $c = -8.$

$x = \frac{-b}{2a} \pm \frac{\sqrt{D}}{2a}$

$x = \frac{-2}{6} \pm \frac{\sqrt{100}}{6}$ The discriminant is 100.

$x = -\frac{2}{6} \pm \frac{10}{6}$ Simplify.

$x = -\frac{1}{3} \pm \frac{5}{3}$

$x = -\frac{1}{3} - \frac{5}{3};$ Write the two roots separately.

$x = -\frac{1}{3} + \frac{5}{3}$

$x = -2;\ x = \frac{4}{3}$ Simplify.

There are two roots; one at $x = -2$ and one at $x = \frac{4}{3}$.

The axis of symmetry is $x = -\frac{1}{3}$.

C. $x^2 - 10x + 22 = 0$

$a = 1, b = -10,$ and $c = 22.$

$x = \frac{-b}{2a} \pm \frac{\sqrt{D}}{2a}$

$x = \frac{10}{2} \pm \frac{\sqrt{12}}{2}$ The discriminant is 12.

$x = 5 \pm \frac{2\sqrt{3}}{2}$ Simplify.

$x = 5 \pm \sqrt{3}$

$x = 5 - \sqrt{3};\ x = 5 + \sqrt{3}$ Write the two roots separately.

There are two roots; one at $x = 5 - \sqrt{3}$ and one at $x = 5 + \sqrt{3}$.

The axis of symmetry is $x = 5$.

D. $3x^2 - x + 2 = 0$

$a = 3, b = -1,$ and $c = 2.$

$x = \frac{-b}{2a} \pm \frac{\sqrt{D}}{2a}$

$= \frac{1}{6} \pm \frac{\sqrt{-23}}{6}$ The discriminant is −23.

There are no real roots because the discriminant is negative.

The axis of symmetry is $x = \frac{1}{6}$.

My Notes

Practice Exercises

Review what you've learned using these practice problems. For practice problems with feedback, try the Coach and Play items in the Practice section online.

1. Which equations represent solutions or ways of finding the solutions to the equation $2x^2 + 7x - 15 = 0$? *Select all that apply.*

   A. $x = -5$

   B. $x = -\frac{3}{2}$

   C. $x = \frac{-7 \pm \sqrt{169}}{4}$

   D. $(2x - 3)(x + 5) = 0$

   E. $x = \frac{7 \pm 13}{2}$

   F. $(2x + 3)(x - 5) = 0$

2. Grant wrote the following equation to find the solutions of a quadratic equation.

   $x = \frac{-10 \pm \sqrt{10^2 - 4(4)(-3)}}{2(4)}$

   What equation could Grant have been solving?

   ________ x^2 + ________ x + ________ = 0

3. The function $f(t) = -16t^2 + 40t + 5$ represents the height, in feet, of a ball t seconds after it is tossed up in the air.

   Which expression represents the time at which the ball will hit the ground in seconds?

   A. $\frac{40 - \sqrt{1{,}920}}{32}$

   B. $\frac{-40 - \sqrt{1{,}920}}{-32}$

   C. $\frac{40 + \sqrt{1{,}280}}{32}$

   D. $\frac{-40 - \sqrt{1{,}280}}{-32}$

4. Use the equation $9x^2 + 24x - 16 = 0$ to answer the questions below.

   What is the value of the discriminant?

   ____

   How many solutions does the equation have?

   ____

5. Use the discriminant to classify each equation below according to its number of real solutions.

   A. $x^2 + 3x + 6 = 0$

   B. $x^2 - 10x + 25 = 0$

   C. $2x^2 - x - 4 = 0$

   D. $-3x^2 + 9x + 7 = 0$

   E. $4x^2 + 4x + 1 = 0$

   F. $5x^2 - 2x + 1 = 0$

   No real solutions: ________

   Exactly one real solution: ________

   Exactly two real solutions: ________

6. The equation $ax^2 - 60x + 36 = 0$ has exactly one solution.

   What is the value of the discriminant?

   ____

   What is the value of a? ________

7. Judy says that the solutions of $6x^2 + 6x = -1$ are rational.

   Which statement explains whether Judy is correct?

   A. Judy is correct because the coefficients of the equation are all positive integers.

   B. Judy is correct because the discriminant is a positive even integer.

   C. Judy is incorrect because the discriminant is not a perfect square integer.

   D. Judy is incorrect because the coefficients of the variable terms are equal and $c \neq 0$.

Practice Exercises *(continued)*

8. Which statement is true about the solutions of $3x^2 - 7x + 8 = 0$?

A. There are no real solutions because the discriminant is negative.

B. There are no real solutions because the discriminant is 0.

C. There is exactly one real solution because the discriminant is 0.

D. There is exactly one real solution because the discriminant is 1.

E. There are exactly two real solutions because the discriminant is positive.

9. Which of these statements are true about the equation $10x - 3x^2 = 3$? *Select all that apply.*

A. There are two real solutions to the equation.

B. The discriminant is equal to 0.

C. All solutions to the equation are rational.

D. All solutions of the equation are positive.

E. All solutions of the equation are integers.

10. Use the equation $2(x - 1)^2 + 4 = 0$ to answer the questions below.

What is the value of the discriminant?

____

How many real solutions does the equation have?

____

11. A quadratic function $f(x) = x^2 - 9x + c$ has x-intercepts at $x = \frac{9 \pm \sqrt{61}}{2}$ and a y-intercept $(0, c)$.

What is the y-intercept of the function?

The y-intercept of the function is (0, __________).

12. A quadratic equation has one root at $x = \frac{5}{2}$. Which of the following quadratic formula representations would result in this root? *Select all that apply.*

A. $x = \frac{-4 \pm \sqrt{36}}{4}$

B. $x = \frac{2 \pm \sqrt{9}}{2}$

C. $x = \frac{-8 \pm \sqrt{4}}{-4}$

D. $x = \frac{-8 \pm \sqrt{4}}{4}$

Apply

How Much Would You Charge for Your App?

When people create and sell an app for a smartphone or tablet, how do they determine the app's price? Watch the video to get started.

photo: Pixabay

App Development: View this video segment to learn how you could use a quadratic function to help you set the price of a new app.

Suppose you have designed and created a new app. Now you need to decide how much to charge for it. You have conducted some market research by surveying a group of people in your target market. The table shows your results.

Survey Results	
Proposed Price	Percent Surveyed Who Would Buy the App
\$0.99	76%
\$1.99	58%

To determine the best price for your app, you will need to do the following:

- Estimate the number of people in your target market. If you make an app for skateboarders, for example, how could you estimate the number of skateboarders in the United States and the number of those who own smartphones or tablets?
- Develop a linear function that predicts the number of downloads of the app you will sell based on the selling price.
- Develop a function that predicts the total amount of revenue, or income, you will make from selling your app based on the selling price.
- Determine the selling price that will maximize your revenue. Assume that the online store where you will sell your app requires that all prices end in the digit 9, such as \$0.99 or \$1.59.

Show what you've learned by completing the other performance tasks in the online Apply section.

Apply *(continued)*

Your answer to Apply will be assessed on the following criteria:

1. Estimating the number of people in your target market and explaining how you obtained that estimate
2. Writing a linear function that models the number of downloads of the app you will sell based on its price and explaining your reasoning
3. Writing a function that models the revenue you will make from your app based on its price and explaining your reasoning
4. Determining the price that will maximize your revenue and the amount of the maximum revenue and showing your work

Scale / Criteria	4 Exceeds Criteria	3 Meets Criteria	2 Progressing to Criteria	1 Below Expectations	0 No Expectation
Target Market	Gives a reasonable estimate for the number of people in the target market and explains how this estimate was determined.	Gives a reasonable estimate for the number of people in the target market, but makes a minor error in explaining how to determine this estimate.	Follows correct procedures for estimating the number of people in the target market, but finds an unreasonable estimate based on calculation errors.	Gives an unreasonable estimate for the number of people in the target market, and shows little or no evidence of how this estimate was determined.	Does not attempt task.
Function for Number of Downloads	Correctly writes a linear function that models the number of downloads that will sell; explains all reasoning.	Makes a minor calculation error in determining a linear function that models the number of downloads that will sell; explains all reasoning.	Makes a major error in determining one parameter of the linear function; explains most reasoning.	Makes major errors when determining both parameters; does not explain reasoning.	Does not attempt task.
Function for Revenue	Correctly writes a function that models the revenue based on the price of the app; explains all reasoning.	Makes a minor error when writing a function that models the revenue; explains all reasoning.	Makes a major error when writing a function that models the revenue; explains most reasoning.	Uses an incorrect function type when writing a function that models the revenue; explains little or no reasoning.	Does not attempt task.
Selling Price	Correctly determines the price that will maximize the revenue; shows all work.	Makes a minor error when determining the price that will maximize the revenue; shows all work.	Makes a major error when determining the price that will maximize the revenue; shows most work.	Makes multiple conceptual or computational errors when determining the price that will maximize the revenue; shows little or no work.	Does not attempt task.

UNIT 12: Graphs of Quadratic Functions

12.1 Analyze Graphs of Quadratic Functions

photo: Getty Images

Lesson Objectives

- Graph quadratic functions and identify the domain and range.
- Explore the effect of replacing $f(x)$ by $f(x) + k$ or by $f(x + k)$ on the graph of quadratic functions.
- Create a quadratic function that describes a relationship between two quantities.
- Compare linear, quadratic, and exponential functions.
- Transform quadratic functions between standard form and vertex form.
- Identify the graph of a quadratic function, and find the vertex and axis of symmetry of a parabola.
- Interpret key features of quadratic functions in context.
- Determine the zeros of a quadratic function using its graph, and graph a quadratic function using its zeros.

Essential Question

- How are the key features of a quadratic graph related to the characteristics of the related algebraic function?

Investigations

Making Predictions from Data

Let's revisit popuplation. Is it linear, exponential, or something else?

Exploring Graphs of Quadratic Functions

Dogs can model sweaters, so you can model real-world situations with quadratic equations.

Exploring Transformations of Quadratic Functions

Get my best side! Sketch graphs of quadratic functions using their key features.

Finding Vertex Form

Model real-world situations in standard and vertex forms. It only requires completing the square.

Modeling with a Quadratic Function

Help a shelter maximize adoptions. Quadratics to the rescue!

Interpreting Forms of Quadratic Functions

Explore parabolas in the Bouncing Ball Lab. Watch for the five key attributes of a quadratic graph.

Key Vocabulary

axis of symmetry, parabola, standard form, vertex form/graphing form of a quadratic

Discover

As you complete Engage and the investigations, record the most important ideas you've learned.

Engage	**Investigation 1**
Investigation 2	**Investigation 3**
Investigation 4	**Investigation 5**

Name ______________________________ Date ______________

Check for Understanding

Analyze Graphs of Quadratic Functions: Investigation 1

Last year, attendance at the community theater play was 200 when the price per ticket was \$8. The group has noticed that for each \$4 increase in the price of a ticket, 20 fewer people attend the performance.

Use this information to answer questions 1–4.

1. Choose the expression that models the number of tickets sold for the number of price increases, x.

A. $200 + 4x$ **B.** $200 + 20x$

C. $200 - 4x$ **D.** $200 - 20x$

2. Choose the expression that models the price of a ticket for the number of price increases, x.

A. $4x - 8$ **B.** $8 + 4x$

C. $4 + 8x$ **D.** $8x - 4$

3. What is the equation that represents the theater group's income for the play, expressed in both standard and vertex forms? *Choose one standard and one vertex form.*

A. $I(x) = -80x^2 + 640x + 1{,}600$ **D.** $I(x) = -160(x - 4.75)^2 + 4{,}410$

B. $I(x) = -16x^2 + 768x + 10{,}816$ **E.** $I(x) = -80(x - 4)^2 + 2{,}880$

C. $I(x) = -160x^2 + 1{,}520x + 800$ **F.** $I(x) = -16(x - 24)^2 - 10{,}816$

4. What are the values that make each statement true? *Fill in the blanks.*

A. The number of ticket price increases that will maximize profit is___________.

B. The number of tickets they expect to sell at the new price is ___________.

C. The maximum income the theater group can earn is ___________.

5. Explain what each of the standard and vertex forms of the theater group's function reveals about its graph and what these features represent within the context.

Name ______________________________ Date ______________

Check for Understanding

Analyze Graphs of Quadratic Functions: Investigation 2

1. Consider the function $f(x) = 2(x - 10)^2 - 8$. Analyze the equation's parameters. *Circle the terms or fill in the blank to correctly complete each statement.*

A. The parabola opens [upward / downward], and has an axis of symmetry at [$x = -2$ / $x = 5$ / $x = 8$ / $x = 10$].

B. The function has a [minimum / maximum] value of y = _______, which occurs at x = _______.

C. The function's y-intercept is (_______, _______), and its x-intercepts are (_______, _______) and (_______, _______).

2. Match the parameters of the function $g(x) = -0.25(x + 2)^2$ to the features of its graph. *Not all options will be used.*

A. x-intercept(s)

B. y-intercept

C. axis of symmetry

D. vertex

E. point symmetrical to $(-4, -1)$

I. $x = -2$

II. $x = 2$

III. $(2, 0)$

IV. $(0, 1)$

V. $(0, -1)$

VI. $(4, -1)$

VII. $(-2, 0)$

3. A suspension footbridge has two 2-meter vertical supports 20 meters apart. The lowest point of the cable connecting them is 1 meter above the path. If the y-axis represents the left support, what is the vertex form of the equation for the parabola describing the suspension cable? *Fill in the blanks.*

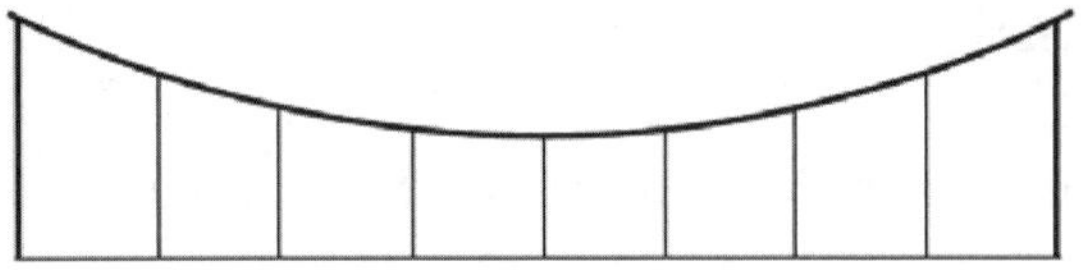

y = _______ $(x -$ _______ $)^2 +$ _______

4. Explain how you can write the vertex form for the equation of a parabola, given its y-intercept and vertex.

Name ______________________________ Date ____________

Check for Understanding

Analyze Graphs of Quadratic Functions: Investigation 3

1. Write each of the following in vertex form by completing the square.

   A. $f(x) = x^2 - 8x + 12 = (x^2 - \_\_\_\_ x) + 12$

   $= (x^2 - \_\_\_\_ x + \_\_\_\_) + 12 - \_\_\_\_$

   $= (x - \_\_\_\_)^2 + \_\_\_\_$

   B. $g(x) = 5x^2 - 30x + 49 = \_\_\_\_(x^2 - \_\_\_\_ x) + 49$

   $= \_\_\_\_(x^2 - \_\_\_\_ x + \_\_\_\_) + 49 - \_\_\_\_$

   $= \_\_\_\_(x - \_\_\_\_)^2 + \_\_\_\_$

   C. $h(x) = -0.5x^2 + 6x + 4 = \_\_\_\_(x^2 + \_\_\_\_ x) + 4$

   $= \_\_\_\_(x^2 - \_\_\_\_ x + \_\_\_\_) + 4 - \_\_\_\_$

   $= \_\_\_\_(x - \_\_\_\_)^2 + \_\_\_\_$

The manager of a horse farm plans to build a training arena alongside the fence of one pasture to use as one length, and she has 400 feet of fencing for the other three sides.

Use the given information to answer questions 2–3.

2. Which of the following represents an expression for the length of the arena in terms of its width, w?

   A. $400 - w$ B. $400 - 3w$ C. $400 - 2w$ D. $200 - w$

3. Determine the width and length for the maximum area the manager can enclose.

   A. Write the function to maximize the area based on the width, w, in feet.

   $A(w) =$ ______________________________

   B. Write the function in vertex form. $A(w) =$ ____________________

   C. For the maximum area of ________ square feet, the width should be ________ feet and the length should be ________ feet.

4. What are a reasonable domain and range for the length, l, and width, w, of a general area function represented by $A(w) = a(w - h)^2 + k$? Explain your reasoning.

Name ______________________________ Date ______________

Check for Understanding

Analyze Graphs of Quadratic Functions: Investigation 4

The table shows the number of dogs that a Retriever Rescue group expects to place for adoption each month based on different adoption fees.

Adoption Fee (\$), x	Likely Number of Dogs Adopted per Month, $A(x)$
44	57
48	54
52	51
56	48

1. Write an expression that models the number of dogs likely to be adopted from the rescue group per month when the adoption cost is x dollars.

__________ $x +$ __________

2. Write the function that models the total amount, A, in dollars that the rescue group expects to collect from adoption fees when it charges x dollars to adopt a retriever.

A. Standard Form $A(x) =$ ______________________________

B. Vertex Form $A(x) =$ ______________________________

C. What is the expected amount of collected fees when the fee is \$64? __________

D. What is the expected number of dogs adopted when the fee is \$64? __________

3. Identify the key features of the graph using one of the function forms.

A. y-intercept: (______, ______); x-intercept(s): (______, ______); (______, ______)

B. The vertex of the parabola is (________, ________).

C. The maximum amount the group can expect to collect is __________, and this amount is expected when the adoption fee is __________. The number of dogs likely to be adopted for this fee is __________.

D. A reasonable domain for the function is [0, __________], and a reasonable range is [0, __________].

4. Explain why the adoption cost that maximizes the collected fees does not also maximize the number of dogs adopted per month. How much would you recommend that the rescue group charge to adopt a dog and why?

Name ______________________________ Date ____________

Check for Understanding

Analyze Graphs of Quadratic Functions: Investigation 5

1. Match each form of a quadratic model for the first bounces of a dropped ball to the information it shows in the feature of its graph. *More than one option may apply.*

A. Factored Form	**I.** Maximum height between bounces
B. Vertex Form	**II.** Initial drop height of the ball
C. Standard Form	**III.** Time at maximum height
	IV. Initial velocity of the ball
	V. Time at each bounce
	VI. Axis of symmetry

2. A ball is dropped from a height of 6 feet, and its behavior between the first two bounces is modeled with the equation in vertex form $H(t) = -16.8(t - 0.95)^2 + 1.8$, and in factored form $H(t) = -16.8(t - 1.28)(t - 0.62)$. *Complete the following statements.*

   A. The maximum height of the bounce is __________ feet at __________ seconds.

   B. The time between bounces is __________ seconds.

   C. The standard form of the equation when it has been shifted to begin the bounce at the origin is ______________________________.

   D. The initial velocity of the ball is __________ ft/sec.

3. Write the given equation for the quadratic function in standard form.

   $g(x) = -\frac{5}{4}(x - 4)^2 + 11$ ____________________

4. Write a quadratic function, $h(x)$, in standard form having a vertex at $(1, -5)$ and passing through the point $(3, 7)$.

   ____

5. Explain how you could use the measures for the time and distance between the first and second bounces of any ball to determine its initial velocity and maximum height.

Summary

Before you attempt the Practice Exercises, review what you've learned.

> The **vertex form** of a quadratic function is $f(x) = a(x - h)^2 + k$, where a, h, and k are real numbers and $a \neq 0$.

For a quadratic function in vertex form,

- The sign of a determines the direction of concavity of the graph. If $a > 0$, the parabola opens upward and the function has a minimum. If $a < 0$, the parabola opens downward and the function has a maximum.
- The axis of symmetry is $x = h$.
- The vertex is (h, k).

Graphing

EXAMPLE: Graph a Quadratic Function in Vertex Form

Graph the function $f(x) = -(x - 1)^2 + 9$ without using technology.

SOLUTION:

For this function, $a = -1$, $h = 1$, and $k = 9$.

Because a is negative, the parabola opens downward.

The axis of symmetry is $x = 1$. The vertex is $(1, 9)$.

To find the y-intercept of the graph, substitute 0 for x and solve for $f(0)$.

$$f(0) = -(0 - 1)^2 + 9$$

$$f(0) = 8$$

The y-intercept is $(0, 8)$.

To find the x-intercepts, if any, substitute 0 for $f(x)$ and solve for x.

$0 = -(x - 1)^2 + 9$	
$-9 = -(x - 1)^2$	Subtract 9 from both sides.
$9 = (x - 1)^2$	Divide both sides by -1.
$\pm 3 = x - 1$	Take the square root of both sides.
$+3 = x - 1$ or $-3 = x - 1$	Write as two equations.
$4 = x$ or $-2 = x$	Solve each equation.

The solutions show that the x-intercepts are $(4, 0)$ and $(-2, 0)$.

Find another point on the graph of the function. For example, substitute 3 for x.

$$f(3) = -(3 - 1)^2 + 9$$

$$f(3) = 5$$

So, the point $(3, 5)$ lies on the graph. This point is 2 units to the right of the axis of symmetry at $x = 1$. Based on the symmetry of a parabola, there must be a point on the graph 2 units to the left of the axis of symmetry with the same y-coordinate. This point is at $(-1, 5)$.

Graph the known points and the axis of symmetry, and connect them by sketching a parabola that opens downward.

My Notes

Summary *(continued)*

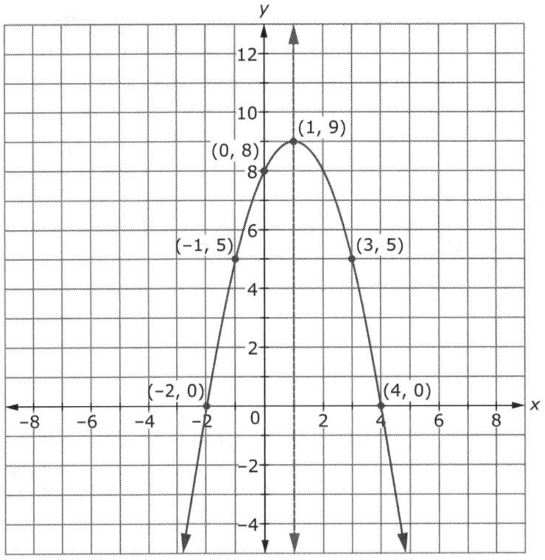

Transformations

The parameters a, h, and k in the vertex form of a quadratic function also indicate transformations compared to the parent function $g(x)= x^2$.

Effects of a, h, and k on the Graphs of Quadratic Functions	
Value of Parameter	Transformation
$k > 0$	A vertical translation up k units
$k < 0$	A vertical translation down $\lvert k \rvert$ units
$h > 0$	A horizontal translation right h units
$h < 0$	A horizontal translation left $\lvert h \rvert$ units
$a < 0$	A reflection across the x-axis
$\lvert a \rvert > 1$	A vertical stretch by a factor of $\lvert a \rvert$
$0 < \lvert a \rvert < 1$	A vertical compression by a factor of $\lvert a \rvert$

EXAMPLE: Use Transformations to Describe a Quadratic Function in Vertex Form

Describe the function $f(x) = -(x - 3)^2 - 1$ as a transformation of $g(x) = x^2$.

SOLUTION:

The parameters of $f(x) = -(x - 3)^2 - 1$ are $a = -1$, $h = 3$, and $k = -1$.

The value of a indicates that $f(x)$ is a reflection of $g(x)$ across the x-axis.

The values of h and k indicate that $f(x)$ is also a translation of $g(x)$ 3 units to the right and 1 unit down.

Check by using the Graphing Calculator.

Converting to Vertex Form

EXAMPLE: Convert the Equation of a Quadratic Function to Vertex Form

Write the quadratic function in vertex form.

$$y = -2x^2 + 8x - 29$$

SOLUTION:

$y = 2x^2 - 8x - 29$

$y = (-2x^2 + 8x) - 29$ Group $-2x^2$ and $8x$.

$y = -2(x^2 - 4x) - 29$ Factor -2 from the parentheses.

$y = -2(x^2 - 4x + \square) - 29 - \square$

Make space for completing the square and compensating for the value that will be added inside the parentheses.

My Notes

Summary *(continued)*

$y = -2(x^2 - 4x + 4) - 29 - (-8)$

Complete the square inside the parentheses. Distribute to the value you added inside the parentheses to determine what change was actually made. Subtract that value outside the parentheses.

$y = -2(x - 2)^2 - 21$ Factor the perfect-square trinomial.

Converting to Standard Form

EXAMPLE: Convert the Equation of a Quadratic Function to Standard Form

Write the quadratic function $y = -3(x - 1)^2 - 9$ in standard form.

SOLUTION:

$y = -3(x - 1)^2 - 9$

$y = -3(x^2 - 2x + 1) - 9$ Expand $(x - 1)^2$.

$y = -3x^2 + 6x - 3 - 9$ Apply the distributive property.

$y = -3x^2 + 6x - 12$ Combine like terms.

Use a Quadratic Function to Model a Data Set

The table shows the number of new cars of a particular model that have been sold each year over a 10-year period.

Year	1	2	3	4	5	6	7	8	9	10
Cars Sold (thousands)	203	245	278	325	341	356	377	371	351	306

Write a function that can be used to predict the number of cars of this model sold in a given year.

SOLUTION:

Start by graphing the data in the table. Let x represent the year number and y represent the number of cars sold, in thousands.

The relationship between x and y appears to be quadratic based on the shape formed by the data points.

The data point (7, 374) appears to be the approximate vertex of the parabola, so let $h = 7$ and $k = 374$ in a quadratic function that models the data.

$y = a(x - h)^2 + k$
$y = a(x - 7)^2 + 374$

Choose a different data point, such as (1,203), and use its coordinates to solve for a.

$$y = a(x - 7)^2 + 374$$
$$203 = a(1 - 7)^2 + 374$$
$$-171 = a(36)$$
$$-4.75 = a$$

So, an equation that models the data is $y = -4.75(x - 7)^2 + 374$. Graph this function to check that it is a good fit for the data.

$y = -4.75(x - 7)^2 + 374$
$y = -4.75(12 - 7)^2 + 374$
$y = 225.25$

In year 12, about 225.25 thousand, or 225,250, cars will be sold.

You can use the quadratic best-fit option in the Graphing Calculator to determine the function. This option gives an equation of

$$y = -4.88x^2 + 67.58x + 131.13.$$

My Notes

Practice Exercises

Review what you've learned using these practice problems. For feedback on other practice problems, try the Coach and Play items in the Practice section online.

1. Using the choices below, classify the functions as represented by the graph of a parabola that opens up, a parabola that opens down, or not a parabola.

A. $f(x) = -x^2$

B. $g(x) = -9 + x^2$

C. $h(x) = 3x^2 - 2x - 5$

D. $j(x) = x^3 - 4x^2$

E. $k(x) = 2x^2 + x^4$

F. $m(x) = 3x - x^2 - 4$

Parabola that opens up: __________

Parabola that opens down: __________

Not a parabola: __________

2. Which of the following statements are true about quadratic functions? *Select all that apply.*

A. All quadratic functions have a domain of all real numbers.

B. If two quadratic functions share a vertex, then they also share a range.

C. Any quadratic function that is represented by a parabola opening down has a maximum value.

D. The minimum value for any quadratic function is 0.

E. If a parabola has no x-intercepts and opens up, then the related quadratic function has a minimum value greater than 0.

3. Franklin needs to create a triangular board for a game. The perpendicular distance between the vertex and longest side of the triangle must be 6 inches less than the base. Let x represent the base of the triangle in inches.

Write a quadratic function of the base length to represent the possible area, in square inches, of the triangle. *Use integers to complete the equation.*

$A(x) = \frac{1}{2}x^2 +$ __________ $x +$ __________

4. The functions $f(x)$ and $g(x)$ are both quadratic functions with the same minimum value. The x-intercepts of $f(x)$ are –2 and 3. The x-intercepts of $g(x)$ are 3 and 8. Which statement could explain the relationship between $f(x)$ and $g(x)$?

A. $g(x) = f(x) + 5$, because each of the x-intercepts of $g(x)$ is 5 units greater than the x-intercepts of $f(x)$.

B. $g(x) = f(x + 5)$, because the functions have the same minimum value.

C. $g(x) = f(x) - 5$, because $g(x)$ shares an x-intercept with $f(x)$.

D. $g(x) = f(x + 5)$, because the points of $g(x)$ is a translation 5 units right of $f(x)$.

5. Given the equation $y = (x - 2)(x + 8)$, what is the vertex of the parabola formed by this equation? *Write the coordinates of the vertex.*

(__________, __________)

Practice Exercises *(continued)*

6. Using the choices below, classify which form of a quadratic function can best be determined from the given information. Assume that the leading coefficient a is known or can be determined from a point that is not a vertex or intercept.

   Standard form: $y = ax^2 + bx + c$

   Vertex form: $y = a(x - h)^2 + k$

   A. the maximum point of the parabola

   B. y-intercept

   C. zeros

   D. the minimum point of the parabola

   E. translation from $y = ax^2$

   Standard form __________

   Vertex form __________

7. A parabola is a translation of $y = \frac{1}{2}x^2$ and has a vertex at (2, −4). Complete two equivalent functions to represent the parabola.

   $y = \frac{1}{2}(x + \underline{\qquad\qquad})^2 + \underline{\qquad\qquad}$

   $y = \frac{1}{2}x^2 + \underline{\qquad\qquad} x + \underline{\qquad\qquad}$

8. A parabola is represented by the equation $y = -x^2 + 4x$. Which of the following are true about the parabola? *Select all that apply.*

   A. The parabola has x-intercepts at (0, 0) and (−4, 0).

   B. The y-intercept of the parabola is the origin.

   C. The parabola can be represented by the equation $y = -(x - 2)^2 + 4$.

   D. The vertex of the parabola is at (−2, 4).

   E. The parabola has an axis of symmetry of $x = 2$.

9. The function $C(x) = 3x^2 - 42x + 187$ represents the estimated monthly cost in dollars of natural gas for 1 year, where x is the number of the month with January represented by $x = 1$.

   During which month is monthly cost expected to be the lowest? *Give the number of the month.*

   Month of lowest cost: $x =$ __________

   What is the estimated minimum monthly cost? *Round to the nearest dollar.*

   Minimum monthly cost: $ __________

10. Using the choices below, identify the quadratic equations that have the same parabola as each of the following functions.

    $f(x) = 2(x + 1)^2 - 3$

    $g(x)$ is a translation of $y = x^2$, such that the vertex is at (−4, 1).

    A. $y = x^2 - 8x - 15$

    B. $y = 2x^2 + 2x - 1$

    C. $y = x^2 + 8x + 17$

    D. $y = 2x^2 + 4x - 1$

    $f(x) =$ __________

    $g(x) =$ __________

11. In the quadratic function $f(x) = 3x^2 + 11$, 3 is the quadratic coefficient.

    What happens to the graph of a quadratic function if its quadratic coefficient is changed from 3 to −7?

    The parabola flips to open down and is __________ vertically, making it __________.

    A. stretched, wider

    B. stretched, narrower

    C. compressed, wider

    D. compressed, narrower

Apply

How Can Math Help You Create a Successful Business?

How can an artist use data to determine how many of each product to make and what prices to charge to yield the highest overall revenue?

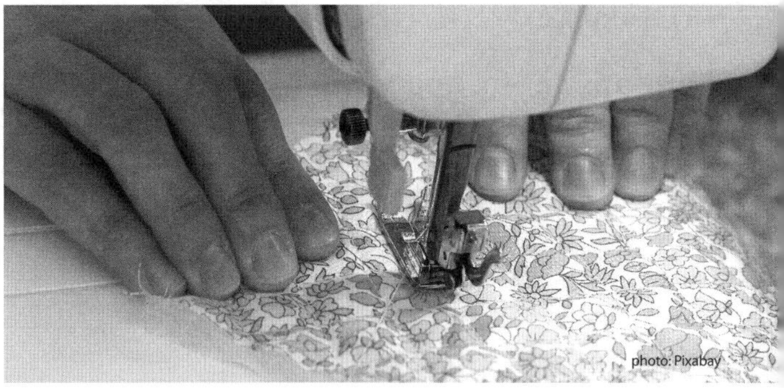

Making a Profit: View this video segment about tracking sales data.

A classmate, Mo Blox, decided to start a business building cell phone wall charger docks made from toy building blocks. The costs will be low, since Mo has lots of blocks left over from childhood. Mo made a questionnaire to post on various social media accounts:

> Don't you hate it when you have to plug your cell phone into a wall charger and there's nowhere to rest the phone? It ends up sitting on the floor – how sad. Now you can end that misery! My new company, eBlox, will be producing awesome cell phone docks that keep your phone right next to the plug. Before I get started, I need to know how much I should charge for each one to make this business work. Please help out by completing this simple questionnaire.
>
> Thanks,
> Mo Blox, Inventor of eBlox
>
> ✂ -
>
> Dude, this eBlox wall dock idea is so cool!
>
> I think you should charge: $______, and I would buy ______ of them.

Mo graphed the results of the survey and found the line of best fit. The function $N(x) = 7.17 - 0.82x$ gives the number of units each person would buy at a given price, x.

Revenue is the same as income; it equals the product of the price and the number of items sold. How many docks should Mo make, and at what price should Mo sell them to earn the most revenue? If Mo's business attracts 20 customers, what will be the revenue?

Show what you've learned by completing the other performance tasks in the online Apply section.

Apply *(continued)*

Your answer to Apply will be assessed on the following criteria:

1. Graphing and analyzing the number function; discussing domain and range
2. Writing, graphing, and analyzing the revenue function; discussing domain and range
3. Describing how the attributes of the revenue function relate to the scenario
4. Determining the price that will maximize revenue and its production implications, then finding the revenue for 20 customers

Scale / Criteria	4	3	2	1	0
	Exceeds Criteria	**Meets Criteria**	**Progressing to Criteria**	**Below Expectations**	**No Expectation**
Number Function	Graphs and analyzes the number function, including domain and range.	Graphs and analyzes the number function, but makes minor errors with the domain and range.	Graphs and analyzes the number function, but graph has errors or domain or range missing.	Graph and analysis are incomplete or incorrect.	Does not attempt task.
Revenue Function	Writes, graphs, and analyzes the revenue function, including domain and range.	Writes, graphs, and analyzes the revenue function; makes minor errors with the domain and range.	Writes, graphs, and analyzes the function, but makes errors; domain and range are missing.	Function, graph, and analysis are incomplete or incorrect.	Does not attempt task.
Context	Correctly describes how the attributes of the revenue function relate to the scenario.	Describes how the attributes of the revenue function relate to the scenario, with minor errors.	Describes how the attributes of the revenue function relate to the scenario, with significant errors.	Incorrectly describes how the attributes of the revenue function relate to the scenario.	Does not attempt task.
Price and Revenue	Determines price to maximize revenue, explains production implications, and finds the revenue for 20 customers.	Determines price to maximize revenue, explains production implications, and finds the revenue for 20 customers, but contains minor errors.	Determines price to maximize revenue, but does not clearly explain production implications, and the revenue for the 20 customers may have an error.	Gives a price that does not maximize revenue, the explanation of the production implications is incomplete, and the revenue for the 20 customers may be missing.	Does not attempt task.

UNIT 13: Recursive, Explicit, and Inverse Functions

13.1 Explore Recursive Functions

photo: Getty Images

Lesson Objectives

- Use functions to model relationships between two quantities, either with an explicit expression or through a recursive process.
- Derive formulas for the sum of arithmetic and geometric series and use the formulas to solve real-world problems.
- Use recursive and explicit functions to model arithmetic sequences and series.

Essential Questions

- How can you represent arithmetic and geometric sequences recursively and explicitly?
- How can you solve problems involving arithmetic and geometric sequences and series?

photo: Getty Images

Investigations

Oil Spills

Clean up an oil spill using patterns and functions. Find the most efficient clean-up method.

Seating Arrangements

Maximize the guests you can invite to a party. Use a recursive formula and be explicit!

Shape Shifter

Play Shape Shifter. Analyze the arithmetic pattern to push your score past 10,000.

Paper Tearing

Tearing paper? It's geometric! Use explicit and recursive formulas to predict the area after each tear.

A New Shape Shifter

Revisit Shape Shifter with a geometric scoring mechanism. Predict your score after *n* turns.

Repeated Triangles

Formulas describe characteristics of a simple fractal. Where do the series diverge or converge?

Key Vocabulary

arithmetic sequence, arithmetic series, common difference, common ratio, geometric sequence, geometric series, iteration, recursive formula, sequence, series

Discover

As you complete Engage and the investigations, record the most important ideas you've learned.

Engage	**Investigation 1**
Investigation 2	**Investigation 3**
Investigation 4	**Investigation 5**

Name ______________________________ Date ______________

Check for Understanding

Explore Recursive Functions: Investigation 1

1. Determine which of the following are arithmetic. *Select all that apply.*

   A. –14, –19, –24, –29 **B.** 2, 4, 8, 16, . . . **C.** 27, 26, 25, 24, . . .

   D. 8.25, 8.75, 9.25, 9.75 **E.** 32, 16, 8, 4, 1 **F.** –3, 4, 12, 21, . . .

2. If the third and fourth terms of a sequence are $a_3 = 2$ and $a_4 = 8$, determine the common difference as well as the first and second terms.

   $d =$ ______ $a_1 =$ ______ $a_2 =$ ______

3. Kia bought a snowboard package including boots and bindings for $525. She contributed $100, and her parents paid the remainder. Kia arranged to pay her parents back in regular monthly payments.

   A. What is the remaining balance after the third payment?

Term	(a_0)	a_1	a_2	a_3
Payment	0	1	2	3
Remaining Balance	425	380	335	

   B. Create an explicit formula to represent the remaining balance after any number of payments.

   ____

4. Use your explicit formula to determine the remaining balance after 8 payments.

   A. 110 **B.** 25 **C.** 20 **D.** 65 **E.** None of these

5. Explain the similarities and differences between the explicit formula for an arithmetic sequence and a linear function. Be sure to include algebraic characteristics and graphic representations.

Name ______________________________ Date ______________

Check for Understanding

Explore Recursive Functions: Investigation 2

1. A 3D theater has seating such that each subsequent row has more seating than the one in front of it, representing an arithmetic sequence. Row A is the first row of seats, followed by row B, etc. If row P has 163 seats, determine the number of seats that are added to each row. Then, use that number to find the number of seats in row B.

Number of seats added to each row: ____________

Row	A	B	...	P
Number of Seats	58		...	163

2. Find the total number of seats in the theater from row A through row P.

A. 1,768 **B.** 3,536 **C.** 840 **D.** 1,989 **E.** None of these

3. If the total number of seats in the entire theater is 3,408 and the last row has 226 seats, how many rows are in the theater?

Total number of rows in the theater = ____________

4. Explain how the formula for a sum of an arithmetic series is created. Use specific vocabulary, in addition to symbols, to explain how the partial sum of an arithmetic series, S_n, can be found. For example, reference S_n as "the sum of any number of terms."

Name ______________________________ Date ____________

Check for Understanding

Explore Recursive Functions: Investigation 3

1. Determine which of the following sequences are geometric. If they are geometric, determine the common ratio.

A. 28, –14, 7, –3.5 Common ratio: ______

B. 9, 18, 27 Common ratio: ______

C. $\frac{81}{4}, \frac{27}{2}, 9, 6$ Common ratio: ______

D. 32, 16, 4, 2 Common ratio: ______

2. If the second and third terms of a geometric sequence are $a_2 = \frac{-3}{4}$ and $a_3 = 3$, determine the common ratio as well as the first and fourth terms.

$r =$ ______ $a_1 =$ ______ $a_4 =$ ______

3. A flu virus affects the attendance at Adams High School, which has a population of 2,460 including teachers and staff. Below are the initial recorded absences. Assume growth through the weekend continues at the same rate.

Date	Monday 11/3	Tuesday 11/4	Wednesday 11/5
Absences	24	36	54

The superintendent closed the school on Tuesday, November 11. What is the rate of growth for absences? How many absentees were expected on the day of closing?

Growth rate: ______ Absences expected on closing day: ______

4. How does a sequence with ratio $r < -1$ compare to a sequence with ratio $r > 1$? How does a sequence with ratio $r < -1$ compare to a sequence where $-1 < r < 1$? Explain.

Name ________________ Date ________

Check for Understanding

Explore Recursive Functions: Investigation 4

1. The Owner of CellMate Cell Phones finds graphic data on new contract sales acquired by a competing company, Fonatics. Use the graph to determine correct functions that represent Fonatics new monthly contract sales. *Select all that apply.*

   A. $f(n) = 256(0.75)^n$

   B. $f(n) = 256(0.75)^{n-1}$

   C. $f(n) = 256\left(\frac{3}{4}\right)^n$

   D. $f(n) = 256\left(1 - \frac{1}{4}\right)^{n-1}$

   E. $f(n) = 256(1.25)^{n-1}$

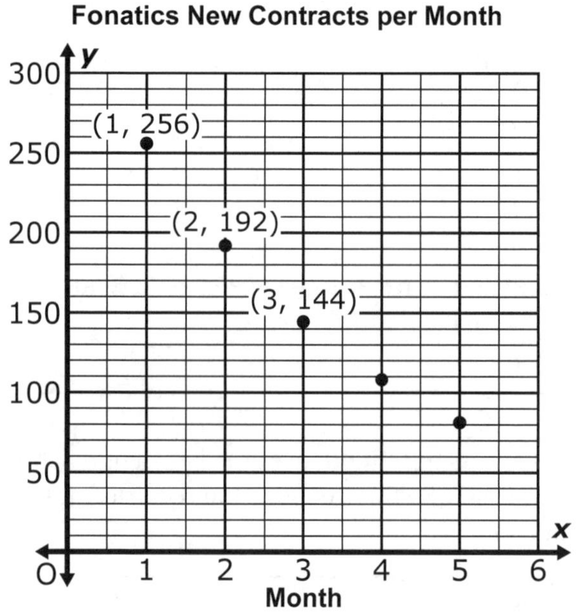

2. Use your function and graph to determine what month the Fonatics new cell contract acquisitions would fall below 10.

   Month __________ Number of new contracts expected __________

3. The CellMate owner also hired a new manager due to stagnant sales. The new manager acquires 81 new contracts in her first month. She believes she can continue to increase the number of new contracts by another third each month, where the second month, 108 new subscriptions would be expected. Use the formula for the sum of a series to approximate the total sum of new contracts the manager would be expected to acquire by the completion of her 10th month.

   A. 1,078 **B.** 4,523 **C.** 4,072 **D.** 3,452 **E.** None of these

4. Explain how the formula for a partial sum of a geometric series is created. Use specific vocabulary, in addition to symbols. For example, reference S_6 as "the sum of 6 terms."

Name ____ Date ____________

Check for Understanding

Explore Recursive Functions: Investigation 5

1. Determine which of the following geometric series are convergent or divergent. Determine the ratio, circle the behavior, and then find the sum, if it exists.

   A. $\frac{4}{25} + \frac{2}{5} + 1 + \ldots$ $r =$ ______   Convergent or Divergent   Sum: ______

   B. $-27 + 18 - 12 + \ldots$ $r =$ ______   Convergent or Divergent   Sum: ______

   C. $-9 - \frac{27}{2} - \frac{81}{4} - \ldots$ $r =$ ______   Convergent or Divergent   Sum: ______

   D. $0.1 + 0.01 + 0.001 + \ldots$ $r =$ ______   Convergent or Divergent   Sum: ______

2. A hand pump vacuum-seal product can be used to preserve food for longer periods of time. The "Seal Me" hand pump vacuum sealer removes 60 milliliters. of air from a large plastic bag on the first pump. The second pump removes 36 milliliters. If the ratio of air being removed remains the same for each successive pump, determine the ratio for air removed and the total amount of air that will be removed if you continue to perform full pumps for an extended period of time, until there is no ability to remove any more air.

   Common Ratio: ____________   Sum: ____________

3. Use the formula for a partial geometric sum, $S_n = \frac{a_1 - a_n \cdot r}{1 - r}$, to explain how the characteristics of the formula change as a convergent geometric series is summed to infinity. Rewrite the new formula for the sum of an infinite geometric series. Be sure to use appropriate vocabulary and terms, in addition to symbols and mathematical notation, in your justification.

Summary

Before you attempt the Practice Exercises, review what you've learned.

Finding Arithmetic Sequences

A **sequence** is a list of ordered numbers. Each number is called a **term** of the sequence. In conventional notation, the first term is a_1, the second term is a_2, and the nth term is a_n.

An **arithmetic sequence** is a sequence in which each term after the first term, a_1, is formed by adding a constant d, called the **common difference**, to the previous term.

EXAMPLE

Find the next 3 terms for the arithmetic sequence: 5, 8, 11, . . .

To find the common difference d, subtract any two consecutive terms.

$8 - 5 = 3$ $11 - 8 = 3$

So the common difference is 3.

My Notes

Using Arithmetic Sequences

A **recursive formula for an arithmetic sequence** defines a term, a_n, using the previous term, a_{n-1}, and the common difference, d. A recursive formula for an arithmetic sequence is: $a_n = a_{n-1} + d$.

An **explicit formula for an arithmetic sequence** allows you to find any term, a_n, if you know the first term and the common difference, d. An explicit formula for an arithmetic sequence is: $a_n = a_1 + (n - 1)d$.

Arithmetic Sequences as Functions

An **arithmetic sequence** belongs to the family of linear functions. By replacing the independent variable with x and the dependent variable with y, the explicit formula is of the same form as a linear function.

$$a_n = a_1 + (n - 1)d$$
$$y = a_1 + (x - 1)d$$
$$y = a_1 + dx - d$$
$$y = dx + (a_1 - d)$$

This is the equation of a line with slope d and y-intercept $a_1 - d$.

Finding the Missing Terms of Arithmetic Sequence

To find the missing terms in an arithmetic sequence given the first and last terms, find the common difference.

Summary *(continued)*

EXAMPLE

An arithmetic sequence with five terms has a first term of 7 and a last term of 23.

Use the explicit formula $a_n = a_1 + (n - 1)d$ to find the value of d.

There are five terms in the sequence, so $n = 5$.

$23 = 7 + (5 - 1)d$

$16 = 4d$

$d = 4$

Now, use a formula to find the missing terms.

$a_1 = 7$

$a_2 = a_1 + 4 = 11$

$a_3 = a_2 + 4 = 15$

$a_4 = a_3 + 4 = 19$

$a_5 = a_4 + 4 = 23$

The three missing terms between 7 and 23 are 11, 15, and 19.

Finding the Sum of an Arithmetic Series

A An **arithmetic series** is the sum of the terms in an arithmetic sequence. The **partial sum** is the sum of a limited or finite number of terms. The nth partial sum, S_n, can be found by finding n times the average of the first and last terms: $S_n = n\left(\frac{a_1 + a_n}{2}\right)$.

EXAMPLE

Find the sum of the first 40 terms for the arithmetic sequence 35, 33, 31, . . .

$a_1 = 35$

$a_{40} = -43$

$S_n = 40\left(\frac{35 + (-43)}{2}\right) = -160$

Finding Geometric Sequences

A A **geometric sequence** is a sequence in which each term after the first term, a_1, is formed by multiplying the previous term by a constant, r, called the **common ratio.**

Using Geometric Sequences

A The **recursive formula for a geometric sequence** defines a term, a_n, using the previous term, a_{n-1}, and the common ratio, r. A recursive formula for a geometric sequence is $a_n = ra_{n-1}$.

The nth term for a geometric sequence with first term, a_1, and common ratio, r, is given by the explicit formula $a_n = a_1(r)^{n-1}$, where n is any positive integer.

EXAMPLE

Find the 10th term of the geometric sequence with $a_1 = 64$ and $r = \frac{1}{2}$.

My Notes

Summary *(continued)*

Use the explicit formula $a_n = a_1(r)^{n-1}$.

Substitute 64 for a_1, substitute 10 for n, and substitute $\frac{1}{2}$ for r.

$$a_{10} = 64\left(\frac{1}{2}\right)^{10-1}$$

$$a_{10} = 64\left(\frac{1}{2}\right)^{9}$$

$$a_{10} = 0.125$$

Geometric Sequences as Functions

The explicit formula shows that the geometric sequence belongs to the family of exponential functions.

Finding the Missing Terms of a Geometric Sequence

To find the missing terms of a geometric sequence given the first and last terms, find the common ratio.

EXAMPLE

A geometric sequence with five terms has a first term of 1 and a last term of 81.

Use the explicit formula $a_n = a_1(r)^{n-1}$ to find the value of r.

There are five terms in the sequence, so $n = 5$.

$$81 = 1(r)^{5-1}$$

$$81 = r^4$$

$$r = \sqrt[4]{81}$$

$$r = 3$$

Now, use a formula to find the missing terms.

$$a_1 = 1$$

$$a_2 = a_1 \cdot 3 = 3$$

$$a_3 = a_2 \cdot 3 = 9$$

$$a_4 = a_3 \cdot 3 = 27$$

$$a_5 = a_4 \cdot 3 = 81$$

The three missing terms between 1 and 81 are 3, 9, and 27.

Finding the Sum of a Geometric Series

A A **geometric series** is the sum of the terms of a geometric sequence. The sum of the first n terms of a geometric sequence with first term, a_1, and common ratio, r, is given by $S_n = a_1\left(\frac{1-r^n}{1-r}\right)$.

EXAMPLE

Find the sum of the first 10 terms of the geometric sequence 64, 32, 16, ...

Substitute 64 for a_1, substitute 10 for n, and substitute $\frac{1}{2}$ for r.

$$S_{10} = 64\left(\frac{1-\left(\frac{1}{2}\right)^{10}}{1-\frac{1}{2}}\right)$$

$$S_{10} = 127.875$$

Finding the Sum of a Geometric Series Given the First and Last Terms

You can also find the sum of a geometric series if you know the first and last terms and the common ratio.

My Notes

Practice Exercises

Review what you've learned using these practice problems. For practice problems with feedback, try the Coach and Play items in the Practice section online.

1. Look at the table.

n	1	2	3	4
a_n	3	5	9	17

Which recursive relationship correctly describes the relationship between n and a_n in the table?

A. $a_0 = 0$
$a_n = a_{n-1} + 4 - n$

B. $a_0 = 1$
$a_n = a_{n-1} + 2$

C. $a_n = 1$
$a_n = 4n - a_{n-1}$

D. $a_0 = 2$
$a_n = 2a_{n-1} - 1$

E. $a_0 = 2$
$a_n = a_{n-1} + n$

2. Kenny grows tomato plants from seeds every year. He uses one seed for each tomato plant he grows.

The first year, he had a small garden and grew only two tomato plants. Each year, he expanded the garden to increase the number of tomato plants by one more than the previous year.

The number of tomato plants in any year, n, can be determined by $f(n) = n + 1$.

If Kenny continues this pattern for 25 years, what is the total number of tomato seeds Kenny will have used over that time?

____________ seeds

3. Look at the geometric sequence.

144, ________, ________, −18

What is the common ratio, r, for this geometric sequence, and what are the values of the missing terms a_2 and a_3?

Be sure to write any fractions in decimal form.

$r =$ ____________

$a_2 =$ ____________

$a_3 =$ ____________

4. Examine the sequence.

n	1	2	3
a_n	−3	−12	−48

What is the value of a_7?

$a_7 =$ ______

5. The following exponential function is an explicit formula that represents a geometric sequence.

$f(n) = 3^{n-1}$

Which statements are accurate? *Select all that apply.*

A. The initial value of the function is 3.

B. The common ratio of the geometric sequence is 3.

C. The recursive rule that models the same sequence is $a_n = 3 \cdot a_{n-1}$.

D. The initial value of the function is 1.

Practice Exercises *(continued)*

6. While Dalton recovers from knee surgery, his trainer recommends that he bikes until he can return to his normal running program. Dalton is told to bike for 12 minutes for his workouts during the first week of recovery, and then steadily increase his biking time each week thereafter. By the 9th week, Dalton should be able to bike for an hour with no swelling and can then transition to running.

   Dalton wants to know what the time commitment for each workout will be during his recovery period.

   Which of the following information is needed to find the missing terms in the sequence? *Select all that apply.*

   A. Dalton's recovery training represents a geometric sequence.

   B. Dalton's recovery training represents an arithmetic sequence.

   C. $a_n = a_1 + (n - 1)d$
   $1 = 12 + (9 - 1)d$
   $-11 = 8d$
   $d = -1.375$

   D. $a_n = a_1 + (n - 1)d$
   $60 = 12 + (9 - 1)d$
   $48 = 8d$
   $d = 6$

   E. $a_1 = 12$

   F. $a_2 = 60$

   G. $S_n = n\left(\frac{a_1 + a_n}{2}\right)$
   $S_n = 9\left(\frac{12 + 60}{2}\right) = 324$

   H. $a_n = a_1(r)^{n-1}$
   $60 = 12(r)^{9-1}$

   I. $a_2 = 12 + d$

   J. $a_4 = 12 + d$

7. The arithmetic series $S_n = (1 + 2 + 3 \ldots + n)$ is equal to $\frac{n(n+1)}{2}$.

   Which of the following is the formula for the series $(2 + 4 + 6 + \ldots + 2n)$?

   A. $\frac{n(n+1)}{2} + n$

   B. $\frac{n^2(n+1)}{2}$

   C. $n(2n + 1)$

   D. $n(n + 1)$

8. The following is a geometric sequence:

   3, 6, 12, 24, 48, 96, . . .

   Which statements are accurate? *Select all that apply.*

   A. The initial value of the sequence is 3.

   B. The common ratio of the sequence is 3.

   C. The recursive rule that models the sequence is $a_n = 6 \cdot a_{n-1}$.

   D. The explicit formula that models the sequence is $f(n) = 3 \cdot 2^{n-1}$.

9. Consider the following geometric sequence.

   3, 9, 27, 81, 243, . . .

   Which formula models the sequence?

   A. $a_n = a_n \cdot 3$

   B. $a_n = a_{n-1} \cdot 3$

   C. $a_n = a_{n+1} \cdot 3$

   D. $a_n = 3n$

10. Consider the arithmetic series with a first term of 5 and seventh term of 23.

    What is sum of these first seven terms?

    The sum is ________________.

Apply

How Many Digits Are Needed in a Telephone Number?

photo: Library of Congress

Identify a pattern between the number of digits and the number of possible phone numbers.

Did you ever wonder why phone numbers in the United States (including area codes) contain 10 digits? As we move through the 21st century, do you think 10 digits will be enough to ensure that everyone has a unique phone number? Watch the video to learn about the history of telephone numbers and where we go from here.

Phone Numbers: View this video segment to see how you would identify a pattern between the number of digits and the number of possible phone numbers.

Do some research to determine the number of phones in the entire world. Further, assume that every phone in the world uses the same phone number rules that are used in the United States: each number contains 10 digits, and the first digit cannot be 0 or 1.

Now think about the following questions:

- Historically, how has the format of phone numbers changed mathematically to meet the need for more phone numbers?
- If phone numbers had just 2 digits, how many phone numbers would be possible? What about 5 digits? What about 50 digits?
- How many digits should a phone number contain?

Telephone Numbers

Use your thinking about the above questions to solve the following problem.

It is estimated that the number of phones in the world will continue to increase by 25 million every month. How many digits are needed right now to accommodate all phones? When would more digits be needed for phone numbers so each user had one unique number?

Justify your answers using sequences and correct mathematical explanations.

Show what you've learned by completing the other performance tasks in the online Apply section.

Apply *(continued)*

Your answer to Apply will be assessed on the following criteria:

1. Researching necessary and accurate information to answer the problem
2. Determining the number of digits needed now and in the future
3. Using appropriate algebraic representations and equations including, but not limited to, relevant sequences and series
4. Correctly explaining why a particular number of digits is needed for each person to have a unique number using arithmetic or geometric sequences and series

Criteria \ Scale	4	3	2	1	0
	Exceeds Criteria	Meets Criteria	Progressing to Criteria	Below Expectations	No Expectation
Adequate Research	Gathers valid and usable information.	Gathers mostly valid and usable information.	Gathers somewhat valid and usable information.	Gathers mostly invalid and unusable information.	Does not perform research.
Correct Computations	Correctly calculates the number of users and digits needed based on collected data and problem statement.	Calculates the number of users and digits needed based on collected data and problem statement with no more than one minor error.	Calculates the number of users and digits needed based on collected data and problem statement with several minor errors.	Calculates the number of users and digits needed based on collected data and problem statement with major errors.	Does not attempt computations.
Algebraic Representations and Equations	Creates correct algebraic models to determine the number of users and digits.	Creates mostly correct algebraic models to determine the number of users and digits.	Creates some correct algebraic models to determine the number of users and digits.	Creates incorrect algebraic models to determine the number of users and digits.	Does not create a model.
Conclusions with Justifications	States a reasonable conclusion and uses the model to provide a correct justification.	States a reasonable conclusion and uses the model to provide a mostly correct justification.	States an unreasonable conclusion or uses the model to provide an incorrect justification.	States an unreasonable conclusion and uses the model to provide an incorrect justification.	Provides no conclusion.

UNIT 13: Recursive, Explicit, and Inverse Functions

13.2 Explore Inverse Functions

photo: Getty Images

Lesson Objectives

- Find the inverse of a function by solving the equation $f(y) = x$ for y, and decide the appropriate domain of the inverse.
- Identify the graphical relationship between a function and its inverse.
- Solve real-world problems that can be modeled by composite functions.

Essential Questions

- How can you identify a function's inverse numerically, graphically, and algebraically?
- How can you restrict the domain of a function to ensure the inverse is also a function?

photo: Getty Images

Investigations

How Do You Get a Better Deal with Discounts?

Combine a dollar-off coupon with a percent-off deal. Does order matter?

What Is a Function?

Relation, function, or both? Do combined functions make a function?

Exploring Composite Functions

Model the dollar-off, percent-off deal with a composite function.

Function Machine

Use the Function Machine to analyze functions and their inverses.

Graphs of Inverse Functions

Look under the microscope to compare details of the graphs of functions and their inverses.

Diving In

Real-world situations have restrictions, as do functions. Restrict an inverse to make it a function.

Key Vocabulary

composite function, inverse function

Discover

As you complete Engage and the investigations, record the most important ideas you've learned.

Engage	**Investigation 1**
Investigation 2	**Investigation 3**
Investigation 4	**Investigation 5**

Name ________________________________ Date ____________

Check for Understanding

Explore Inverse Functions: Investigation 1

1. Removing which ordered pair would make the relation $R = \{(1, 1), (2, 1), (2, 2)\}$ a function?

A. only (1, 1)

B. only (2, 1)

C. (1, 1) or (2, 1)

D. (2, 1) or (2, 2)

2. For the relation represented by each of the equations $y = \pm\sqrt{-5x}$, $y = -\sqrt{-5x}$, and $y = \sqrt{-5x}$, consider the domain and range. *Circle the correct words in the sentences.*

The domains of the three relations are [the same / different], and the ranges are [the same / different]. The relation $y = \pm\sqrt{\;5x}$ [is / is not] a function, the relation $y = -\sqrt{-5x}$ [is / is not] a function, and the relation $y = -\sqrt{-5x}$ [is / is not] a function.

3. Elon is trying to determine if the relation represented by the following table is a function.

x	−8	−6	−4	−2	0	2	4	6	8
y	0	1	−2	3	−4	5	−6	7	−8

He plotted the points in the table and drew lines through the points as shown.

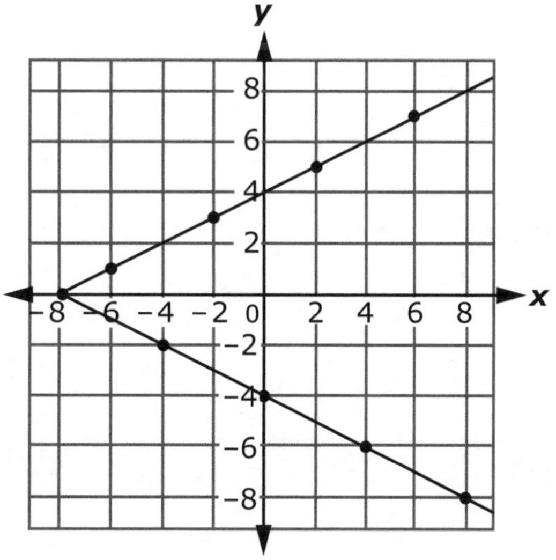

Elon says that the graph shows that the relation can be represented by the equation $\frac{1}{2}(x + 8) = |y|$, and since a vertical line can be drawn that intersects the graph more than once, the relation represented by the table is not a function. Is he correct? *Explain.*

Name ______________________________ Date ______________

Check for Understanding

Explore Inverse Functions: Investigation 2

1. A cell phone company has a Monthly Unlimited Family Plan where they charge a flat fee of \$45 in addition to \$25 per phone, represented as $M(x) = 25x + 45$. If anyone in the family works for the cell company, the entire bill will be given a 15% discount, $D(x) = 0.15x$. Which of the following composite functions represents this discounted bill? *Select all that apply.*

A. $D \cdot M(x)$

B. $M \cdot D(x)$

C. $0.15(25x + 45)$

D. $25(0.15x) + 45$

E. $25x + 45(0.15)$

F. $M(D(x))$

G. $D(M(x))$

2. Determine the composite functions if $f(x) = 6 - 2x$, $g(x) = \frac{x}{2}$, and $h(x) = -x^2$. Rewrite your composite in expanded form.

A. $f(g(x)) =$ ____________ **B.** $g(f(x)) =$ ____________

C. $h(f(x)) =$ ____________ **D.** $g(g(x)) =$ ____________

E. $f(f(x)) =$ ____________ **F.** $f(h(x)) =$ ____________

3. Jayden found the composite function $k \circ p(x)$ where $k(x) = 4 - \frac{x}{2}$ and $p(x) = 2x + 10$. Describe the errors she made, and explain the mistakes she made in her reasoning.

$k(p(x)) = 4 - \frac{2x + 10}{2}$

$k(p(x)) = 4 - x + 10$

$k(p(x)) = 14 - x$

Name ____ Date ____________

Check for Understanding

Explore Inverse Functions: Investigation 3

Max challenges Maria to a number game. He asks her to pick a number between 1 and 10, confident that he can tell her this original number after she completes the following steps. Maria picks a number and Max then tells her to do the following operations:

- Add four
- Multiply by 3
- Subtract 10

Max then asks Maria for her new value. Maria says "29." Max replies:

1. "OK! Your original number is: ____________!"

2. What are the operation steps that must be made to "undo the operations" that Maria made in order to change the result of 29 back to the original number chosen?

    Step 1: ____________

    Step 2: ____________

    Step 3: ____________

3. Write an algebraic equation that reflects the operations Maria made on n, the original number she chose to obtain the resulting number r.

    Maria's Equation: $r =$ ____________

4. Write an algebraic expression that reflects the operations Max made on the resulting number r, to obtain the original number n.

    Max's Equation: $n =$ ____________

5. Explain how the domains and ranges among the functions that make up a composite relationship are the same or different from the relationships between the domains and ranges of a function and its inverse.

Name ____________________ Date __________

Check for Understanding

Explore Inverse Functions: Investigation 4

1. Use the graph of $f(x) = -\frac{1}{3}x + 2$ to find at least 4 points of the inverse function, listing them in the table, (a). Graph the inverse function on the same axis, (b).

A.

x	$f^{-1}(x)$

B.

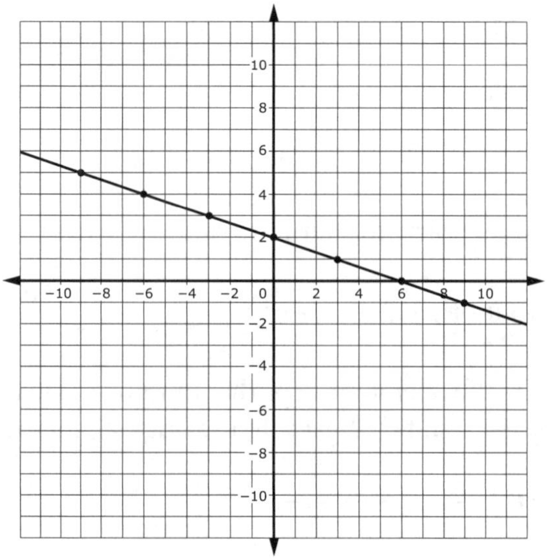

2. Use the data in your table and the graph to create the algebraic inverse function.

   Inverse of $f(x) = -\frac{1}{3}x + 2$: ____________

3. Ty was assigned the algebraic function: $g(x) = -4x - 8$ to create the algebraic inverse function. Analyze his justification and process, correcting errors he has made or gaps in his reasoning, attending to precision in your response.

   If $g(x) = -4x - 8$, then $y = -4x - 8$. All inverses reflect over $y = x$ so the inverse can be written: $x = -4y - 8$. Simplify: $g^{-1}(x) = \frac{(x+8)}{4}$; rewritten as $g^{-1}(x) = x + 2$.

Name ______________________________ Date ____________

Check for Understanding

Explore Inverse Functions: Investigation 5

Darin has taken 4 tests in Algebra II scoring 86%, 81%, 72%, and 78%.

1. Choose the functions that will describe the average of his scores, A, upon taking the fifth test, t. *Select all that apply.*

   A. $A(t) = \frac{86 + 81 + 72 + 78 + t}{4}$

   B. $A(t) = \frac{t - (86 + 81 + 72 + 78)}{5}$

   C. $A(t) = \frac{86 + 81 + 72 + 78 + t}{5}$

   D. $A(t) = \frac{317 + t}{5}$

   E. $A(t) = \frac{317 - t}{5}$

   F. $A(t) = \frac{317 + t}{4}$

2. Create the inverse function that will compute the value of the fifth test when the average score goal is entered as the domain value.

   Inverse Function: ______________________________

3. Darin wants to be sure to obtain an average of at least 80.0% for the 5-test average. What is the lowest percentage he must earn on the fifth test?

   Minimum possible fifth test percentage for 80.0% average: __________

4. Darin is motivated to earn at least 83.0% for the 5-test average. What score on the fifth test would make this possible?

   Minimum percentage on the fifth test for 83.0% average: __________

5. Define the domain and range values for the original function to find Darin's average percentage. Then, explain how the domain and range of the average percentage affect the related inverse function's domain and range, when finding the fifth test score. Note: There is no option for extra credit on the fifth test.

Summary

Before you attempt the Practice Exercises, review what you've learned.

Functions can be combined using composition of functions. The composition of functions f and g is written $f \circ g$ and is defined as $(f \circ g)(x) = f(g(x))$. For $f(g(x))$, the function g is performed first, and the resulting output becomes the input for the function f. The domain of $f \circ g$ consists of the x values in the domain of g for which $g(x)$ is in the domain of f.

EXAMPLE

Find $(f \circ g)(x)$ and $(g \circ f)(x)$ for $f(x) = x - 2$ *and* $g(x) = x^2 + 1$.

$$(f \circ g)(x) = f[g(x)]$$
$$= f(x^2 + 1)$$
$$= (x^2 + 1) - 2$$
$$= x^2 - 1$$
$$(g \circ f)(x) = g[f(x)]$$
$$= g(x - 2)$$
$$= (x - 2)^2 + 1$$

For a function and its inverse, $(f^{-1} \circ f)(x) = x$ and $(f \circ f^{-1})(x) = x$.

My Notes

Finding Inverse Functions

Functions that undo each other are called inverse functions. The notation for the inverse of f is f^{-1}, read "the inverse of f." The inverse of a function can be found algebraically or graphically.

EXAMPLE

Find the inverse of the function $y = 3x - 9$ algebraically.

$y = 3x - 9$	Original equation
$x = 3y - 9$	Switch the x- and y-variables.
$3y = x + 9$	Isolate the y term using inverse operations.
$y = \frac{1}{3}x + 3$	Solve for y.

EXAMPLE

Find the inverse of the function $y = x^2 + 2$ graphically.

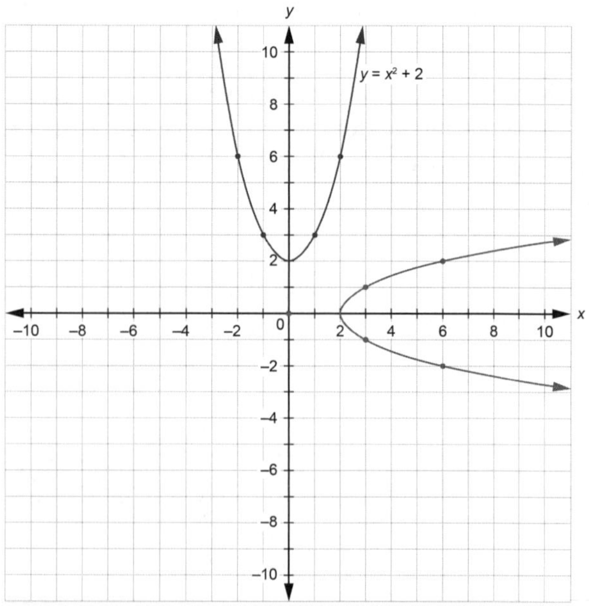

Inverses

To graph the inverse of a function, plot the points formed by switching the x and y values of the function. Connect the points to form the graph.

Practice Exercises

Review what you've learned using these practice problems. For practice problems with feedback, try the Coach and Play items in the Practice section online.

1. What is the domain of the inverse of $y = 4x - 3$?

   A. all real numbers x

   B. all real numbers x, where $x > 3$

   C. all real numbers x, where $x < -3$

   D. all real numbers x, where $x \neq \frac{3}{4}$

   E. all real numbers x, where $x \neq 0$

2. Which statements are true about the graph of the inverse function of $y = 3x + 1$? *Select all that apply.*

   A. The slope of the inverse function will be –3.

   B. The slope of the inverse function will be 3.

   C. The slope of the inverse function will be $\frac{1}{3}$.

   D. The inverse function will have a y-intercept at $\left(0, -\frac{1}{3}\right)$

   E. The inverse function will have a y-intercept at (0, –1).

   F. The inverse function will not be a straight line.

   G. The inverse function will have a vertical asymptote at $x = -\frac{1}{3}$.

3. What will be the axis of symmetry between the graphs of $y = 4x - 2$ and its inverse?

   A. $y = \frac{1}{4}x + \frac{1}{2}$

   B. $y = x$

   C. $y = -x$

   D. $y = 4x$

   E. $y = -\frac{1}{4}x$

4. The table shows the values of x and y for a certain function.

x	0	2	3	4	5
y	2	3	4	5	6

   What will be the value of y for the inverse of this function when $x = 3$?

   A. 0

   B. 2

   C. 3

   D. 4

   E. It cannot be determined from the given information.

5. Which of the following functions is the inverse of the function $h(t) = 2t - b$? *Select all that apply.*

   A. $h^{-1}(t) = \frac{t+b}{2}$

   B. $h^{-1}(t) = \frac{t}{2} + b$

   C. $h^{-1}(t) = t + \frac{b}{2}$

   D. $h^{-1}(t) = \frac{t-b}{2}$

   E. $h^{-1}(t) = \frac{t}{2} - b$

   F. $h^{-1}(t) = \frac{t}{2} + \frac{b}{2}$

6. Match the function $f(x)$ with its inverse, $f^{-1}(x)$.

$f(x)$	$f^{-1}(x)$
$y = \frac{2}{5}x + 10$	______
$y = -\frac{5}{2}x + 10$	______
$2x + 5y = 10$	______

   A. $y = -\frac{5}{2}x + 5$

   B. $y = -\frac{2}{5}x + 4$

   C. $y = \frac{5}{2}x - 25$

Practice Exercises *(continued)*

7. Joey sells fresh chicken eggs. He charges customers a \$5.00 flat delivery fee plus \$2.50 per dozen eggs. The equation $A = 2.50d + 5$ can be used to calculate the amount A (in dollars) that Joey earns for selling d dozen eggs to a customer.

Derive an equation Joey can use to determine how many dozen eggs, d, he needs to sell to a customer to earn A dollars. *Write the entire equation.*

$d =$ ____________ $A +$ ____________

8. The table lists some of the resulting values of given inputs for $f(x)$.

x	−4	−3	−2	−1	0	1	2	3	4
$f(x)$	5	3	7	2	9	6	4	5	1

The function $h(x) = 3x + 5$ can be used to define a new function, $g(x)$, as $g(x) = -2f(h(x)) - 4$.

What is $g(-2)$?

$g(-2) =$ ____________

9. The table represents a function $f(t)$. The function f returns a temperature in degrees Fahrenheit given a temperature, t, measured in degrees Celsius.

t	10	20	30	35	50	95
$f(t)$	50	68	86	95	122	203

Fill in the missing values in the table below for $f^{-1}(t)$, the inverse of $f(t)$.

t	$f^{-1}(t)$
50	
	50
95	

Apply

Does How Fast You Drive Affect How Much Fuel You Use?

What is a race car's optimal performance level?

A race car's performance depends, at least in part, on its fuel consumption. Watch the video to learn more.

The table below shows how one race car's gas mileage in miles per gallon (mpg) changes with its speed in miles per hour (mph).

Race Car Performance: View this video segment to see how to determine a race car's optimal performance level.

Use the data in the table to complete the questions.

Speed (mph)	Gas Mileage (mpg)
60	7.1
80	8.2
100	8.8
120	8.4
140	7.5

Gas Mileage

Write an equation that models the car's gas mileage as a function of its speed. Identify a reasonable domain for your function. Then, graph its inverse, and explain what the inverse represents.

- At what speed is the car's fuel efficiency the greatest?
- Using your model, how much fuel would the car use during a 500-mile NASCAR race?

Use your equation, graph, and precise mathematical language to justify your answers.

Show what you've learned by completing the other performance tasks in the online Apply section.

Apply *(continued)*

Your answer to Apply will be assessed on the following criteria:

1. Writing an equation of a function that models the race car's gas mileage given its speed and explaining how you determined this equation
2. Deciding on a reasonable domain for the function and explaining your reasoning
3. Graphing the inverse of your function and explaining what the inverse represents
4. Answering the questions about the race car and showing your work or explaining how you determined your answers

Criteria \ Scale	4	3	2	1	0
	Exceeds Criteria	**Meets Criteria**	**Progressing to Criteria**	**Below Expectations**	**No Expectation**
Data Modeling	Writes the equation of a function that reasonably models the data (likely a quadratic function) and clearly explains how this equation was determined.	Writes the equation of a function that reasonably models the data set, but the explanation of how this model was determined is incomplete or unclear.	Writes the equation of a function to model the data, and correctly explains how this model was determined; however, the type of function chosen is not appropriate for the data.	Attempts to write a function that models the data, but the response shows little understanding of how to go about this process.	Does not attempt to model the data with a function.
Reasonable Domain	Identifies a reasonable domain for the function and clearly explains the reasoning for the decision.	Identifies a reasonable domain for the function, but the explanation for the decision is incomplete or unclear.	Identifies a reasonable upper bound or lower bound for the domain, but not both; explains reasoning for the decision.	Identifies the mathematical domain of the function, but does not take the real-world situation into account.	Does not attempt to identify a reasonable domain for the function; or identifies a clearly unreasonable domain.
Inverse Function	Uses a valid procedure to correctly graph the inverse of the function, and clearly explains the steps used.	Uses a correct procedure to graph the inverse of the function, but makes minor mistakes when following this procedure.	Demonstrates some understanding of how to graph the inverse of a function, but the response contains errors.	Attempts to graph the inverse of the function, but the response shows little understanding of functions and their inverses.	Does not attempt to graph the inverse of the function.
Function Applications	Correctly applies the function and its inverse to determine the speed of maximum fuel efficiency and the amount of fuel used in a race; shows all work.	Applies the function and its inverse to answer the questions, but makes minor computation errors; shows all work.	Shows some understanding of how to apply the function and its inverse to answer the questions; shows at least some work.	Provides answers to the questions, but shows no work and offers no mathematical justification for the answers.	Does not attempt to answer the questions.

photo: Discovery Education

UNIT 13: Recursive, Explicit, and Inverse Functions

13.3 Explore Function Transformations

Lesson Objective

- Graph functions by applying the behaviors of parent functions when stretched, skewed, or translated by a constant.

Essential Question

- How do numeric values within an algebraic function affect the transformation of the parent function?

photo: Getty Images

Investigations

Modeling a Scenario with a Graph

Dive into an algebraic and graphic model of pricing. Then watch the results when pricing changes.

Exploring Transformations

Consider three more pricing scenarios. How are the functions related to the original function?

Identifying the Effects of Transformations

Discover the algebraic connection between a transformed function and its parent. It's all in the behavior.

Exploring Multiple Transformations

Build advanced equations by combining transformations.

Key Vocabulary

reflection, transformation, translation

Discover

As you complete Engage and the investigations, record the most important ideas you've learned.

Engage	Investigation 1
Investigation 2	Investigation 3

Name ______________________________ **Date** ______________

Check for Understanding

Explore Function Transformations: Investigation 1

Alex started his own yard work business. His first year, he charged each customer \$15 per hour for labor.

Create an equation to model Alex's hourly fee for the first year of his business, then reference this as you answer the scenarios posed in questions 1–3. Match each scenario with the appropriate graph, and describe the transformations with direction and magnitude relative to the first year of his business.

Graph A

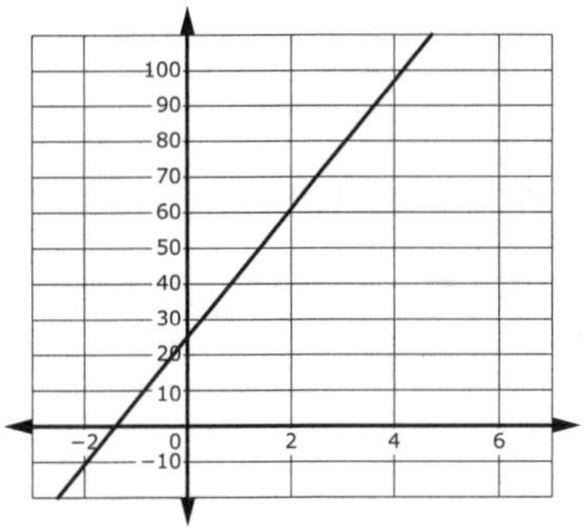

Graph B

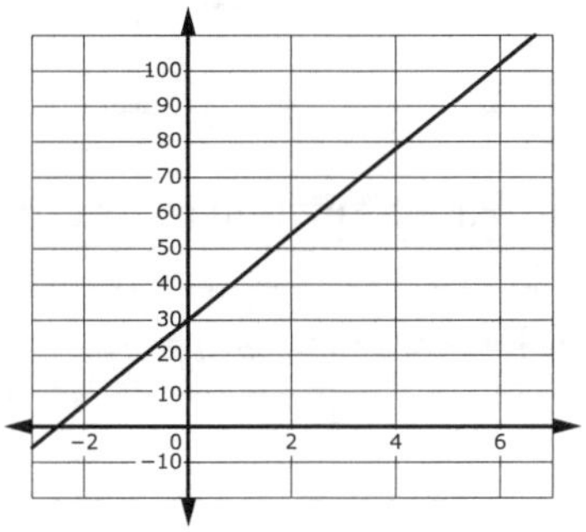

Graph C

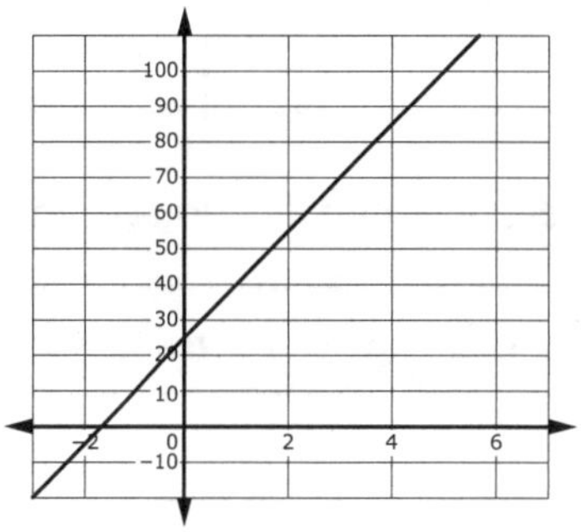

1. Year 2: Alex bought some of his own equipment and started driving to further locations, so he added a \$25 flat fee to his \$15 per hour labor.

Equation: ____________ Graph: ____________

Dilation: ____________ Translation: ____________

2. Year 3: Business was booming! Alex increased the amount charged per hour by 20% and continued to charge the flat fee of \$25.

Equation: ____________ Graph: ____________

Dilation: ____________ Translation: ____________

3. Competition: Alex's friend Jay started the same business and only charges \$12 per hour but charges a flat fee of \$30. How does Jay's profit function relate to Alex's first year?

Equation: ____________ Graph: ____________

Dilation: ____________ Translation: ____________

4. Describe a business scenario that would model a dilation and a vertical shift down if the original model is $y = 28x$. Also, include the supporting equation and graph.

Name ______________________________ Date ____________

Check for Understanding

Explore Function Transformations: Investigation 2

Analyze the characteristics of each function to describe how the graph differs from the related parent function.

1. Describe the translation behavior for each function.

$f(x) = x^2 - 6$	$g(x) = \|x - 6\|$	$h(x) = \sqrt{x} + 6$	$k(x) = \sqrt{x + 6}$	$p(x) = (x - 6)^2$

2. Describe the reflection compared to the parent function as "over *x*-axis" or "over *y*-axis."

$f(x) = -x^2$	$f(x) = (-x)^2$	$p(x) = -\|x\|$	$g(x) = \|-x\|$	$h(x) = -\sqrt{x}$	$k(x) = \sqrt{-x}$

3. Describe the dilation for each function as horizontal or vertical, and as a stretch or a compression.

$f(x) = 4x^2$	$g(x) = \|5x\|$	$h(x) = 0.5\sqrt{x}$	$k(x) = \sqrt{0.5x}$	$p(x) = 0.5\|x\|$
$q(x) = \left(\frac{1}{4}x\right)^2$	$g(x) = \frac{3}{2}\|x\|$	$t(x) = \frac{1}{4}x^2$	$h(x) = \sqrt{\frac{3x}{2}}$	$f(x) = (4x)^2$

4. Explain how a horizontal dilation compares to a vertical dilation of the same factor.

Name ______________________________ Date ____________

Check for Understanding

Explore Function Transformations: Investigation 3

Determine what words you can use to describe each of the functions regarding translation, dilation, or reflection on the related parent function. If there is no change, write "none."

1. Describe $f(x) = 2 - 3x^2$.

Parent Function	Vertical Translation	Horizontal Translation	Dilation	Reflection

2. Describe $g(x) = 0.5|4 - x| - 6$.

Parent Function	Vertical Translation	Horizontal Translation	Dilation	Reflection

3. Write a function that fits the given description: ____________

Parent Function	Vertical Translation	Horizontal Translation	Dilation	Reflection
$h(x) = \sqrt{x}$	Down 4	Right 2	Stretch: 1.5	Over y-axis

4. Write a function that fits the given description: ____________

Parent Function	Vertical Translation	Horizontal Translation	Dilation	Reflection
$k(x) = |x|$	Down 3	Left 5	Stretch: 2	Over x-axis

5. Explain the difference in the reflection of $g(x) = -f(x)$ and $h(x) = f(-x)$. You may wish to reference specific values as part of your explanation.

Summary

Before you attempt the Practice Exercises, review what you've learned.

Graphs can be transformed using **translations**, **reflections**, and **horizontal** or **vertical dilations**.

Translations

A vertical translation transforms a function $f(x) \rightarrow f(x) + k$, where each point on the graph of $f(x)$ is shifted k units up if k is positive and k units down if k is negative. If the point (x, y) is on the parent graph, then $(x, y + k)$ is on the transformed graph. As shown, (2, 4) is on the original graph, so (2, 4 + 5) is on the transformed graph.

EXAMPLE: Vertical Translation

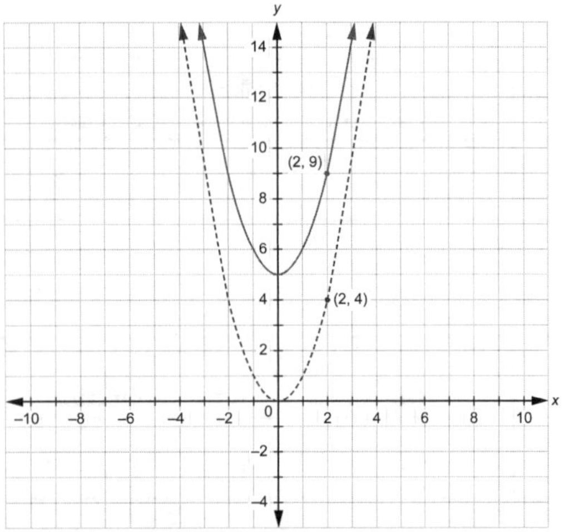

A reflection across the x-axis $f(x) \rightarrow f(x) = -\sqrt{x}$ will reflect the function across the x-axis. If the point (x, y) is on the parent graph, then $(x, -y)$ is on the transformed graph. Note that (4, 2) on the parent graph has been transformed to (4, −2) on the reflected graph.

EXAMPLE: Reflection across the *x*-axis

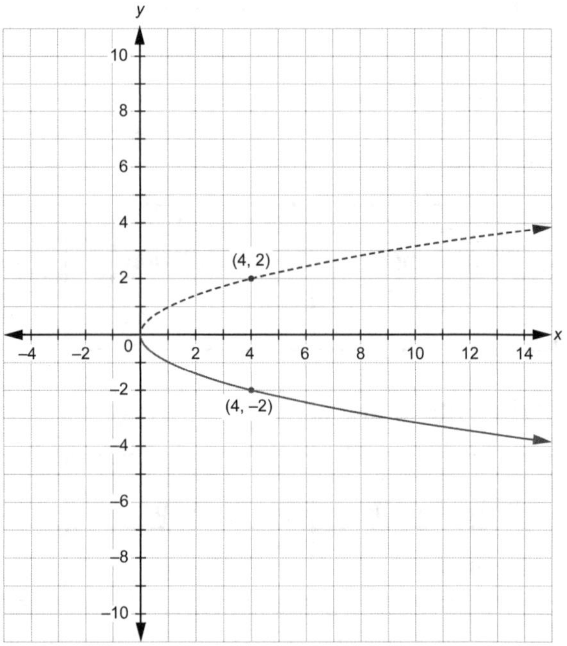

A vertical stretch or compression occurs when $f(x)$ is replaced by $kf(x)$. The transformation of $y = |x|$ to $y = 3|x|$ is a vertical stretch by a factor of 3. Note that every point on the expanded graph has a y-coordinate that is 3 times the y-coordinate of the original function. So, (x, y) will be stretched to $(x, 3y)$ on the transformed graph.

My Notes

Summary *(continued)*

EXAMPLE: Vertical Stretch

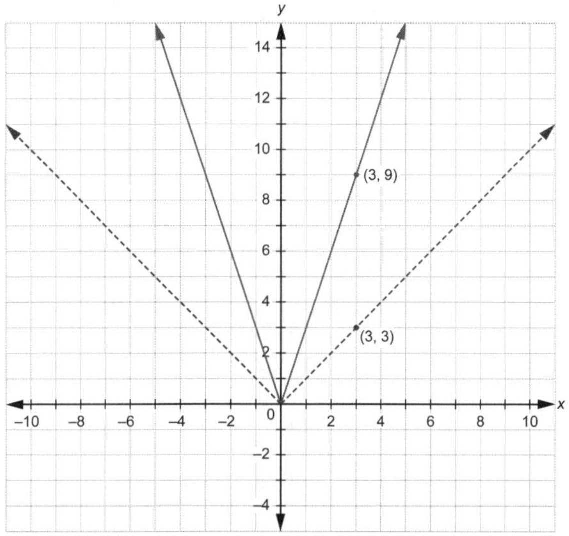

If the value of k is greater than 1 in $y = |kx|$, then the result is a horizontal compression. For the function $y = |2x|$, the graph shows a horizontal compression of the parent graph by a factor of $\frac{1}{2}$. The x-coordinates are multiplied by a factor of $\frac{1}{2}$.

EXAMPLE: Horizontal Compression

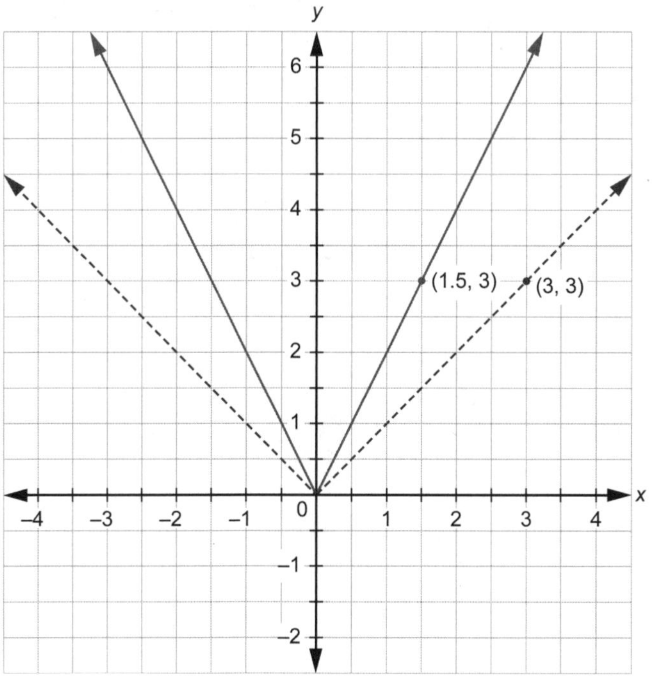

Multiple Transformations

When graphing multiple transformations, it is important to consider the sequence. You can often determine algebraically whether the sequence is important.

Consider the following:

a. Shift the graph of $f(x) = x^2$ five units up and then reflect over the x-axis.

b. Reflect the graph of $f(x) = x^2$ over the x-axis and then shift it 5 units up.

a) and b) are *not* the same because a) results in $f(x) = -(x^2 + 5)$ while b) results in $f(x) = -x^2 + 5$.

My Notes

Summary *(continued)*

To graph a), first shift the graph, and then reflect over the x-axis.

EXAMPLE: Shift and then Reflect

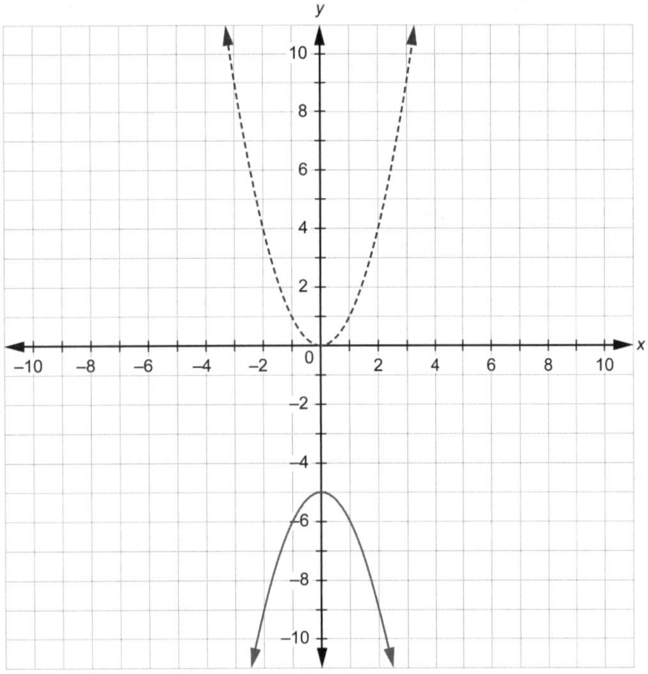

To graph b), first reflect the graph over the x-axis, and then shift it 5 units up.

EXAMPLE: Reflect and then Shift

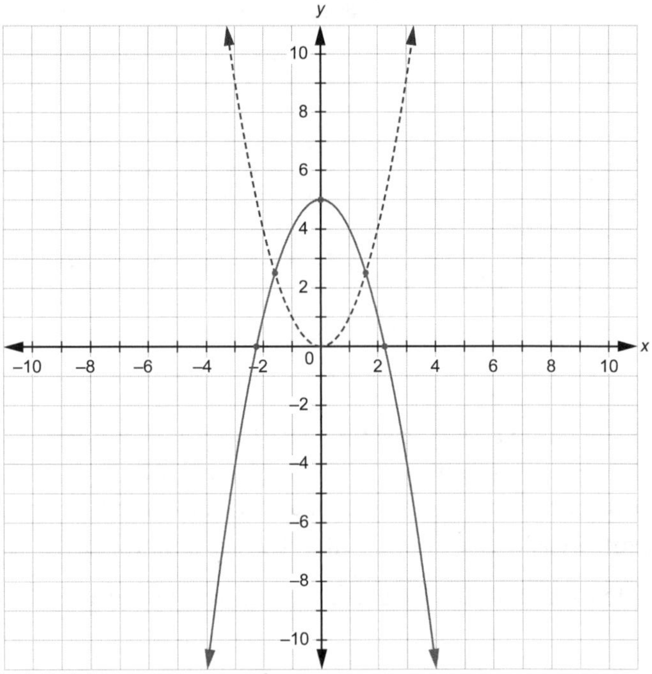

My Notes

Practice Exercises

Review what you've learned using these practice problems. For practice problems with feedback, try the Coach and Play items in the Practice section online.

1. Which graph in the diagram is the graph of $f(x) = \frac{1}{2}\sqrt{x}$?

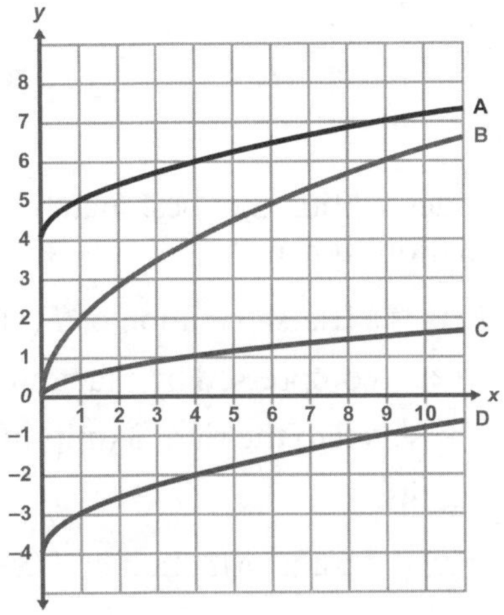

A. graph A **B.** graph B

C. graph C **D.** graph D

2. There are two functions, $f(x) = \frac{1}{2}(x - 4)^2$ and $g(x) = 3(x + 1)^2$.

Which transformation of the graph of f will coincide with the graph of g?

A. $g(x) = \frac{1}{6}f(x - 5)$ **B.** $g(x) = 6f(x + 5)$

C. $g(x) = 6f(x - 5)$ **D.** $g(x) = \frac{1}{6}f(x + 5)$

3. In which of the following ways can the line $f(x) = 3x + 6$ be represented as the transformation of the parent function $y = x$? *Select all that apply.*

A. a shift to the left of 2 units followed by a vertical stretch by a factor of 3

B. a vertical stretch by a factor of 3 followed by a shift up of 6 units

C. a horizontal compression by a factor of 3 followed by a shift upward of 6 units

D. a vertical stretch by a factor of 3 followed by a shift to the left of 2 units

E. a shift left of 6 units followed by a horizontal compression by a factor of 3

4. Draw a line that connects the parent function $f(x) = \sqrt{x}$ to its transformation on the matching graph.

$f(x) = \sqrt{x}$ $f(x) = \sqrt{-2x}$ $f(x) = \sqrt{-\frac{1}{2}x}$

$f(x) = \sqrt{2x}$ $f(x) = -\sqrt{2x}$

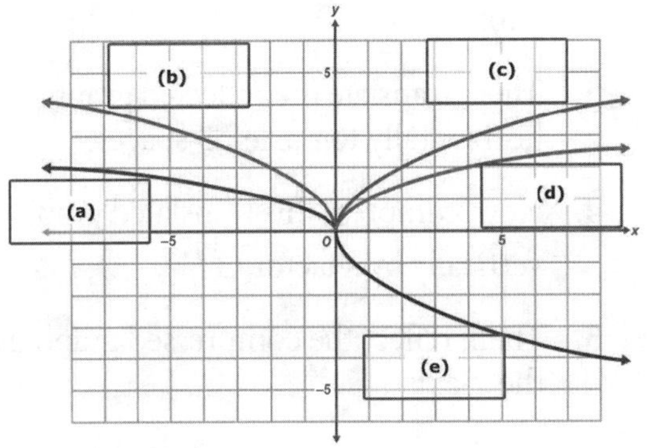

5. Jamaal's class is playing a transformation game. Each student has a shape on a coordinate plane and tracks a particular point on that shape. The teacher calls out transformations to perform.

When they start, the teacher calls out, "Reflect your shape across the *y*-axis," then "shift it 3 units left and 2 units up," and then "rotate 90° clockwise about the origin." The point that Jamaal is tracking ends up at (5, 2) in the new image after the four transformations.

Assuming that all transformations were done correctly, the original location of Jamaal's point was (______, ______).

6. The graph of $f(x) = x$ has been transformed to the graph of $g(x) = x + 2$.

Which of the transformations have occurred? *Select all that apply.*

A. a translation 2 units upward

B. a translation 2 units downward

C. a translation 2 units to the right

D. a translation 2 units to the left

Practice Exercises *(continued)*

7. On a recent assessment, Kelvin was instructed to describe the function $h(x) = -\frac{1}{4}|-x+3|+7$ as a transformation beginning with the parent function. Kelvin's description was marked as incorrect.

   1. Begin by graphing the parent function $y = |x|$.
   2. Next, reflect the parent graph over the y-axis.
   3. Then, translate the reflected graph horizontally to the left 3 spaces.
   4. Now, compress this translated graph vertically by a factor of $\frac{1}{4}$.
   5. Then, reflect the compressed graph over the x-axis.
   6. Finally, translate this reflected graph vertically up 7 spaces.

   Analyze Kelvin's description. Where is Kelvin's error?

   A. Kelvin's error is in line 4. It should read, "Now, compress the translated graph vertically by a factor of 4."

   B. Kelvin's error is in lines 2 and 5. They should read, "Next, reflect the parent graph over the x-axis. Then, follow this reflection with a reflection over the y-axis." In line 6, change the word "compressed" to "translated."

   C. Kelvin's error is in lines 2 and 3. They should read, "Next, translate the parent graph horizontally to the left 3 spaces. Then, reflect this translated graph over the y-axis." In line 4, change the word "translated" to "reflected."

   D. Kelvin's error is in lines 4 and 5. Line 4 should read, "Now, vertically stretch the reflected graph from line 3 by a factor of $-\frac{1}{4}$." Then line 5 should be deleted.

   E. Kelvin's error is in line 3. It should read, "Then, translate the reflected graph horizontally to the right 3 spaces."

8. In the figure, $P(x) = aT(x-h)+k$.

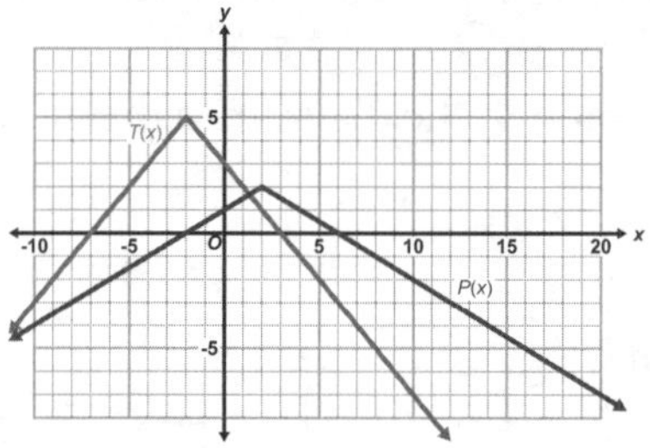

   Select the description that best matches the transformation shown.

   A. $P(x)$ is a transformation of $T(x)$ so that $T(x)$ is compressed by a factor of $\frac{1}{2}$ and translated to the right 3 units and down 4 units.

   B. $P(x)$ is a transformation of $T(x)$ so that $T(x)$ is compressed by a factor of $\frac{1}{2}$ and translated to the right 4 units and down 3 units.

   C. $P(x)$ is a transformation of $T(x)$ so that $T(x)$ is compressed by a factor of 2 and translated to the right 3 units and down 4 units.

   D. $P(x)$ is a transformation of $T(x)$ so that $T(x)$ is compressed by a factor of 2 and translated to the right 4 units and down 3 units.

9. A person is walking along a lake shoreline. He sees himself in the water. What type of transformation is this?

   A. translation **B.** rotation

   C. reflection **D.** glide reflection

10. The vertex of $f(x) = |x|$ is at (0, 0).

    Where is the vertex of $f(x) = 3|x-5|+1$?

    ____

Apply

Can You Launch the Bird through the Hoop?

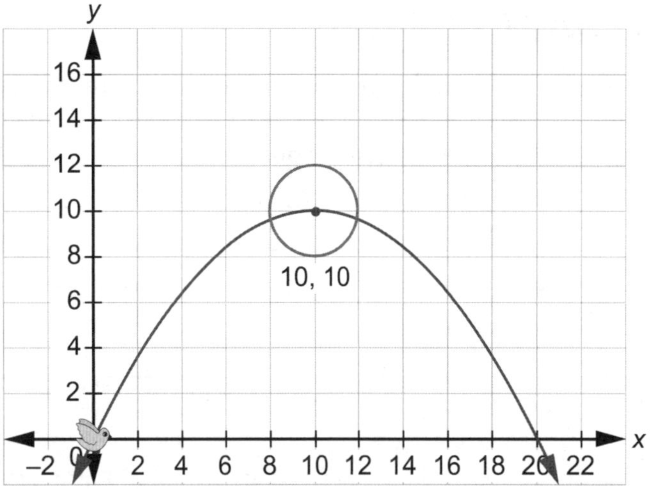

photo: Discovery Education

To play a game, you click a button to release a bird. The path of the bird is based on a quadratic equation, and the goal is to get the bird through the circular hoop of radius 2 units. An example is shown below when the bird is released from (0, 0) and flies through a hoop with center at (10, 10).

If the bird is released from the origin along the curve of the function $f(x) = -110(x - 10)\ 2 + 10$, the bird will go through the hoop and intersect the center of the hoop at (10, 10).

How can you write quadratic equations so the birds will do the following?

1. Go through the center of a hoop centered at (18, 9) with a red bird located at (0, 0).

2. Go through the center of a hoop centered at (40, 30) with a blue bird located at (0, 30).

3. Go through the center of a hoop centered at (35, 35) with a yellow bird located at (5, 20).

Bird through Hoop

In your summary, answer the questions:

How does the equation that guides each bird into the hoop compare to the equation in the original launch from (0, 0) through (10, 10)?

How do the equations change based on the initial position of each bird?

How do the equations change based on the location of the hoop?

Show what you've learned by completing the other performance tasks in the online Apply section.

Apply *(continued)*

Your answer to Apply will be assessed on the following criteria:

1. Writing correct equations
2. Describing the attributes of each equation
3. Comparing each equation to the original equation
4. Comparing each equation to the placement of the bird and hoop

Criteria \ Scale	4 Exceeds Criteria	3 Meets Criteria	2 Progressing to Criteria	1 Below Expectations	0 No Expectation
Correct Equations	Correctly identifies equations that move all three birds through the hoop.	Correctly identifies equations that move two birds through the hoop.	Identifies an equation that moves only one bird through the hoop.	Does not identify an equation that moves any of the three birds through the hoop.	Does not attempt the task.
Attributes of Functions	Correctly identifies transformation type, slope, and intercepts of each function.	Correctly identifies transformation type, slope, and intercepts of each function with no more than two minor errors.	Identifies transformation type, slope, and intercepts of each function with several minor errors or a major error.	Incorrectly identifies transformation type, slope, and intercepts for each function.	Does not attempt the task.
Comparing to Original Function	Compares the intercepts, vertex, and vertical stretch of each function compared to the original function.	Compares the intercepts, vertex, and vertical stretch of two functions compared to the original function.	Compares the intercepts, vertex, and vertical stretch of one function compared to the original function.	Incorrectly compares the intercepts, vertex, and vertical stretch of each function compared to the original function.	Does not attempt the task.
Comparing to Bird and Hoop Location	States that each function passes through the point where the bird starts and the location of the hoop as points on the graph of the function.	States that two functions pass through the point where the bird starts and the location of the hoop as points on the graph of the function.	States that one function passes through the point where the bird starts and the location of the hoop as points on the graph of the function.	Incorrectly states that the functions pass through the point where the bird starts and the location of the hoop as points on the graph of the function.	Does not attempt the task.

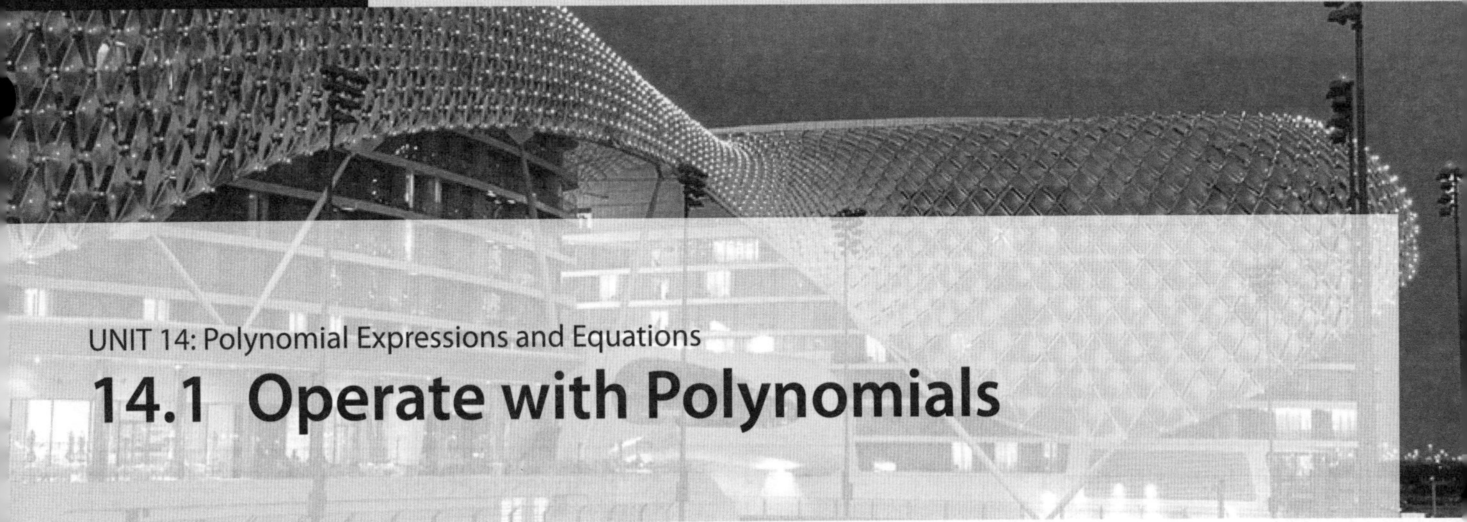

photo: Getty Images

UNIT 14: Polynomial Expressions and Equations

14.1 Operate with Polynomials

Lesson Objectives

- Apply knowledge of operations on polynomials to factor complex polynomial expressions.
- Verify polynomial identities.
- Identify difference of square and difference of cube relationships in higher degree polynomials, and use quadratic factoring techniques to find factors and roots of higher degree $P(x)$.
- Understand the meaning of single, double, and conjugate factors in large degree polynomials and how the factors relate to the roots and graphic behavior of $P(x)$.

Essential Questions

- What meaning can you derive from writing polynomials in different forms?
- What are the similarities and differences among a polynomial's roots, zeros, and solutions?

Investigations

Examining Roots of Polynomials

Rewind. Take another look at quadratic and cubic functions. Find functions with specific solutions.

Model with Polynomials

How do polynomials express themselves? How do their characteristics affect a graph?

Factoring with Quadratic Techniques

Factor and solve complex polynomials to get to the root of real-world problems.

Factoring Cubes

Sums and differences of cubes lead the way to polynomial roots.

Building Pascal's Triangle

What's in a pattern? Pascal's pattern of triangles has polynomial power.

Key Vocabulary

combination, degree of a monomial, degree of a polynomial, permutation, trinomial, zero of a function

Discover

As you complete Engage and the investigations, record the most important ideas you've learned.

Engage	
Investigation 1	**Investigation 2**
Investigation 3	**Investigation 4**

Name ______________________________ Date ____________

Check for Understanding

Operate with Polynomials: Investigation 1

1. Match each polynomial expression on the left with an equivalent expression on the right.

A. $3x(1 + 16x + 2x^2)$

B. $-24 + 48x^2 - 3x + 6x^3$

C. $(4x^3 + 48x^2 + 3x) + (2x^3 + x + 24)$

D. $-4x + 6x^3 - 24 + 48x^2$

E. $3(2x^2 + 1)(x + 8)$

F. $(7x^3 + 48x^2 - 24) - (x^3 - 4x - 24)$

I. $6x^3 + 48x^2 + 4x$

II. $6x^3 + 48x^2 + 3x + 24$

III. $6x^3 + 48x^2 - 4x - 24$

IV. $6x^3 + 48x^2 + 3x$

V. $6x^3 + 48x^2 - 3x - 24$

VI. $6x^3 + 48x^2 + 4x + 24$

2. Which of these polynomial expressions have a degree of 2? *Select all that apply.*

A. $7x^2 + 5x - 1$

B. $3x^3 - 2$

C. $x^4 + 2x^2 + 2$

D. $2x^3 - x + 11$

E. $x^2 + 9x - 10$

F. $2x^4 - 35$

3. Consider the polynomial expression $4x^4 + 40x^2 + 100$. *Fill in the blanks below.*

The polynomial expression has ____________ terms, so it is a ____________.

4. The graph of the polynomial function $p(x) = (x)(x + 1)(x + 1)(x - 3)$ has x-intercepts at which of these points? *Select all that apply.*

A. $(-3, 0)$

B. $(-1, 0)$

C. $(0, 0)$

D. $(1, 0)$

E. $(2, 0)$

F. $(3, 0)$

5. Two cargo containers are being loaded onto a ship. The first container is a cube with an edge length of x feet. The second container has a length that is 5 feet less than that of the first, a width that is 2 feet less, and a height that is 3 feet less. How many cubic feet greater is the volume of the first container than that of the second container? Explain.

Name ______________________________ Date ______________

Check for Understanding

Operate with Polynomials: Investigation 2

1. What is the completely factored form of the polynomial $x^4 - 21x^2 - 100$?

A. $(x - 2)(x + 2)(x^2 + 25)$

B. $(x - 5)(x + 5)(x^2 + 4)$

C. $(x - 2)(x + 2)(x - 5)(x + 5)$

D. $(x^2 + 4)(x^2 + 25)$

2. Which of the following polynomials has a factor of $x + 3$? *Select all that apply.*

A. $x^4 - 13x^2 - 48$

B. $x^4 - x^2 - 72$

C. $x^4 - 28x^2 + 75$

D. $x^4 + 8x^2 - 9$

E. $x^4 + 12x^2 + 27$

F. $x^4 - 19x^2 + 90$

3. For each of the equations shown in the table below, give the number of rational roots, the number of irrational roots, and the number of non-real roots.

Equation	Number of Rational Roots	Number of Irrational Roots	Number of Non-Real Roots
$x^4 - 256 = 0$			
$x^4 + 11x^2 + 18 = 0$			
$x^4 - 10x^2 + 24 = 0$			
$x^4 - 25 = 0$			
$x^4 - 50x^2 + 49 = 0$			
$x^4 - 13x^2 + 30 = 0$			

4. Stocks trade on the New York Stock Exchange from 9:30 a.m. to 4:00 p.m. The price of a certain stock was at or above its opening price all day during a particular trading day. The number of dollars, $d(x)$, the stock was above its opening price during the day can be modeled by the function $d(x) = \frac{1}{12}x^4 - x^3 + 3x^2$, where x represents the number of hours since the opening. Was the stock at its opening price at any time during the day other than the open? If so, what time? Explain how you got your answer.

Name ______________________________ Date ______________

Check for Understanding

Operate with Polynomials: Investigation 3

1. Which of these are factors of the binomial $x^3 - 125$? *Select all that apply.*

A. $x + 5$ **B.** $x^2 + 5x + 25$ **C.** $x - 25$

D. $x^2 - 5x + 25$ **E.** $x + 25$ **F.** $x - 5$

2. Consider the function $f(x) = x^3 - k^3$, where k is an integer. *Fill in each of the blanks in the sentence below with the appropriate number of x-intercepts.*

If k is negative, the graph of $f(x)$ has ____________ x-intercept(s), if k is 0, the graph of $f(x)$ has ____________ x-intercept(s), and if k is positive, the graph of $f(x)$ has ____________ x-intercept(s).

3. Which of these are factors of the binomial $x^3 + 216$? *Select all that apply.*

A. $x + 6$ **B.** $x^2 + 6x + 36$ **C.** $x - 36$

D. $x^2 - 6x + 36$ **E.** $x + 36$ **F.** $x - 6$

4. The graph of which of these functions coincides with the graph of $g(x) = 3x^5 + 81x^2$?

A. $m(x) = (3x^2)(x - 3)(x^2 - 3x + 9)$ **B.** $n(x) = (3x^2)(x - 3)(x^2 + 3x + 9)$

C. $p(x) = (3x^2)(x + 3)(x^2 - 3x + 9)$ **D.** $q(x) = (3x^2)(x + 3)(x^2 + 3x + 9)$

5. Dexter and Lisa both factored the binomial $x^6 - 64$. Dexter thought of it as a difference of squares, first getting $(x^3 - 8)(x^3 + 8)$, and then using both the difference of cubes formula and the sum of cubes formula to get $(x - 2)(x^2 + 2x + 4)(x + 2)(x^2 - 2x + 4)$, or $(x - 2)(x + 2)(x^2 - 2x + 4)(x^2 + 2x + 4)$. Lisa thought of it as a difference of cubes, first using the difference of cubes formula to get $(x^2 - 4)(x^4 + 4x^2 + 16)$, and then factoring the difference of squares to get $(x - 2)(x + 2)(x^4 + 4x^2 + 16)$. Why didn't Dexter and Lisa get the same result? Did one of them make a mistake? Explain your answer.

Name ______________________________ Date ______________

Check for Understanding

Operate with Polynomials: Investigation 4

1. When expanding which of these binomial expressions would you use the same row of Pascal's Triangle as when expanding the expression $(4x + 9)^3$? *Select all that apply.*

A. $(4x + 7)^4$ **B.** $(3x + 9)^2$ **C.** $(2x + 7)^3$

D. $(3x + 5)^3$ **E.** $(2x + 9)^4$ **F.** $(4x + 5)^2$

2. Consider the expanded form of the binomial expression $(x - 6)^5$. *Fill in the blanks.*

The first, ____________, and ____________ terms of the expanded form of the expression are positive, and the ____________, ____________, and ____________ terms are negative.

3. Write each of the binomial expressions given in the table below in expanded form.

Binomial Expression	Expanded Form
$(4x^2 + 3y)^3$	
$(2x - y^3)^4$	
$(3x^4 - 2y)^3$	
$(x^3 + 5y)^4$	
$(5x - 4y^2)^3$	

4. Joe answered a question on a test that asked him for the fourth term of the expanded form of the binomial expression $(x - 2y^2)^6$. He went to the row of Pascal's Triangle with six numbers in it, found the fourth number from the left, which is 10, and multiplied it by the product of x^{6-4} and $(-2y^2)^4$ to get $(10)(x^{6-4})(-2y^2)^4 = (10)(x^2)(16y^8) = 160x^2y^8$. Joe got the question wrong. What mistake(s) did he make? Explain your answer.

Summary

Before you attempt the Practice Exercises, review what you've learned.

Vocabulary

There are many terms related to working with polynomials. Defining and understanding these terms is necessary for you to be able to successfully manipulate polynomials.

> **Zero of a function** is the root of a function, where a function equals the value of zero.

EXAMPLE

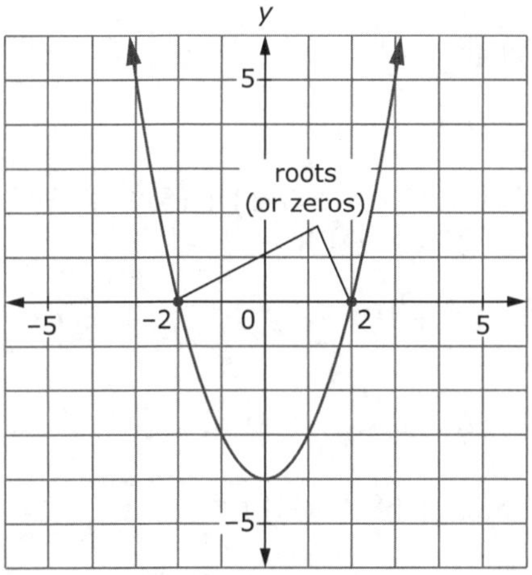

- **Monomial** – A polynomial with one term. Example: $4x^5$ or 3.
- **Polynomial** – An expression that can have constants, variables, and exponents, but has no division by a variable; the variable exponents are integers; it cannot have an infinite number of terms. Example: $3x^2 + 4x + 6$.
- **Degree of a monomial** – The largest exponent of the variable. Example: The degree of $4x^5$ is 5.
- **Degree of a polynomial** – If there is only one exponent, the degree is the largest exponent of that variable. If there is more than one exponent, the degree is the largest sum of the exponents in any term. Example: The degree of $3x^2 + 4x + 6$ is 2. The degree of $4xy^2 + 3x - 5$ is 3. Polynomials are named by their degree. 1 = linear, 2 = quadratic, 3 = cubic, 4 = quartic, and so on.
- **Polynomial function** – A function such as a quadratic, a cubic, a quartic, and so on, involving only non-negative integer powers of x. In general terms, $f(x) = a_nx^n + a_{n-1}x^{n-1} + \ldots + a_2x^2 + a_1x + a_0$.
- For all arguments x, where n is a non-negative integer and $a_0, a_1, a_2, \ldots, a_n$ are constant coefficients.
- **Polynomial equation** – An equation that can be written as a polynomial and set equal to a value: $a_nx^n + a_{n-1}x^{n-1} + \ldots + a_2x^2 + a_1x + a_0 = 0$.
  Example: $x^2 - 4 = 0$.
- **Expanded form of a polynomial** – Writing the polynomial with the highest degree term first. Example: $x^5 - 9x^2 + 2x + 1$.
- **Leading coefficient** – The coefficient of a polynomial's highest degree term. Example: The leading coefficient of $4xy^2 + 3x - 5$ is 4.
- **Root of an equation** – The solution of an equation. The number of roots equals the degree of the polynomial. Example: The degree of $x^2 - 4 = 0$ is 2, so there are 2 roots. The roots for $x^2 - 4 = 0$ are $x = 2$ or $x = -2$.

My Notes

Summary *(continued)*

Expanded Form of a Polynomial

The expanded form of a polynomial has no parentheses, like terms are combined, and the terms are written in descending degrees.

EXAMPLE

This polynomial is in factored form. Multiply the factors, and write the polynomial in expanded form.

$$(x - 1)(x + 2)(x - 5) = 0$$

You can choose either pair of factors to multiply first.

$$(x^2 + x - 2)(x - 5) = 0 \text{ or } (x - 1)(x^2 - 3x - 10) = 0$$

Multiply out, group like terms, and put them in order.

$$x^3 - 4x^2 - 7x + 10 = 0$$

Model with Polynomials

You can use polynomials to model real-world ideas, such as volume and surface area.

EXAMPLE

Find the volume and surface area of this box.

Volume:

$$V = [(12x + 1)(35x^2 + 5x)] \text{ ft}^3$$

$$V = [(420x^3 + 95x^2 + 5x)] \text{ ft}^3$$

Surface Area:

$$A = 2(wl + hl + hw)$$

$$A = [2(7x + 1) \cdot (12x + 1) + (5x) \cdot (12x + 1) + (5x) \cdot (7x + 1)] \text{ ft}^2$$

$$A = [2(84x^2 + 19x + 1) + (60x^2 + 5x) + (35x^2 + 5x)] \text{ ft}^2$$

$$A = [2(179x^2 + 29x + 1)] \text{ ft}^2$$

$$A = [(358x^2 + 58x + 2)] \text{ ft}^2$$

My Notes

EXAMPLE

If $x = 2$, determine what would be the exact surface area of the block. Use the surface area to determine how many quarts of paint you need to paint all exterior sides, if a quart of paint covers 100 ft^2.

$$A = 358(2)^2 + 58(2) + 2 \text{ ft}^2$$

$$A = 1550 \text{ ft}^2$$

$$\frac{1550}{100} = 15.5$$

So you would need 16 quarts.

Summary *(continued)*

Factoring Second-Degree Polynomials

EXAMPLE

Use the difference of squares $a^2 - b^2 = (a + b)(a - b)$ to factor $4y^2 - 36y^6$.

$$4y^2 - 36y^6$$
$$= 4y^2 (1 - 9y^4)$$
$$= 4y^2 (1 + 3y^2)(1 - 3y^2)$$

EXAMPLE

Use the perfect square trinomial $(a + b)2 = a2 + 2ab + b2$ or $(a - b)2 = a2 - 2ab + b2$ to factor $9x2 - 6x + 1$.

$$9x^2 - 6x + 1$$
$$= (3x - 1)(3x - 1)$$
$$= (3x - 1)^2$$

Sum and Difference of Cubes

EXAMPLE

Use the difference of cubes $a^3 - b^3 = (a - b)(a^2 + ab + b^2)$ to factor $x^6 - 64$.

First, note that $a = x^2$ and $b = 4$.

$$x^6 - 64$$
$$= (x^2)^3 - 4^3$$
$$=(x^2 - 4)((x^2)^2 + (4)(x^2) + (x^2))$$
$$=(x + 2)(x - 2)(x^4 + 4x^2 + 16)$$

EXAMPLE

Use the sum of cubes $a^3 + b^3 = (a + b)(a^2 - ab + b^2)$ to factor $27x^3 + 1$.

First, note that $a = 3x$ and $b = 1$.

$$27x^3 + 1$$
$$= (3x)^3 + 1^3$$
$$= (3x + 1)((3x)^2 - (3x)(1) + 1^2$$
$$= (3x + 1)(9x^2 - 3x + 1)$$

Number and Types of Roots

EXAMPLE

Find the roots of $x^3 - 3x^2 - 2x + 6$.

First, set the polynomial equal to zero, and factor.

$$x^3 - 3x^2 - 2x + 6 = 0$$
$$(x - 3)(x^2 - 2) = 0$$

Set each factor equal to zero, and solve for x.

$$(x - 3) = 0$$
$$(x^2 - 2) = 0$$

So, $x^3 - 3x^2 - 2x + 6$ has one rational root and two irrational roots.

EXAMPLE

Find the roots of $x^5 - x^4 + x^3 - x^2 - 12x + 12$.

$$x^5 - x^4 + x^3 - x^2 - 12x + 12 = 0$$
$$(x - 1)(x^4 + x^2 - 12) = 0$$
$$(x - 1)(x^2 - 3)(x^2 + 4) = 0$$
$$(x - 1)(x + \sqrt{3})(x - \sqrt{3})(x^2 + 4) = 0$$
$$(x - 1)(x + \sqrt{3})(x - \sqrt{3})(x + 2i)(x - 2i) = 0$$

So, $x^5 - x^4 + x^3 - x^2 - 12x + 12 = 0$ has one real, two irrational, and two complex roots.

My Notes

Summary *(continued)*

EXAMPLE

Find the roots of $4x^3 - 12x^2 + 16$.

$$4x^3 - 12x^2 + 16 = 0$$
$$(x + 1)(4x^2 - 16x + 16) = 0$$
$$(x + 1)(2x - 4)^2 = 0$$
$$(x + 1) = 0, (2x - 4) = 0, (2x - 4) = 0$$
$$x = -1, x = 2, x = 2$$

Since two of the roots are the same, we say that there are three roots but the polynomial has a duplicity at 2. Look at the graph to see what happens at a repeated root.

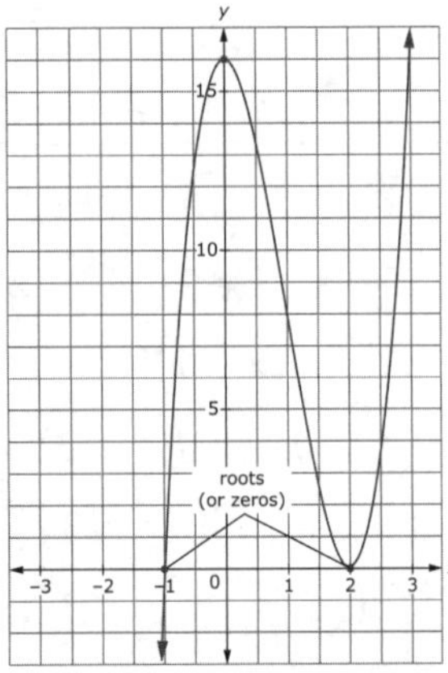

The graph is tangent to the x-axis at the repeated root.

My Notes

The Binomial Theorem

The binomial theorem is a way to expand a binomial expression raised to a power without having to multiply out all of the binomials. The binomial theorem is illustrated in Pascal's triangle.

Pascal's triangle is made so that each number is the sum of the two diagonal numbers above it. Each row begins and ends with 1. The triangle is symmetric and has many patterns buried in it. One pattern shows the coefficients for an expanded binomial term.

1	$(x+1)^0$
1 1	$(x+1)^1$
1 2 1	$(x+1)^2$
1 3 3 1	$(x+1)^3$
1 4 6 4 1	$(x+1)^4$
1 5 10 10 5 1	$(x+1)^5$
1 6 15 20 15 6 1	$(x+1)^6$
1 7 21 35 35 21 7 1	$(x+1)^7$

EXAMPLE

Using Pascal's triangle, expand $(x + 1)^4$.

The coefficients for row $(x + 1)^4$ are 1, 4, 6, 4, and 1. Start with a variable of degree equal to the row, and decrease the power as you move to the right.

$$1x^4 + 4x^3 + 6x^2 + 4x + 1$$

EXAMPLE

Using Pascal's triangle, expand $(3x + 2)^5$.

The coefficients for row $(x + 1)^5$ are 1, 5, 10, 10, 5, 1. Since the coefficient of x is 3 and not 1, raise $(3x)$ to the power equal to the row, and decrease the power as you move to the right. Since the constant is 2 instead of 1, start at the right with 2^5 and decrease the power as you move to the left. The sum of the exponents in each term equals the degree of the row.

$$(3x + 2)^5 =$$
$$(1)(3x)^5(2)^0 + (5)(3x)^4(2)^1 + (10)(3x)^3(2)^2 +$$
$$(10)(3x)^2(2)^3 + (5)(3x)^1(2)^4 + (1)(3x)^0(2)^5 =$$
$$243x^5 + 810x^4 + 1080x^3 + 720x^2 + 240x + 32$$

Practice Exercises

Review what you've learned using these practice problems. For practice problems with feedback, try the Coach and Play items in the Practice section online

1. Rewrite the expression. *Combine all like terms. Your final expression should not contain any parentheses.*

$3(x^2 + 2) + (x^3 - 3x^2 - 4) - x(x^2 - 3x) =$

____

2. Factor the expression completely.

$44x^5 - 40x^4 - 4x^3$

A. $4x^3\,(11x + 1)(x - 1)$

B. $4x^3\,(11x - 1)(x - 1)$

C. $4x^3\,(11x - 1)(x + 1)$

D. $(-4x^3\,)(11x + 1)(x + 1)$

3. Factor the expression completely.

$(x^4 - 81)$

A. $(x^2 - 9)(x^2 - 9)$

B. $(x - 3)(x + 3)(x - 3)(x + 3)$

C. $(x - 3)(x + 3)(x^2 + 9)$

D. $(x - 3)(x + 3)(x + 9)(x - 9)$

E. $(x^2 + 9)(x^2 - 9)$

4. Factor the expression completely.

$x^6 - 133x^3 + 1{,}000$

A. $(x^2 + 8)(x^4 - 125)$

B. $(x + 2)(x^2 - 2x + 4)(x + 5)(x^2 - 5x + 25)$

C. $(x - 2)(x^2 + 2x + 4)(x - 5)(x^2 + 5x + 25)$

D. $(x - 2)(x^2 - 2x - 4)(x - 5)(x^2 - 5x - 25)$

5. Factor the expression completely.

$6x^5 + 27x^4 + 21x^3$

A. $(3x^3)(2x^2 + 5x + 7)$

B. $(3x^2)(2x + 7)(x + 1)$

C. $(3x^3 + 3x)(2x^2 + 7x^2)$

D. $(3x^3)(2x^2 + 7)$

6. Expand the binomial.

$(2x + 3y)^4$

A. $8x^4 + 72x^3y + 144x^2y^3 + 72xy^3 + 12y^3$

B. $16x^4 + 81y^3$

C. $16x^4 + 528x^9y^6 + 81y^3$

D. $16x^4 + 96x^3y + 216x^2y^2 + 216xy^3 + 81y^4$

7. Use the binomial theorem to simplify.

$(2x - 3)^4$

A. $2x^4 - 6x^3 + 24x^2 - 24x + 81$

B. $4x^4 + 8x^3 + 16x^2 + 12x + 9$

C. $16x^4 + 32x^3 + 96x^2 + 216x + 81$

D. $16x^4 - 96x^3 + 216x^2 - 216x + 81$

8. Determine the missing factors of the polynomial expression.

$x^6 + 35x^3 + 216$

(____________)$(x^2 - 3x + 9)$

(____________)$(x^2 - 2x + 4)$

Practice Exercises (*continued*)

9. Apply your understanding of numeric and algebraic operations to determine whether each equation is true or false:

A. $4t^2 + 36 = (2t + 6)^2$

B. $(15x - 3x)^2 = 12^2\,x^2$

C. $(2n - 5)^3 = (2n - 5)(4n^2 - 20n + 25)$

D. $-5^2 = -25$

E. $\left(-\frac{3z^4}{5y}\right)^2 = \frac{9z^8}{5y^2}$

F. $\frac{6k^3 - 9m}{-2} = -3k^3 + 9m$

True: ____________

False: ____________

10. Supply the missing term in the binomial expansion.

$(x + y)^6 = x^6 + 6x^5y + 15x^4y^2 +$ ____________ $+ 15x^2y^4 + 6xy^5 + y^6$

11. Determine the missing factor for the expression $30x^4 + 65x^3 + 30x^2$.

$(5x^2)(2x + 3)$ (____________)

12. Expand the expression.

$(6t - 3n^2)(6t + 3n^2)$

Combine all like terms. Your final answer should not contain any parentheses.

____

Apply

Have You Seen the Lizards?

Biologists are tracking a species of lizard on a Japanese island. Their population grew over the first six years from the initial observation of 144 lizards. At some point during the sixth year, after reaching a maximum population of more than 1500 lizards, the population took a turn and began to decline. Watch the video to learn more.

What is the best way to monitor, model, and predict changes in animal populations?

Where Did All the Lizards Go?: View this video segment about lizard populations.

The biologists modeled the lizard population with the following function:
$f(x) = 0.1x^5 - 1.9x^4 + 2.6x^3 + 68.2x^2 + 4.8x + 144$

Write the function in factored form, and find the zeros. Discuss the number and types of zeros, including the real-world meaning of each.

Then, make a graph and a table of values far enough out to predict the future of the species. State the practical domain and range, and describe what will happen to the lizard population according to this model.

Show what you've learned by completing the other performance tasks in the online Apply section.

Apply *(continued)*

Your answer to Apply will be assessed on the following criteria:

1. Writing the function in factored form and identifying the zeros
2. Graphing the function and making a table of values
3. Stating the domain and range of the function
4. Predicting what will happen to the population based on the findings

Scale / Criteria	4	3	2	1	0
	Exceeds Criteria	**Meets Criteria**	**Progressing to Criteria**	**Below Expectations**	**No Expectation**
Factored Form	Correctly writes the function in factored form and identifies the zeros; classifies and interprets them properly.	Writes the function in factored form and identifies the zeros with a minor error; classifies and interprets them properly.	Writes the function in factored form and identifies the zeros with a minor error; classifies and interprets them improperly.	Writes the function in a factored form that is not equivalent and identifies zeros incorrectly; does not classify or interpret them.	Does not attempt task.
Graph and Table	Correctly graphs the function and makes an adequate table of values for a prediction.	Graphs the function and makes an adequate table of values for a prediction with minor errors.	Graphs the function and makes an adequate table of values for a prediction but has significant errors.	Graphs the function incorrectly and makes an inadequate table of values for a prediction.	Does not attempt task.
Domain and Range	States the domain and range appropriately.	States the domain and range of the function with minor errors.	Correctly states either the domain or range of the function.	Incorrectly states both the domain and range.	Does not attempt task.
Prediction	Predicts what will happen to the population based on the findings and clearly explains the reasoning.	Predicts what will happen to the population based on the findings, but the explanation is unclear.	Predicts what will happen to the population based on incorrect reasoning.	Incorrectly predicts what will happen to the population and the prediction is not based on the findings.	Does not attempt task.

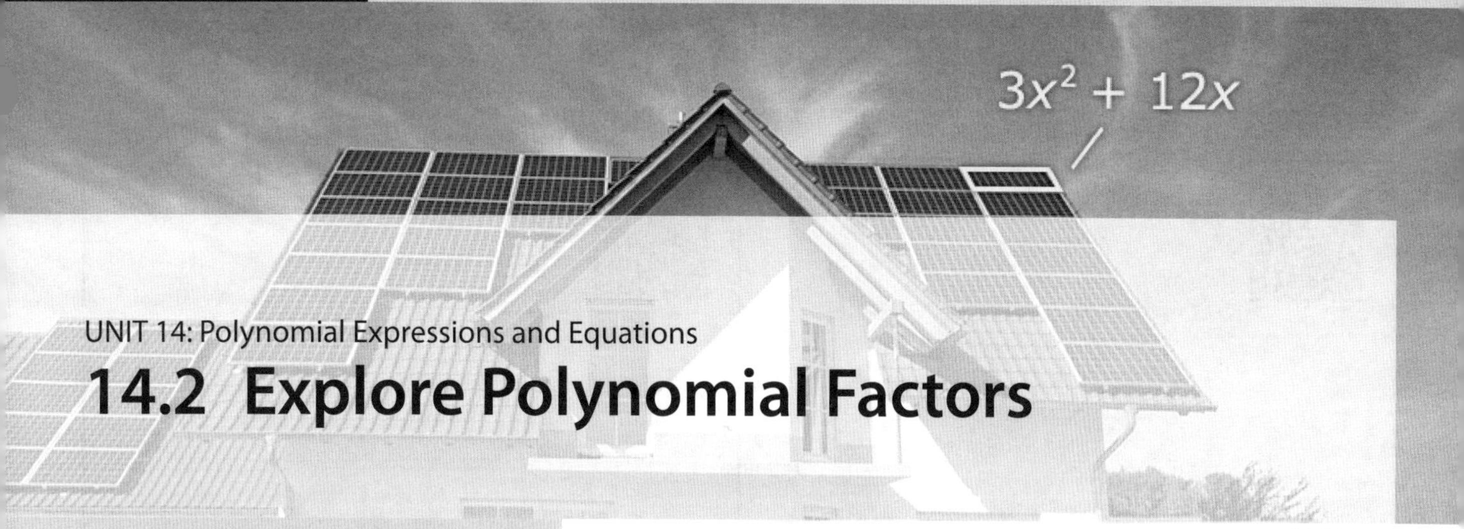

UNIT 14: Polynomial Expressions and Equations

14.2 Explore Polynomial Factors

photo: Getty Images

Lesson Objectives

- Find all zeros of a polynomial by factoring and interpret their meaning as solutions to polynomial equations.
- Use the rational root theorem to find the possible rational roots of polynomials.
- Translate between expanded and factored forms of polynomials to simplify expressions and garner insight.

Essential Question

- How can you find the roots of a polynomial equation, and what do these roots mean in the context of a function and its graph?

photo: Getty Images

Investigations

Quadratic Long Division

Dividing polynomials may look complex. It's just elementary division taken to the algebraic level.

Polynomial Long Division

Find factors of polynomials using division. Is it a factor if there is a remainder?

Using Long Division to Find Roots

Dig for the roots of a polynomial with division.

Synthetic Division

Take the synthetic shortcut to long division. How can factors help you find x-intercepts?

Exploring the Rational Root Theorem

Can rational roots be relatives of leading coefficients and constants of a function?

Finding Roots and Factors

Roots can be rational, irrational, and complex. Find them all!

Key Vocabulary

remainder theorem

Discover

As you complete Engage and the investigations, record the most important ideas you've learned.

Engage	**Investigation 1**
Investigation 2	**Investigation 3**
Investigation 4	**Investigation 5**

Name ______________________________ Date ______________

Check for Understanding

Explore Polynomial Factors: Investigation 1

1. Which of the following division problems give a remainder? *Select all that apply.*

A. $4x - 3\overline{)8x^2 + 22x - 21}$ B. $3x + 5\overline{)6x^2 - 17x - 34}$ C. $5x - 1\overline{)10x^2 + 43x - 9}$

D. $5x + 2\overline{)15x^2 - 19x - 10}$ E. $4x - 7\overline{)12x^2 - 13x - 12}$ F. $3x + 8\overline{)9x^2 + 21x - 8}$

2. Relate the factoring of a polynomial to polynomial division. *Then fill in the blanks in the sentence shown below.*

When there is no remainder in polynomial division, the ____________ and the ____________ are factors of the ____________.

3. Match each of the division problems on the left with the correct quotient on the right.

A. $x + 7\overline{)2x^2 + 5x - 60}$ I. $2x - 5 + \frac{2}{x+7}$

B. $x + 7\overline{)2x^2 + 9x - 33}$ II. $2x - 11 + \frac{9}{x+7}$

C. $x + 7\overline{)2x^2 + 11x - 13}$ III. $2x - 13 + \frac{7}{x+7}$

D. $x + 7\overline{)2x^2 + 3x - 68}$ IV. $2x - 7 + \frac{4}{x+7}$

E. $x + 7\overline{)2x^2 + 7x - 45}$ V. $2x - 9 + \frac{3}{x+7}$

F. $x + 7\overline{)2x^2 + x - 84}$ VI. $2x - 3 + \frac{8}{x+7}$

4. Determine the answer to the division problem $x^4 + 1\overline{)x^8 - 2x^5 - 2x - 1}$. *Then circle the correct words in the sentence shown below.*

The answer to the division problem [does / does not] include a remainder, so the binomial $x^4 + 1$ [is / is not] a factor of the polynomial $x^8 - 2x^5 - 2x - 1$.

5. Madeline said she divided a polynomial by a polynomial and got a negative remainder. Is this possible, or did she make a mistake? Give evidence to support your answer.

Name ______________________________ Date ______________

Check for Understanding

Explore Polynomial Factors: Investigation 2

1. One of the factors of the polynomial $3x^3 + 10x^2 - 51x - 18$ is $x - 3$. Which of these are other factors of the polynomial? *Select all that apply.*

   A. $x - 6$ **B.** $3x + 1$ **C.** $x + 3$

   D. $3x + 2$ **E.** $x + 6$ **F.** $3x - 1$

2. Each of the equations below has a root of $x = 5$. Which also has two irrational roots?

   A. $2x^3 - 5x^2 - 37x + 60 = 0$ **B.** $3x^3 - x^2 - 75x + 25 = 0$

   C. $2x^3 - 9x^2 - 14x + 45 = 0$ **D.** $3x^3 - 5x^2 - 58x + 40 = 0$

3. The graph of $p(x) = x^4 - 3x^3 - 3x^2 + 12x - 4$ is shown. Use the graph and polynomial long division to help fill in the blanks in the sentence that follows with the exact values of the roots of the equation $x^4 - 3x^3 - 3x^2 + 12x - 4 = 0$ from least to greatest.

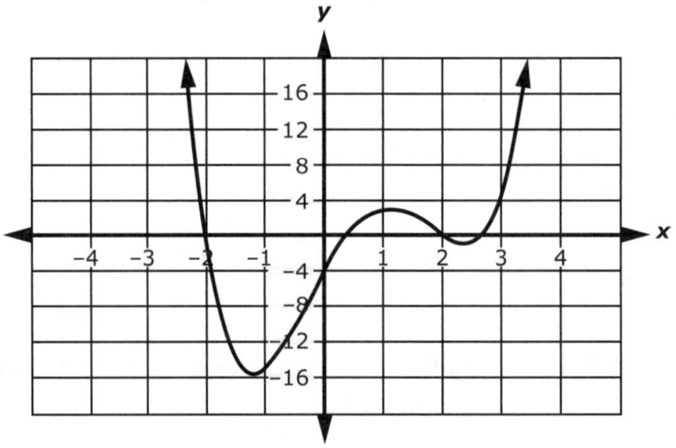

The equation's roots are $x =$ ______, $x =$ ______, $x =$ ______, and $x =$ ______.

4. Suppose you know the polynomial equation $4x^4 + 11x^3 + 13x^2 + 44x - 12 = 0$ has two non-real, complex roots of $x = \pm 2i$. Explain how you can use polynomial long division to help find the equation's other roots. What are the other roots of the equation?

Name ______________________________ Date ______________

Check for Understanding

Explore Polynomial Factors: Investigation 3

1. Which of the following synthetic division problems show that the binomial $x + 1$ is a factor of the trinomial into which it is being divided? *Select all that apply.*

A. $$\begin{array}{r|rrr} 1 & 1 & -2 & -3 \\ & & 1 & -1 \\ \hline & 1 & -1 & -4 \end{array}$$

B. $$\begin{array}{r|rrr} -1 & 1 & -6 & -7 \\ & & -1 & 7 \\ \hline & 1 & -7 & 0 \end{array}$$

C. $$\begin{array}{r|rrr} 1 & 1 & 4 & -5 \\ & & 1 & 5 \\ \hline & 1 & 5 & 0 \end{array}$$

D. $$\begin{array}{r|rrr} 1 & 1 & 7 & -8 \\ & & 1 & 8 \\ \hline & 1 & 8 & 0 \end{array}$$

E. $$\begin{array}{r|rrr} -1 & 1 & 3 & -4 \\ & & -1 & -2 \\ \hline & 1 & 2 & -6 \end{array}$$

F. $$\begin{array}{r|rrr} -1 & 1 & -5 & -6 \\ & & -1 & 6 \\ \hline & 1 & -6 & 0 \end{array}$$

2. Tamara used synthetic division to divide the polynomial $P(x)$ by a binomial as shown.

$$\begin{array}{r|rrrr} 3 & 1 & -4 & -17 & 60 \\ & & 3 & -3 & 60 \\ \hline & 1 & -1 & -20 & 0 \end{array}$$

If she graphed $P(x)$ on a coordinate plane, which point must the graph pass through?

A. (–3, 0) **B.** (0, –3) **C.** (0, 3) **D.** (3, 0)

3. Examine the synthetic division problem below, and then fill in the blanks in the sentence that follows. When a variable is necessary in an answer, use x.

$$\begin{array}{r|rrrr} 4 & 1 & 0 & 0 & -64 \\ & & 4 & 16 & 64 \\ \hline & 1 & 4 & 16 & 0 \end{array}$$

The synthetic division problem shows that the resulting depressed polynomial, or quotient, when ______ is divided by ______ is ______, so this validates the formula for the factoring of the ______ of ______.

4. Is it possible to use synthetic division to show that the binomial $x^2 - 9$ is a factor of the polynomial $x^3 + 6x^2 - 9x - 54$? If so, how would you do it? Explain your reasoning.

Name ______________________________ Date ______________

Check for Understanding

Explore Polynomial Factors: Investigation 4

1. Consider the equation $2x^2 + 11x - 21 = 0$. *Fill in the blank in the sentence below.*

   The number of possible rational roots given by the rational root theorem is ______.

2. Which of these are possible rational roots of the equation $4x^2 + 15x - 54 = 0$ according to the rational root theorem? *Select all that apply.*

   A. $x = \frac{4}{9}$ **B.** $x = -\frac{1}{4}$ **C.** $x = \frac{27}{2}$ **D.** $x = -\frac{2}{3}$

   E. $x = -\frac{1}{6}$ **F.** $x = \frac{9}{2}$ **G.** $x = -\frac{4}{3}$ **H.** $x = \frac{1}{9}$

3. Darryl used the rational roots theorem to analyze the equation $8x^2 - 18x - 35 = 0$. First, he defined the factors of the constant as set P and the factors of the leading coefficient as set Q. Then, he listed the numbers in each set as shown in the table.

Set P	±1	±5	±7	±35
Set Q	±1	±2	±4	±8

   Next, he wrote the equation's possible rational roots as follows: $\pm1, \pm5, \pm7, \pm35, \pm\frac{1}{2}, \pm\frac{5}{2}, \pm\frac{7}{2}, \pm\frac{35}{2}, \pm\frac{1}{4}, \pm\frac{5}{4}, \pm\frac{7}{4}, \pm\frac{35}{4}, \pm\frac{1}{8}, \pm\frac{5}{8}, \pm\frac{7}{8}, \pm\frac{35}{8}$. Finally, he factored the trinomial, set both factors equal to 0, and found the roots to be $x = -\frac{7}{2}$ and $x = \frac{5}{4}$. Darryl's teacher said that some of the values he gave were wrong. Which were they?

   A. the numbers in set P

   B. the numbers in set Q

   C. the equation's possible rational roots

   D. the equation's roots

4. According to the rational root theorem, what are the possible rational roots of the equation $6x^4 - 13x^2 - 15 = 0$? How many of these are roots of the equation? Explain.

Name ______________________________ Date ______________

Check for Understanding

Explore Polynomial Factors: Investigation 5

1. Which are roots of the equation $2x^3 - 3x^2 - 17x - 12 = 0$? *Select all that apply.*

   A. $x = -4$ B. $x = \frac{3}{2}$ C. $x = 1$ D. $x = -\frac{2}{3}$

   E. $x = -\frac{3}{2}$ F. $x = 4$ G. $x = \frac{2}{3}$ H. $x = -1$

2. What is the fully factored form of the equation $6x^3 + 25x^2 - 17x - 36 = 0$?

   A. $(x - 1)(2x - 9)(3x - 4) = 0$ B. $(x - 1)(2x + 9)(3x + 4) = 0$

   C. $(x + 1)(2x - 9)(3x + 4) = 0$ D. $(x + 1)(2x + 9)(3x - 4) = 0$

3. Consider the equation $4x^5 + 7x^3 - 2x = 0$. *Fill in the blanks with the equation's roots.*

   Rational roots: $x =$ ______ and $x = \pm$______ Complex, non-real roots: $x = \pm$______

4. Edwin was trying to find the roots of the quartic equation $8x^4 + 48x^3 + 73x^2 + 6x + 9 = 0$. First, he listed all the possible rational roots as follows:

   $x = \pm 1, \pm\frac{1}{2}, \pm\frac{1}{4}, \pm\frac{1}{8}, \pm 3, \pm\frac{3}{2}, \pm\frac{3}{4}, \pm\frac{3}{8}, \pm 9, \pm\frac{9}{2}, \pm\frac{9}{4}, \pm\frac{9}{8}$. Then, he tested each of the possible rational roots, identifying only $x = -3$ as a rational root by using synthetic division as shown.

$$\begin{array}{r|rrrrr} -3 & 8 & 48 & 73 & 6 & 9 \\ & & -24 & -72 & -3 & -9 \\ \hline & 8 & 24 & 1 & 3 & 0 \end{array}$$

   Since Edwin already tested each of the possible rational roots, he concluded that the equation has no more rational roots and that the fully factored form of the equation is $(x + 3)(8x^3 + 24x^2 + x + 3) = 0$. However, his teacher said that this is not the fully factored form. What did Edwin do wrong? Explain, giving the fully factored form in your answer.

Summary

Before you attempt the Practice Exercises, review what you've learned.

Dividing Polynomials

You can divide polynomials by polynomials. If there is no remainder, then the divisor and the quotient are factors of the dividend. You can use polynomial division to factor polynomials and to identify whether a polynomial is a factor. Two ways to divide polynomials are **long division** and **synthetic division**.

Long Division

EXAMPLE

Divide $x^2 - 9x - 10$ by $x + 1$.

SOLUTION:

Set up the division problem the same way you would for numeric division.

$$\begin{array}{r} x - 10 \\ x + 1 \overline{)\ x^2 - 9x - 10} \\ -(x^2 + x) \\ \hline -10x - 10 \\ -(-10x - 10) \\ \hline 0 \end{array}$$

Since there is no remainder, $x^2 - 9x - 10 = (x + 1)(x - 10)$.

My Notes

EXAMPLE

Divide $3x^3 - 5x^2 + 10x - 3$ by $3x + 1$.

SOLUTION:

$$\begin{array}{r} x^2 - 2x + 4 \\ 3x + 1 \overline{)3x^3 - 5x^2 + 10x - 3} \\ -(3x^3 + 1x^2) \\ \hline -6x^2 + 10x - 3 \\ -(6x^2 - 2x) \\ \hline 12x - 3 \\ -(12x + 4) \\ \hline -7 \end{array}$$

Since there is a remainder, $3x + 1$ is not a factor of $3x^3 - 5x^2 + 10x - 3$. The remainder is written as a fraction, so the quotient is $x^2 - 2x + 4 + \frac{-7}{3x + 1}$.

EXAMPLE

Divide $2x^3 - 9x^2 + 15$ by $2x - 5$.

SOLUTION:

When one degree is not represented within a polynomial, an easy way to handle it is to put a coefficient of 0 in front of the term and include it when dividing. The term with the 0 coefficient acts as a placeholder. In this case, the x term is not represented.

$$\begin{array}{r} x^2 - 2x - 5 \\ 2x - 5 \overline{)2x^3 - 9x^2 + 0x + 15} \\ -(2x^3 - 5x^2) \\ \hline -4x^2 + 0x + 15 \\ -(-4x^2 + 10x) \\ \hline -10x + 15 \\ -(-10x + 25) \\ \hline -10 \end{array}$$

Since there is a remainder, $2x - 5$ is not a factor of $2x^3 - 9x^2 + 15$. The quotient is $x^2 - 2x + 5 + \frac{-10}{2x - 5}$.

Summary *(continued)*

Synthetic Division

Another type of polynomial division is called synthetic division. It is essentially a shortcut way of doing long division. However, synthetic division can only be performed with a linear factor.

EXAMPLE

Divide $x^2 + 5x + 6$ by $x + 3$.

First, determine that $x + 3$ gives you a root of -3.

Then, set up the division with just the coefficients:

$$\begin{array}{r|rrr} -3 & 1 & 5 & 6 \\ & & -3 & -6 \\ \hline & 1 & 2 & 0 \end{array}$$

Bring the first number down.

Then, multiply the result by the divisor (-3) and put it above the line.

Add the second column and bring down the sum, 2.

Multiply the 2 by the divisor (-3) and put it below the 6.

Add the third column. The 0 shows that there is no remainder.

The numbers below the horizontal bar are the coefficients of the quotient. So, the quotient is $x + 2$.

Since there is no remainder, -3 is a root and $x^2 + 5x + 6 = (x + 3)(x + 2)$.

Rational Root Theorem

The rational root theorem is a generalized method of how to find the rational roots of a polynomial. Using the factors of the leading coefficient and the constant term, you can determine the possible rational roots of an equation, which you can test to determine the actual rational roots. The theorem will only identify possible rational roots, not irrational or complex roots.

The rational root theorem states that all rational roots of a polynomial are within the set of all factors of the constant term divided by the factors of the leading coefficient. If all the factors of the constant term are possible values of p and all factors of the leading coefficient are possible values of q, then all $\frac{p}{q}$ are possible roots of the polynomial. All rational roots of $P(x)$ are all values such that $P\left(\frac{p}{q}\right) = 0$.

EXAMPLE

Find all the rational roots of $P(x) = 2x^4 + x^3 - 19x^2 - 9x + 9$.

Factors of the constant term: $\pm1, \pm3, \pm9$

Factors of leading coefficient: $\pm1, \pm2$

Possible values of $\frac{p}{q}$: $\pm1, \pm\frac{1}{2}, \pm3, \pm\frac{3}{2}, \pm9, \pm\frac{9}{2}$.

You can use long division or synthetic division to check possible roots to determine which is an actual rational root. Once you have found a root, continue dividing or using other factoring methods until the polynomial is completely factored.

My Notes

Summary *(continued)*

The rational roots are $-3, -1, \frac{1}{2}$, and 3. Since the polynomial has a degree of 4, you know that there will be exactly 4 roots. Since there are 4 rational roots, there cannot be any irrational or complex roots.

So, the factored form of the polynomial is $P(x) = (x + 1)(x + 3)(x - 3)(2x - 1)$.

EXAMPLE

Find all the roots of $2x^5 + 3x^4 - 30x^3 - 57x^2 - 2x + 24 = 0$.

Use the rational root theorem to determine the possible rational roots:

$$\pm\frac{1}{2}, \pm1, \pm\frac{3}{2}, \pm2, \pm3, \pm4, \pm6, \pm8, \pm12, \pm24$$

Use polynomial division and factoring to factor the expression. The following factors result in no remainder:

$$(x + 1), (2x + 3), \text{ and } (x - 4)$$

Based on the factors, the actual rational roots are:

$$x = -1, -\frac{3}{2}, 4$$

This polynomial has a degree of 5, so there are 5 roots. Since 3 are rational, the other 2 must be irrational or complex.

To find the irrational and complex roots, use long or synthetic division to factor out all of the rational factors you found one at a time, and then factor the remainder.

$$(2x^5 + 3x^4 - 30x^3 - 57x^2 - 2x + 24) \div (x + 1) = 2x^4 + x^3 - 31x^2 - 26x + 24$$

$$(2x^4 + x^3 - 31x^2 - 26x + 24) \div (x - 4) = 2x^3 + 9x^2 + 5x - 6$$

$$(2x^3 + 9x^2 + 5x - 6) \div (2x + 3) = 2x^2 + 6x - 4$$

Factoring out $(x + 1)$, $(2x + 3)$, and $(x - 4)$ results in $2x^2 + 6x - 4$.

Use the quadratic formula to solve for x to find the remaining roots:

$$x = \frac{-3 \pm \sqrt{17}}{2}$$

The roots of $2x^5 + 3x^4 - 30x^3 - 57x^2 - 2x + 24$ are $x = -1, \frac{-3 \pm \sqrt{17}}{2}, -\frac{3}{2}, 4$.

My Notes

Practice Exercises

Review what you've learned using these practice problems. For practice problems with feedback, try the Coach and Play items in the Practice section online.

1. Which of the following statements describe the graph of $y = -x(x + 5)(x - 1)^2$? *Select all that apply.*

   A. The graph has a y-intercept of 0.

   B. The graph has an x-intercept of 5.

   C. The graph has a double root on the x-axis at $x = 1$.

   D. The graph has 3 x-intercepts.

   E. The graph has a zero at $x = -1$.

2. The polynomial $x^4 + 2x^3 - 8x^2 - 18x - 9$ has a double root.

   Identify the double root and the other roots of the polynomial.

   A. −9 **B.** −3

   C. −1 **D.** 1

   E. 3 **F.** 9

   Double root: ______

   Other roots: ______

3. The remainder when the polynomial $R(x) = 5x^4 + 18x^3 + kx + 15$ is divided by $(x + 3)$ is −21. Use this information to determine the value of the coefficient k.

   A. $k = -309$ **B.** $k = -111$

   C. $k = -39$ **D.** $k = -21$

   E. $k = -15$ **F.** $k = -3$

   G. $k = 3$ **H.** $k = 15$

   I. $k = 21$ **J.** $k = 39$

   K. $k = 111$ **L.** $k = 309$

4. Determine the standard form by expanding the function $f(x) = (2x - 3)(x + 4)(x - 3)$.

   Your final answer should not have any parentheses.

   $f(x) =$ ____________

5. A polynomial equation is defined as $3x^5 + a_1x^3 + a_2x - 6 = 0$, where a_1 and a_2 are integers. Identify the set of all the possible integer roots and rational roots.

   A. ± 1 **B.** ± 2

   C. ± 3 **D.** ± 4

   E. ± 6 **F.** $\pm\frac{1}{2}$

   G. $\pm\frac{1}{3}$ **H.** $\pm\frac{1}{6}$

   I. $\pm\frac{2}{3}$ **J.** $\pm\frac{3}{2}$

   Integer roots: ____________

   Other rational roots: ____________

6. The roots of a polynomial equation $ax^2 - b = 0$ are defined as $\pm\sqrt{\frac{b}{a}}$, where a and b are nonzero integers.

   Which is true about these roots according to the rational root theorem?

   A. The roots must be rational because a and b are nonzero integers.

   B. The roots must be rational because a and b are rational numbers.

   C. The roots are rational if and only if a and b are both perfect squares.

   D. The roots cannot be rational because all square roots are irrational numbers.

Practice Exercises *(continued)*

7. Rewrite the polynomial $p(x) = 2x^4 + 4x^3 - 8x^2 - 20x - 10$ as a product of its factors, then interpret the graphic behavior.

   Fill in the blanks to complete a factorization of p(x).

   $p(x) =$ ______ $(x^2 +$ ______ $)(x +$ ______ $)^2$

   Determine the coordinates for points A, B, and C, the x-intercepts of $p(x)$.

   If necessary, round coordinates to the hundredths place.

   A(______, ______)

   B(______, ______)

   C(______, ______)

8. The binomial $(x + 1)$ is one of the factors of polynomial $T(x) = x^4 + 5x^3 - 20x - 16$.

   Use division strategies and quadratic factoring techniques to write $T(x)$ as a product of its factors.

   $T(x) =$ ____________

   A. $(x - 1)$ **B.** $(x + 1)$
   C. $(x - 2)$ **D.** $(x + 2)$
   E. $(x - 4)$ **F.** $(x + 4)$
   G. $(x - 8)$ **H.** $(x + 8)$
   I. $(x - 16)$ **J.** $(x + 16)$

9. Apply the rational root theorem and quadratic factoring techniques to write $P(x) = 8x^3 + 30x^2 - 11x - 12$ as a product of its factors.

   Choose the factors from the options below.

   $P(x) =$ ____________

   A. $(x - 1)$ **B.** $(x + 1)$
   C. $(x - 2)$ **D.** $(x + 2)$
   E. $(x - 3)$ **F.** $(x + 3)$
   G. $(x - 4)$ **H.** $(x + 4)$
   I. $(2x - 1)$ **J.** $(2x + 1)$
   K. $(4x - 1)$ **L.** $(4x + 1)$
   M. $(8x - 1)$ **N.** $(8x + 1)$
   O. $(2x - 3)$ **P.** $(2x + 3)$
   Q. $(4x - 3)$ **R.** $(4x + 3)$
   S. $(8x - 3)$ **T.** $(8x + 3)$

10. The binomial $(x + 2)$ is one of the factors of polynomial $K(x) = x^4 + 5x^3 + 2x^2 - 20x - 24$. Use division strategies and quadratic factoring techniques to write $K(x)$ as a product of its factors.

    $K(x) =$ ____________

Apply

Will You Go with the Flow?

Have you ever followed a list of instructions to complete a task? If so, you have followed an algorithm.

In mathematics, computer programing, and in many tasks in life, we complete tasks by following step-by-step procedures, or algorithms. Watch the video to learn about an important algorithm related to your own studies.

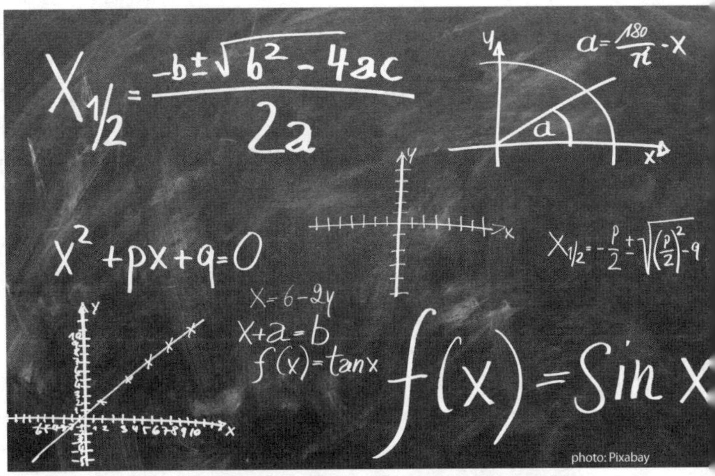

Italian mathematician Paolo Ruffini developed a method of dividing polynomials.

Paolo and the Polynomials: View this video segment about Italian mathematician Paolo Ruffini.

One way that computer programmers plan out an algorithm is by creating a flowchart.

Research how to create a flowchart and plan an algorithm for Paolo Ruffini's synthetic division. Be sure to include special cases in your flowchart. Then, test and revise your algorithm as needed by dividing the following expressions:

A. $(x^2 - 3x + 2) \div (x - 1)$

B. $(x^3 - 27) \div (x - 3)$

C. $(x^4 - 3x + 2x^3 - 6) \div (x - 2)$

D. $(4x^3 - 8x^2 - x + 5) \div (2x - 1)$

Show what you've learned by completing the other performance tasks in the online Apply section.

Apply *(continued)*

Your answer to Apply will be assessed on the following criteria:

1. Creating a flowchart (by hand or with technology) for the steps of Ruffini's synthetic division, using standard notation for flowcharts
2. Describing and analyzing the steps of the flowchart
3. Executing the algorithm for four division problems
4. Justifying your work and reasoning

Criteria \ Scale	4 Exceeds Criteria	3 Meets Criteria	2 Progressing to Criteria	1 Below Expectations	0 No Expectation
Flowchart Creation	Correctly and thoroughly lists Ruffini's synthetic division steps in flowchart form.	Lists Ruffini's synthetic division steps in flowchart form with minor errors in flowchart.	Lists Ruffini's synthetic division steps in flowchart form with minor errors in flowchart and synthetic division steps.	Lists Ruffini's synthetic division steps in flowchart form with major errors in flowchart and division steps.	Does not create flowchart.
Analysis of Flowchart	Provides an accurate analysis of each step in the flowchart.	Provides analysis of most steps in the flowchart.	Provides analysis of most steps in the flowchart with minor errors.	Provides analysis of some steps in the flowchart with major errors.	Does not describe flowchart.
Run the Algorithm	Correctly runs all four of the division problems through the algorithm to determine the quotient.	Correctly runs three of the four division problems through the algorithm to determine the quotient.	Correctly runs two of the four division problems through the algorithm to determine the quotient.	Correctly runs one of the four division problems through the algorithm to determine the quotient.	Does not run algorithm.
Reasoning	Completely and accurately justifies work for the flowchart, its analysis, and the four division problems.	Completely justifies work for the flowchart, its analysis, and the four division problems with minor errors.	Completely justifies work for two of the three: flowchart, its analysis, and the division problems.	Completely justifies work for one of the three: flowchart, its analysis, and the division problems.	Does not give reasoning.

UNIT 14: Polynomial Expressions and Equations

14.3 Analyze Polynomial Functions

photo: Getty Images

Lesson Objectives

- Identify key features of polynomial functions.
- Build and interpret graphs by finding points, using technology, or changing parameters.
- Determine and interpret the rate of change of a polynomial function over a specified interval.
- Model real-world situations and solve problems using polynomial functions.

Essential Questions

- How can you determine the features of a polynomial function based on the equation or graph, and how can you explain their meanings in a real-world context?
- How can you interpret the rates of change of a polynomial function both algebraically and in a real-world context?

Investigations

Use a Polynomial to Build a Fence

Write and analyze a fence function. Use the key features to help you understand the situation.

Roller Coasters

Ride the hills and valleys of a polynomial. Investigate key graphic features for character indications.

Write the Equation of a Polynomial

Tailor polynomial equations to specific characteristics.

Graph That Box

Maximize box volume within given constraints. It's a function of height.

Examine and Graph Polynomials

Identify the features of complex polynomials without graphing. You have the tools.

Discover

As you complete Engage and the investigations, record the most important ideas you've learned.

Engage	
Investigation 1	**Investigation 2**
Investigation 3	**Investigation 4**

Name ______________________________ Date ____________

Check for Understanding

Analyze Polynomial Functions: Investigation 1

1. The average rate of change of which of these functions is positive in the interval [0.5, 1.5]? *Select all that apply.*

   A. $f(x) = 2x^3 - 8x^2 - 3x + 15$

   B. $g(x) = -x^4 + x^3 - 4x + 10$

   C. $h(x) = x^4 + x^3 + 2x^2 + x - 6$

   D. $k(x) = x^3 + 12x^2 - x + 5$

   E. $m(x) = -x^3 + 9x^2 + 5x + 1$

   F. $n(x) = 2x^4 - 7x^2 - 11x - 13$

2. The graph of a polynomial function is shown. *Circle the correct words in the sentence.*

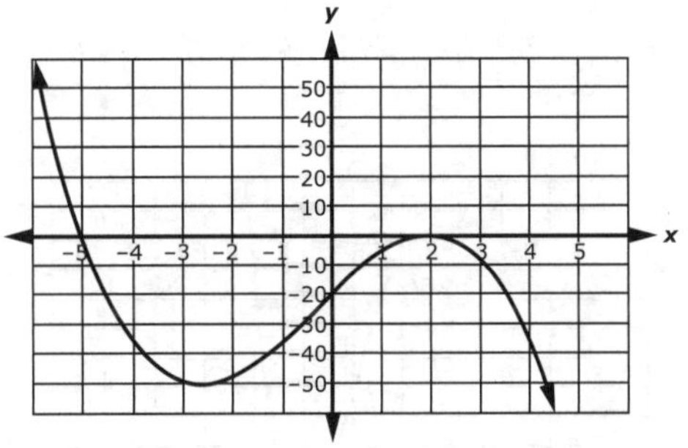

   The function's degree is [even / odd], and its leading coefficient is [negative / positive].

3. Consider the function $p(x) = 3x^4 - x^3 - 5x^2 - 9x - 7$. *Fill in the blanks below.*

   In the function's graph, as $x \to -\infty$, $p(x) \to$ ________, and as $x \to \infty$, $p(x) \to$ ________.

4. Theo analyzed the function $q(x) = x^3 - 4x^2 - 2x + 25$ without graphing. First, he found the average rate of change within the interval [1.5, 2.5] to be −5.75. Then, he found the average rate of change within the interval [2.5, 3.5] to be 1.25. He noticed that within the first interval, the average rate of change is negative, and within the second interval, it is positive. He concluded that since the intervals are joined by 2.5, the function must have a minimum at this value of x. Is Theo's reasoning correct? Explain your answer.

Name ______________________________ Date ______________

Check for Understanding

Analyze Polynomial Functions: Investigation 2

1. Which functions have graphs that pass through the x-axis? *Select all that apply.*

A. $y = (x + 8)^{15}$ **B.** $y = (x + 5)^{12}$ **C.** $y = (x + 9)^{18}$

D. $y = (x + 4)^{16}$ **E.** $y = (x + 3)^{11}$ **F.** $y = (x + 6)^{19}$

2. A polynomial function has a zero of $x = -1$ with a multiplicity of 3, a zero of $x = 6$ with a multiplicity of 1, and a leading coefficient of 2. What is the function's equation in expanded form?

A. $y = 2x^4 - 3x^3 - 15x^2 - 17x - 6$

B. $y = 2x^4 - 3x^3 + 15x^2 - 17x - 6$

C. $y = 2x^4 - 6x^3 - 30x^2 - 34x - 12$

D. $y = 2x^4 - 6x^3 + 30x^2 - 34x - 12$

3. The graph of a cubic function is shown. *Fill in the blanks in the sentence that follows.*

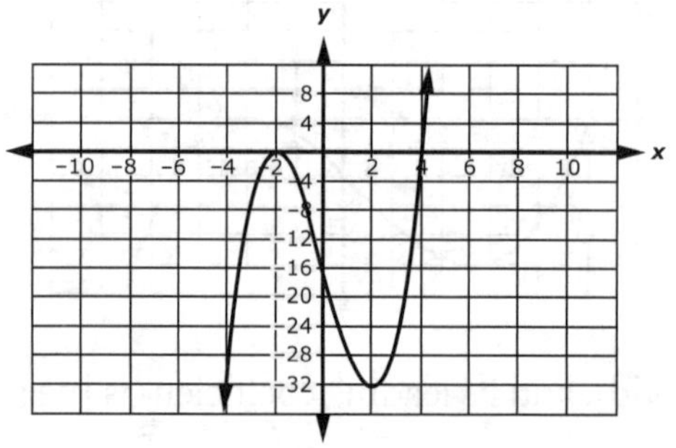

The function has a zero of $x =$ ______ with a multiplicity of 1, a zero of $x =$ ______ with a multiplicity of 2, and a y-intercept at the point (______, ______) so its equation in expanded form must be $y =$ ____________________.

4. Barbara was asked to write the equation of a quartic function in expanded form that has a leading coefficient of 1, zeros of $x = -5$ and $x = 3$, and a graph that has a positive local maximum. What equation(s) could she write? Explain how you got your answer.

Name ______________________________ Date ______________

Check for Understanding

Analyze Polynomial Functions: Investigation 3

1. Suppose an open-top box is created from a 24-unit by 16-unit rectangular piece of grid paper by cutting equal-sized squares with side lengths of x units out of the corners and folding. If the box's volume is defined as a function of x, what is a reasonable domain?

   A. (0, 8) **B.** (0, 12)

   C. (0, 16) **D.** (0, 24)

2. By cutting equal-sized squares out of the corners of a piece of rectangular construction paper with a length of 40 centimeters and a width of 25 centimeters and folding, an open-top box will be created. Which statements are correct? *Select all that apply.*

   A. The box can have a height of 10 cm.

   B. The box can have a height of 15 cm.

   C. The box's maximum volume is 2,250 cm^3.

   D. The box's maximum volume occurs when its height is 5 cm.

   E. The box's maximum volume occurs when its width is 20 cm.

   F. The box's maximum volume occurs when its length is 35 cm.

3. Liam will create an open-top box from a rectangular piece of cardboard by cutting equal-sized squares out of the corners and folding. He defined x as the side length of each of the squares in centimeters and y as the volume of the box in cubic centimeters. Then he wrote a function and graphed it as shown, forgetting to restrict the function's domain.

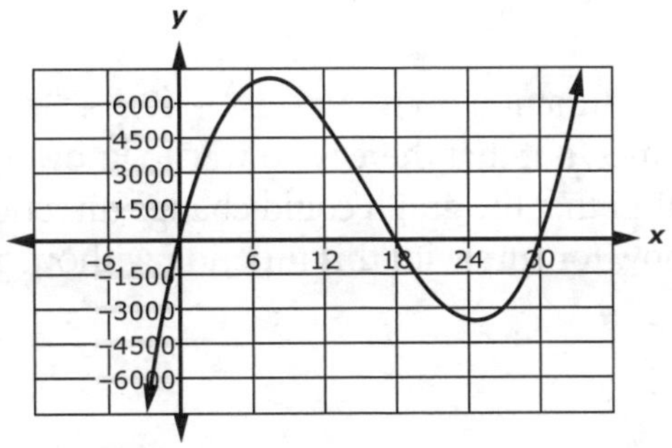

Based on the graph, what are the dimensions of the piece of cardboard? Explain.

Name ____________________ Date ____________

Check for Understanding

Analyze Polynomial Functions: Investigation 4

1. Which statement is correct about the graph of the function $f(x) = x^4 - 11x^2 - 18x - 8$?

   A. It has an x-intercept of –8.
   B. It has a y-intercept of –8.
   C. It has an x-intercept of 8.
   D. It has a y-intercept of 8.

2. For the graphs of each of the functions listed in the table, give the end behavior.

Function	End Behavior as $x \to -\infty$	End Behavior as $x \to +\infty$
$g(x) = x^5 + x^4 - 9x - 9$	$g(x) \to$	$g(x) \to$
$h(x) = -x^4 - 2x^3 + 3x^2 + 8x + 4$	$h(x) \to$	$h(x) \to$
$k(x) = x^6 - 3x^5 + x - 5$	$k(x) \to$	$k(x) \to$
$m(x) = -x^5 + 5x^4 - 2x + 10$	$m(x) \to$	$m(x) \to$
$n(x) = x^4 + 9x^3 + 14x^2 - x - 7$	$n(x) \to$	$n(x) \to$
$p(x) = -x^6 + 6x^5 - x^2 + 8x - 12$	$p(x) \to$	$p(x) \to$

3. When the function $q(x) = x^4 - 13x^3 + 21x^2 + 85x + 50$ is written in factored form, which of these is one of the factors? *Select all that apply.*

   A. $x + 2$
   B. $x - 1$
   C. $x + 10$
   D. $x - 5$
   E. $x - 50$
   F. $x + 5$
   G. $x - 25$
   H. $x + 1$
   I. $x + 25$
   J. $x - 10$
   K. $x + 50$
   L. $x - 2$

4. Quinn claims that the graph of the function $r(x) = x^4 - 2x^3 - 19x^2 + 32x + 48$ could possibly change direction a maximum of three times, but that there is no way to know for sure without graphing the function. Norah agrees with Quinn that the graph could change direction a maximum of three times, but she says that there is a way to know for sure with this function without graphing. Who is right? Explain your answer.

Summary

Before you attempt the Practice Exercises, review what you've learned.

Polynomial Behavior

How do polynomials behave? Is it possible to look at a polynomial function and make predictions about what its graph will look?

Points on the graph can describe the **maxima** (or peaks) and the **minima** (or troughs). The maximum and minimum values are boundaries of when the function changes from increasing to decreasing, and vice versa.

EXAMPLE

For $y = 2x^3 - 4x^2 - 22x + 24$ the maximum is in the interval $-4 < x < 0$, and the minimum is in the interval $0 < x < 4$.

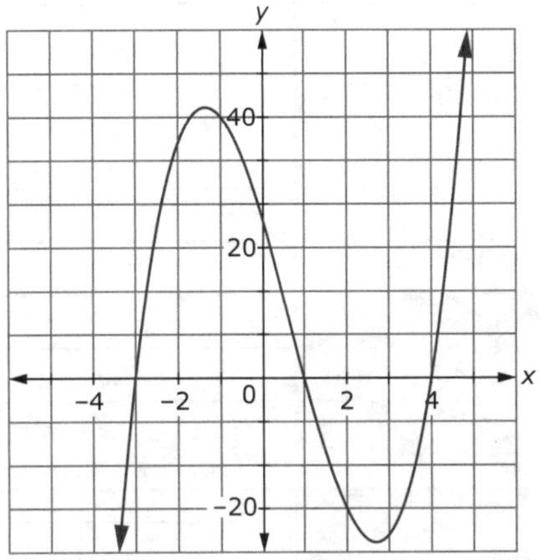

The leading coefficient of a polynomial and the degree of the polynomial describe the **end behavior.**

My Notes

Summary *(continued)*

This is summarized in the following table.

Equation	$y = x^2$
Degree	Even
Leading Coefficient	Positive
End Behavior	$f(x) \to +\infty$, as $x \to -\infty$ $f(x) \to +\infty$, as $x \to +\infty$
Example	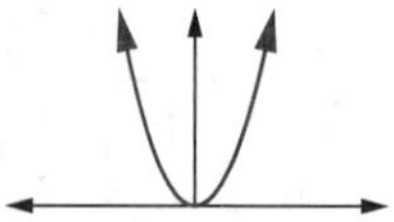

Equation	$y = -x^2$
Degree	Even
Leading Coefficient	Negative
End Behavior	$f(x) \to -\infty$, as $x \to -\infty$ $f(x) \to -\infty$, as $x \to +\infty$
Example	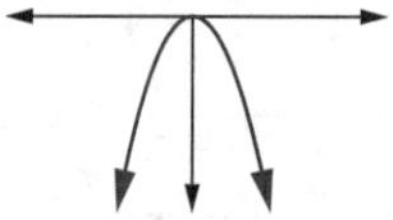

Equation	$y = x^3$
Degree	Odd
Leading Coefficient	Positive
End Behavior	$f(x) \to -\infty$, as $x \to -\infty$ $f(x) \to +\infty$, as $x \to +\infty$
Example	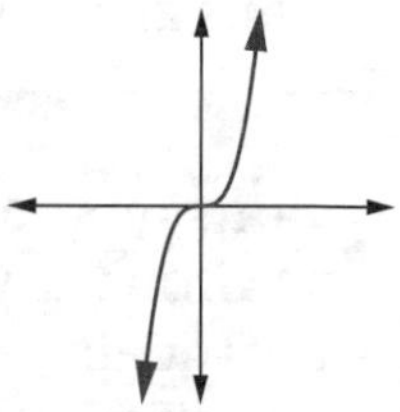

Equation	$y = -x^3$
Degree	Odd
Leading Coefficient	Negative
End Behavior	$f(x) \to +\infty$, as $x \to -\infty$ $f(x) \to -\infty$, as $x \to +\infty$
Example	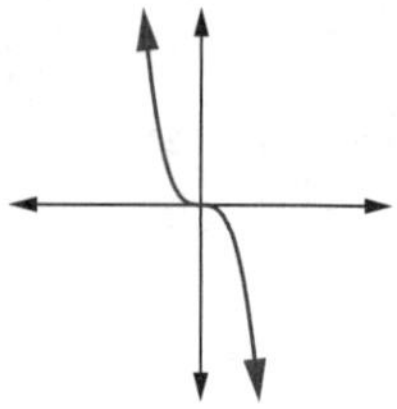

My Notes

EXAMPLE

For $y = 2x^3 - 4x^2 - 22x + 24$, you see that the polynomial is of degree 3 and has a positive leading coefficient. This tells us that as x decreases, y decreases. In function notation, this is written as $f(x) \to -\infty$, as $x \to -\infty$. As x increases, y increases, or $f(x) \to +\infty$, as $x \to +\infty$. We now know that when we graph the function, the curve will continue up and to the right and down and to the left.

You can determine the **roots** of a polynomial.

> These **roots** give specific information about where the graph crosses the axes and where significant turning points are occurring.

These values can be used to help graph the function.

Practice Exercises

Review what you've learned using these practice problems. For practice problems with feedback, try the Coach and Play items in the Practice section online.

1. The traffic at an intersection peaks two times during the course of a day.

   Which of the following statements best explains the best polynomial function to model the number of cars passing through the intersection at a given point in time?

   A. A linear function will best model the traffic because the total traffic that passes through the intersection will increase over time.

   B. A polynomial function of degree 2 will best model the traffic because the traffic data will show two maximums.

   C. A polynomial function of degree 3 will best model the traffic because the data will have 2 intervals of increasing traffic and 1 interval of decreasing traffic.

   D. A polynomial function of degree 4 will best model the traffic because the graph can have two maximums.

2. The standard form of a quartic function is $f(x) = ax^4 + bx^3 + cx^2 + dx + e$.

   If $a > 0$ and $b < c < d < e < 0$, what will be true of the graph?

   A. As $x \to -\infty$, the function will approach $-\infty$ because most of the coefficients are negative.

   B. Because $a > 0$, the graph will open upward. The function will approach $+\infty$ on both the left and the right.

   C. As $x \to +\infty$, the function will approach $-\infty$ because the constant term of the function is negative.

   D. Because $e < 0$, the function will open downward, and the function will approach $-\infty$ on both the left and the right.

3. Analyze the end behavior for each polynomial function shown below. Consider, as $x \to \pm\infty$, what each function is tending to.

   For each polynomial below, choose the two answer choices that best describe what each function is tending to.

   A. $f(x) \to -\infty$ as $x \to -\infty$

   B. $f(x) \to +\infty$ as $x \to -\infty$

   C. $f(x) \to -\infty$ as $x \to +\infty$

   D. $f(x) \to +\infty$ as $x \to +\infty$

   $f(x) = -x^5 + 2x^4 + 4x^3 + 5$: ____________

   $f(x) = x^5 - 2x^4 - 4x^3 - 5$: ____________

   $f(x) = 2x^4 + 4x^3 + 5$: ____________

4. Match each of these polynomial functions with the category that describes its end behaviors.

   A. As $x \to -\infty$, $p(x) \to -\infty$
   As $x \to +\infty$, $p(x) \to -\infty$

   B. As $x \to -\infty$, $p(x) \to +\infty$
   As $x \to +\infty$, $p(x) \to +\infty$

   C. As $x \to -\infty$, $p(x) \to +\infty$
   As $x \to +\infty$, $p(x) \to -\infty$

   D. As $x \to -\infty$, $p(x) \to -\infty$
   As $x \to +\infty$, $p(x) \to +\infty$

   $p(x) = -4x^4 + 4x^2 - 3$ ____________

   $p(x) = 3x^2 - 7$ ____________

   $p(x) = -3x^3 + 2x^2 - x - 1$ ____________

   $p(x) = 2x^5 + 8$ ____________

Practice Exercises *(continued)*

5. The table shows Miranda's distance from home after she leaves to pick up her brother from practice.

Time (min)	0	2	4	6	8	10	12	14	16
Distance (miles)	0	1.2	3.0	3.8	4.1	3.7	2.8	1.0	0

Which of the following polynomials best models Miranda's distance in miles as a function of x, the time in minutes?

Use a graphing calculator and the best fit curve feature to help determine the polynomial that models this data.

A. $-0.1x^2 - x - 0.2$ **B.** $-0.1x^2 + x - 0.2$

C. $-x^2 + x - 2$ **D.** $0.1x^2 + 2x - 0.4$

6. A polynomial function, $k(x)$, is degree 4 with four real zeros. One of the zeros has a multiplicity of 2.

Which of the following statements are always true about the graph of $k(x)$? *Select all that apply.*

A. The graph could have the shape of a parabola.

B. The graph could have the general shape of an M or a W.

C. The graph is tangent to the x-axis.

D. The graph has four x-intercepts.

E. The graph passes through the origin.

7. The table represents the gas mileage of a car at different speeds.

Speed (mph)	40	45	50	55	60	65	70	75	80
Gas Mileage (mpg)	25	26	27	28	27	26	23	22	20

Use a graphing calculator to determine which of the following functions is the best estimate of the gas mileage of a car traveling at a speed of v miles per hour.

A. $M = -0.01v^2 + 1.22v - 5.65$

B. $M = 0.01v^2 - 1.22v + 5.65$

C. $M = -0.14v^2 + 33.29$

D. $M = 0.14v^2 - 33.29$

8. The graph shows a quartic function.

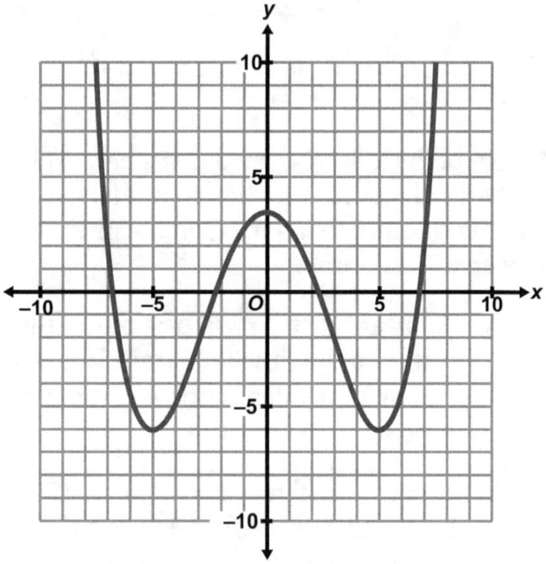

Classify each interval as having negative, zero, or positive average rates of change.

A. $[-2, 2]$ **B.** $[-7, -2]$

C. $[-4, -1]$ **D.** $[1, 4]$

E. $[3, 5]$ **F.** $[5, 7]$

Negative: ____________

Zero: ____________

Positive: ____________

9. Consider a polynomial $f(x) = ax^n + bx^2 + c$, where a, b, c, and n represent constants. Determine values for each constant so that:

- $(1, 0)$ is an x-intercept of $f(x)$.
- $(0, 2)$ is a y-intercept of $f(x)$.
- As $x \to -\infty, f(x) \to +\infty$ and as $x \to +\infty$, $f(x) \to -\infty$.

$a =$ ____________

$b =$ ____________

$c =$ ____________

$n =$ ____________

Apply

How Did the Function Cross the Bridge?

photo: Discovery Education

In the world of architecture, polynomial functions are everywhere you look! Find the dimensions of the Sheikh Zayed Bridge, pictured below.

Use this information to determine a polynomial or a piecewise polynomial function that could be used to describe the bridge's curved support structure. Then, use the Graphing Calculator to graph the function or functions that define the Sheikh Zayed Bridge.

Graphing Calculator: Use the Graphing Calculator in the Apply problem to graph the function or functions that define the Sheikh Zayed Bridge.

Answer

How many pieces of functions could be used to define it? What are they? What are the domain and range of each one?

Show what you've learned by completing the other performance tasks in the online Apply section.

Apply *(continued)*

Your answer to Apply will be assessed on the following criteria:

1. Collecting and graphing data that can be used to determine a function that models the bridge
2. Creating and graphing the function, including restrictions on the domain
3. Asking a question that can be answered by your function
4. Providing a solution for your question, and an explanation

Scale / Criteria	4	3	2	1	0
	Exceeds Criteria	Meets Criteria	Progressing to Criteria	Below Expectations	No Expectation
Collecting and Graphing Data	Collects and graphs enough relevant and reliable data to accurately model the bridge.	Collects and graphs almost enough relevant and reliable data to model the bridge.	Collects and graphs some relevant and reliable data to model the bridge.	Collects and graphs inadequate data to model the bridge.	Does not attempt task.
Creating a Function	Correctly identifies and graphs the function, including restrictions on the domain.	Identifies and graphs the function, including restrictions on the domain, with minor errors.	Identifies and graphs the function, without including restrictions on the domain, or with significant errors.	Gives a graph and function, but they do not come close to modeling the bridge.	Does not attempt task.
Posing a Question	Poses a clear question that can be answered by the function.	Poses a question that can be answered by the function but is vague or unclear.	Poses a question that has to do with the bridge but cannot be answered completely by the function.	Poses an irrelevant question.	Does not attempt task.
Providing a Solution	Provides a correct solution for the question and explains it thoroughly.	Provides a solution that contains minor errors and gives a partial explanation.	Provides a solution that contains significant errors and does not explain.	Provides an irrelevant solution with no explanation.	Does not attempt task.

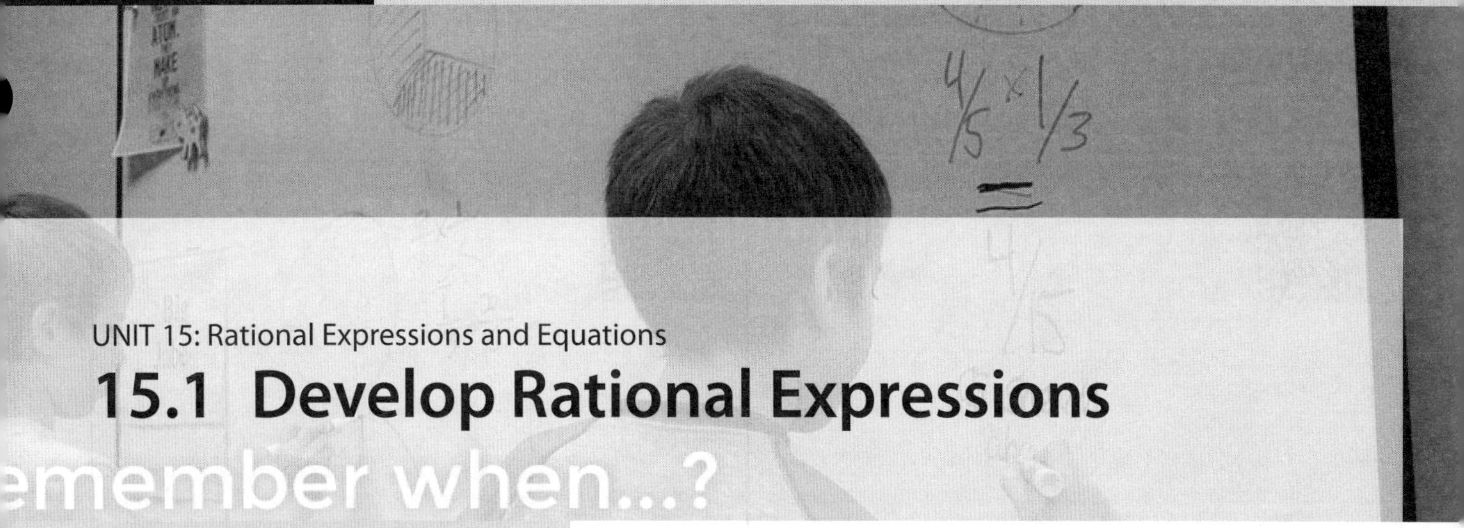

UNIT 15: Rational Expressions and Equations

15.1 Develop Rational Expressions

photo: Discovery Education

Lesson Objectives

- Apply the division algorithm to express rational expressions as a quotient and remainder.
- Perform arithmetic operations on rational expressions using rules for rational number arithmetic.
- Understand that rational expressions can be combined arithmetically and form a group under addition, subtraction, multiplication, and division.

Essential Question

- How does simplifying rational expressions compare to performing operations with rational numbers?

photo: Getty Images

Investigations

Those Were the Days!

What happens as the denominator of a fraction gets smaller? Some say it leads to a black hole.

Multiply and Divide Rational Expressions

Compare multiplication and division of rational numbers to rational variable expressions.

Add and Subtract Rational Expressions

Compare finding the LCD of rational numbers to finding LCD for rational variable expressions.

Expression Remix

Go on an error hunt! Look for common errors lurking in complex fractions.

Polynomial Division By a Binomial

Divide polynomials. How hard can it be? You have accumulated all the tools you need.

Equivalent Expressions

Remember finding equivalent and simplest form fractions? Explore, only with variables.

Key Vocabulary

complex fraction, denominator, divisor, fraction, least common denominator (LCD), numerator, rational number, reciprocal, term (in an expression)

Discover

As you complete Engage and the investigations, record the most important ideas you've learned.

Engage	**Investigation 1**
Investigation 2	**Investigation 3**
Investigation 4	**Investigation 5**

Name ____________________ Date __________

Check for Understanding

Develop Rational Expressions: Investigation 1

1. Match each of the expressions on the left with its simplified form on the right.

A. $\frac{3x^4y}{8y^6} \cdot \frac{6xy^4}{15x^3}$ I. $\frac{3x^8}{y^3}$

B. $\frac{(2x^7)^2}{y^8} \div \frac{12x^9}{y^{10}}$ II. $\frac{3}{x^3y^4}$

C. $\frac{4y^5}{18x^2y^2} \cdot \frac{6x^{11}y}{x^3y^6}$ III. $\frac{x^5y^2}{3}$

D. $\frac{9x^5y^6}{6y^8} \div \frac{3xy}{6x^4}$ IV. $\frac{16x^8}{3y^6}$

E. $\frac{15x^{12}}{(x^5y^3)^3} \cdot \frac{y^5}{5}$ V. $\frac{3x^2}{20y}$

F. $\frac{32x^8}{4xy^7} \div \frac{3x^2y}{2x^3y^2}$ VI. $\frac{4x^6}{3y^2}$

2. Which of the following expressions can be simplified to -1? *Select all that apply.*

A. $(x-5)\cdot(5-x)^{-1}$ B. $\frac{x+8}{x-8}$ C. $\frac{-3+x}{3+x}$

D. $\frac{x-4}{-4+x}$ E. $(x+9)^{-1}\cdot(x+9)$ F. $\frac{7-x}{x-7}$

3. What is the completely simplified form of the expression $\frac{x^2-2xy+y^2}{xy-y} \div \frac{5x-5y}{x-x^2}$? Are there any values that the variables x and/or y cannot equal? Explain your answer.

Name ______________________________ Date ______________

Check for Understanding

Develop Rational Expressions: Investigation 2

1. What is the simplified form of the rational expression $\frac{2x}{x^2 - 4x - 5} - \frac{1}{x^2 - 5x}$?

A. $\frac{2x - 1}{x(x - 5)}$

B. $\frac{2x + 1}{x(x - 5)}$

C. $\frac{(x - 1)(2x - 1)}{x(x - 5)(x + 1)}$

D. $\frac{(x - 1)(2x + 1)}{x(x - 5)(x + 1)}$

2. Consider the rational expression $\frac{x}{x^2 + 2x - 48} + \frac{2}{x^2 - 18x + 72}$. Which values of x make the simplified form of the expression undefined? *Select all that apply.*

A. –12 **B.** –8 **C.** –6 **D.** –2

E. 2 **F.** 6 **G.** 8 **H.** 12

3. On a test, Gabe got the right answer when simplifying a rational expression as shown.

$$\frac{2}{n - 4} - \frac{16}{n^2 - 16} + \frac{2}{n + 4}$$

$$\frac{2}{n - 4} - \frac{16}{(n - 4)(n + 4)} + \frac{2}{n + 4}$$

$$\frac{2(n - 4)}{(n - 4)(n + 4)} - \frac{16}{(n - 4)(n + 4)} + \frac{2(n + 4)}{(n - 4)(n + 4)}$$

$$\frac{2n - 8}{(n - 4)(n + 4)} - \frac{16}{(n - 4)(n + 4)} + \frac{2n + 8}{(n - 4)(n + 4)}$$

$$\frac{2n - 8 - 16 + 2n + 8}{(n - 4)(n + 4)}$$

$$\frac{4n - 16}{(n - 4)(n + 4)}$$

$$\frac{4(n - 4)}{(n - 4)(n + 4)}$$

$$\frac{4}{n + 4}$$

Even though his answer was correct, he didn't get full credit. Explain why this might be.

Name ____________________ Date ____________

Check for Understanding

Develop Rational Expressions: Investigation 3

1. As a first step in simplifying the rational expression $\frac{\frac{2}{x}+4}{\frac{6}{x}+8}$, would it be an error to multiply the 4 in the numerator of the greater fraction by $\frac{1}{x}$?

A. Yes, because this would change the value of the expression.

B. No, because this wouldn't change the value of the expression.

C. No, because $\frac{2}{x}$ and the product of 4 and $\frac{1}{x}$ have a common denominator.

D. Yes, because $\frac{2}{x}$ and the product of 4 and $\frac{1}{x}$ don't have a common denominator.

2. Consider the rational expression $\frac{25x-\frac{1}{x}}{\frac{1}{x}+5}$. *Fill in the blank in the sentence below.*

The simplest form of the rational expression is ____________.

3. When attempting to simplify the rational expression $\frac{3-\frac{1}{x+1}}{3+\frac{1}{x-1}}$, Jonah got an answer of $\frac{(3x+2)(x-1)}{(x+1)(3x-2)}$.

To check his work, he substituted 1 for x in both the original expression and the simplified expression. He found the value of the original expression to be undefined, since the fraction $\frac{1}{x-1}$ in the denominator of the greater fraction cannot have a 0 in its denominator. For the simplified expression, he found a value of 0. Because the two expressions had different values when x was 1, he concluded that he must have made a mistake when simplifying the original expression. Did he? Explain.

Name ______________________ Date ____________

Check for Understanding

Develop Rational Expressions: Investigation 4

1. Match each of the rational expressions on the left with its simplified form on the right.

A. $\frac{6x^2 - 5x - 6}{3x + 2}$ I. $3x + 2$

B. $\frac{8x^3 + 27}{2x + 3}$ II. $4x^2 - x + 9$

C. $\frac{8x^3 - 10x^2 + 15x - 27}{2x - 3}$ III. $2x + 3$

D. $\frac{8x^3 - 27}{2x - 3}$ IV. $4x^2 - 6x + 9$

E. $\frac{6x^2 + 5x - 6}{3x - 2}$ V. $2x - 3$

F. $\frac{6x^2 - 13x + 6}{2x - 3}$ VI. $4x^2 + 6x + 9$

G. $\frac{8x^3 + 10x^2 + 15x + 27}{2x + 3}$ VII. $3x - 2$

H. $\frac{6x^2 + 13x + 6}{2x + 3}$ VIII. $4x^2 + x + 9$

2. Which of the following rational expressions can be simplified by factoring and reducing? *Select all that apply.*

A. $\frac{5x^2 + 28x - 18}{x + 6}$ B. $\frac{2x^2 + x - 12}{x + 2}$ C. $\frac{4x^2 - 29x - 24}{x - 8}$ D. $\frac{x^2 + 10x + 27}{x + 3}$

E. $\frac{4x^2 - 19x - 10}{x - 5}$ F. $\frac{5x^2 - 16x - 16}{x - 4}$ G. $\frac{x^2 + 5x - 45}{x + 9}$ H. $\frac{2x^2 - 19x + 28}{x - 7}$

3. What is the simplified form of the rational expression $\frac{3x^4 + 12x^3 - 5x - 20}{x + 4}$? Can long division, synthetic division, and factoring and reducing all be used to simplify? Explain.

Name ______________________________ Date ______________

Check for Understanding

Develop Rational Expressions: Investigation 5

1. Selma's teacher asked her to write an expression equivalent to $\frac{4x-11}{11-4x}$. He didn't say whether it should be fully simplified. How many possible expressions can Selma write?

   A. Only one, whether the expression has to be fully simplified or not.

   B. An infinite number, whether the expression has to be fully simplified or not.

   C. Only one if the expression has to be fully simplified, and an infinite number if the expression does not have to be fully simplified.

   D. An infinite number if the expression has to be fully simplified, and only one if the expression does not have to be fully simplified.

2. Which of the following pairs of expressions are equivalent? *Select all that apply.*

   A. $\frac{(x+7)^2}{x^2+x-56} \cdot \frac{(x+8)^2}{x^2+6x-91}$ and $\frac{(x+7)(x+8)}{x+13}$

   B. $\frac{3x}{x^3} + 4$ and $\frac{4x^3+3x}{x^3}$

   C. $\frac{10}{x} - \frac{x}{10} + \frac{10}{x^2}$ and $\frac{-x^2+100x+100}{10x^2}$

   D. $\frac{5x^2-5}{8} \div \frac{x+1}{x-1}$ and $\frac{5(x+1)^2}{8}$

   E. $\frac{x^2-x-12}{x^2-9x+14} \cdot \frac{x^2-5x-14}{x^2+9x+18}$ and $\frac{x^2-2x-8}{x^2+4x-12}$

   F. $\frac{x+9}{x} - \frac{12}{2x}$ and $\frac{x+3}{x}$

3. Suppose your friend challenged you to come up with an expression in the form $\frac{x+a}{x+b} \div \frac{x-c}{x-d}$, where a, b, c, and d are constants, that is equivalent to the expression $\frac{x}{x-6} + \frac{2}{x+9}$. Is it possible? If so, what are the values of a, b, c, and d? Explain.

Summary

Before you attempt the Practice Exercises, review what you've learned.

In this lesson, you added, subtracted, multiplied, divided, and simplified rational expressions.

When you simplify an algebraic expression, the result is equivalent to the original expression. If you replace the variable with a value in both the original and simplified expressions, the resulting numeric values will be equal. Remember to note what values make the original expression undefined.

Multiplying Rational Expressions

Multiply rational expressions in the same way as numerical fractions. First, reduce any common factors. Then multiply, just like when you multiply fractions.

$$\frac{6x^2}{2} \cdot \frac{8}{3x} = 8x$$

Then, check for any restrictions on the variable (any values of x that make the expression undefined). Set the original denominator that contains a variable equal to 0:

$$3x = 0$$
$$x = 0$$

The expression is undefined when $x = 0$, so the restriction on the variable is $x \neq 0$.

My Notes

Dividing Rational Expressions

To divide by an algebraic rational expression, multiply by the reciprocal of the divisor, just like when you divide by a fraction.

$$\frac{x^2 - 9}{x^3} \div \frac{x^2 + 3x}{x^5} = \frac{x^2 - 9}{x^3} \cdot \frac{x^5}{x^2 + 3x}$$

First, factor the expression to help you simplify it.

$$\frac{(x+3)(x-3)}{x^3} \cdot \frac{x^5}{x(x+3)}$$

Then, identify common factors to simplify it.

$$(x - 3) \cdot x$$
$$x^2 - 3x$$

Identify the restrictions on the variable:

$$x \neq 0$$
$$x \neq -3$$

Adding and Subtracting Rational Expressions

When you add and subtract algebraic rational expressions, you need to find a common denominator, just like when you add and subtract numerical fractions.

$$\frac{1}{x} + \frac{2}{x^2} - \frac{3}{2x}$$

A common denominator is $2x^2$.

Rewrite each expression with the common denominator, combine, and simplify:

$$\frac{1}{x}\left(\frac{2x}{2x}\right) + \frac{2}{x^2}\left(\frac{2}{2}\right) - \frac{3}{2x}\left(\frac{x}{x}\right)$$
$$\frac{2x}{2x^2} + \frac{4}{2x^2} - \frac{3x}{2x^2}$$
$$\frac{2x + 4 - 3x}{2x^2}$$
$$\frac{-x + 4}{2x^2}$$

Looking back at the original expressions, $x \neq 0$.

Simplifying Complex Algebraic Fractions

You can apply the same ideas to simplify algebraic fractions.

Practice Exercises

Review what you've learned using these practice problems. For practice problems with feedback, try the Coach and Play items in the Practice section online.

1. Ms. Hillard wrote this synthetic division on the board to represent a rational expression, $\frac{p(x)}{q(x)}$:

$$\begin{array}{r|rrrr} -4 & 1 & 0 & 0 & 16 \\ & & -4 & 16 & -64 \\ \hline & 1 & -4 & 16 & -48 \end{array}$$

Which of the following statements are true about the rational expression $\frac{p(x)}{q(x)}$? *Select all that apply.*

   A. Both $p(x)$ and $q(x)$ are binomial expressions.

   B. The numerator, $p(x)$, is a cubic polynomial.

   C. The denominator is $q(x) = x - 4$.

   D. The denominator $q(x)$ is a factor of the numerator $p(x)$.

   E. The numerator is $p(x) = x^3 + 16$.

2. Given $h(t) = \frac{4}{t}$ and $k(t) = \frac{t+5}{t-1}$, determine the expressions that are equal to these combinations of $h(t)$ and $k(t)$.

   A. $\frac{t^2+t+4}{t^2+9t-4}$

   B. $-\frac{t^2+9t-4}{t^2+t+4}$

   C. $\frac{t^2+4t+4}{t^2-1}$

   D. $-\frac{t^2+t+4}{t^2-t}$

   E. $\frac{-t^2+4t-6}{t^2-1}$

   F. $\frac{t^2+9t-4}{t^2-t}$

$h(t) + k(t)$ ____________

$h(t) - k(t)$ ____________

$\frac{h(t)+k(t)}{h(t)-k(t)}$ ____________

3. Which of the following is the quotient of $\frac{2x^3+9x^2+5}{2x+1}$?

   A. $x^2 + 4 + \frac{1}{2x+1}$

   B. $2x^2 + 7x - 7 + \frac{12}{2x+1}$

   C. $x^2 + 8x + 4 + \frac{5}{2x+1}$

   D. $x^2 + 4x - 2 + \frac{7}{2x+1}$

   E. $2x^2 + 9x + 4 + \frac{4}{2x+1}$

4. The polynomial function $p(x) = 2x^3 + 5x^2 - 18x - 45$ is being divided by various binomial factors. Classify each quotient as being evenly divisible, or as not being evenly divisible and therefore having a remainder.

   A. $\frac{2x^3+5x^2-18x-45}{x+3}$

   B. $\frac{2x^3+5x^2-18x-45}{x-3}$

   C. $\frac{2x^3+5x^2-18x-45}{x-9}$

   D. $\frac{2x^3+5x^2-18x-45}{2x+5}$

   E. $\frac{2x^3+5x^2-18x-45}{2x-5}$

   F. $\frac{2x^3+5x^2-18x-45}{2x-9}$

Evenly divisible: ____________

Not evenly divisible: ____________

5. Determine values of m, n, p, and q so that:

$$\frac{x^2+mx+n}{9-x^2} \div \frac{x^2+6x-16}{x^2+px+q} = \frac{-x+3}{x+8}, \text{ for all } x > 3.$$

$m =$ ____________ $n =$ ____________

$p =$ ____________ $q =$ ____________

Practice Exercises *(continued)*

6. The incomplete work below represents $(2x^3 - x^2 + 4x + 8) \div (x + 2)$

$$\begin{array}{r|rrrr} -2 & 2 & -1 & 4 & 8 \\ & & \boxed{A} & 10 & \boxed{B} \\ \hline & 2 & \boxed{C} & 14 & \boxed{D} \end{array}$$

Each letter represents a missing value from synthetic division. Fill in the values below.

A. = ____________

B. = ____________

C. = ____________

D. = ____________

7. Match each rational expression with its equivalent rational expression.

A. $\frac{2x+1}{x+2} - \frac{x+3}{x+3}$

B. $\frac{2x+2}{x+2} - \frac{x+5}{x+3}$

C. $\frac{2x+4}{x+2} - \frac{x+7}{x+3}$

$\frac{x-1}{x+2}$ ____________

$\frac{x-1}{x+3}$ ____________

$\frac{x^2+x-4}{x^2+5x+6}$ ____________

8. Simplify the complex algebraic fraction: $\dfrac{\frac{1}{t+2} + \frac{1}{t-2}}{2t - \frac{2t^2}{t+2}}$

A. $\frac{t}{2(t-2)}$

B. $\frac{1}{2t-4}$

C. $\frac{2t+4}{2t^2}$

D. $\frac{6t}{t+2}$

E. $\frac{2}{t-2}$

9. The reciprocal of the total resistance in an electrical circuit is the sum of the reciprocals of the individual resistances for n resistors in the circuit.

$$\frac{1}{T_n} = \frac{1}{R_1} + \frac{1}{R_2} + \frac{1}{R_3} + \cdots \frac{1}{R_n}$$

Create the expressions for total resistance of a circuit with two resistors, to complete the equations below.

A. $\frac{R_1 + R_2}{R_1R_2}$

B. $\frac{2R_1 + 2R_2}{R_1 + R_2}$

C. $\frac{R_1R_2}{R_1 + R_2}$

D. $\frac{R_1 + R_2}{2(R_1 + R_2)}$

$\frac{1}{T_2}$ = ____________

T_2 = ____________

10. A commuter plane makes daily flights between two cities. The distance between the two airports is 250 miles. The average wind velocity is v miles per hour. The average speed of the airplane is r miles per hour.

Which of the following expressions represents how much faster, in hours, it is to travel going in the same direction as the wind?

A. $\frac{250}{2r - 2v}$

B. $\frac{500}{r^2 + v}$

C. $\frac{250v}{r^2 + v^2}$

D. $\frac{500v}{r^2 - v^2}$

Apply

What Does It Take to Be a Top Triathlete?

photo: iStock

A long-distance triathlon race is an intense test of fitness and endurance. Research the events in such a race, the distances involved in each event, and the time limits for each part of the race. Any participant who meets the time limits is declared a champion.

Suppose a participant can bike 13 mph faster than they swim, and can run 4 times faster than they swim.

Write a rational function that gives the participant's time for the race based on the swim speed. What is the minimum swim speed for the participant to be declared a champion? At that speed, what is the participant's finishing time for the entire race?

Show what you've learned by completing the other performance tasks in the online Apply section.

Apply *(continued)*

Your answer to Apply will be assessed on the following criteria:

1. Researching and citing appropriate long-distance triathlon race information
2. Writing a rational function based on the participant's relative speeds for each part of the race
3. Finding the minimum swim speed that would satisfy the swim, bike, and run times, with the given relative speed information
4. Answering the final question and explaining your answer

Scale / Criteria	4 Exceeds Criteria	3 Meets Criteria	2 Progressing to Criteria	1 Below Expectations	0 No Expectation
Race Information	Correctly cites appropriate long-distance triathlon race information.	Cites appropriate long-distance triathlon race information, with some lack of clarity.	Cites mostly accurate long-distance triathlon race information.	Race information is incomplete or incorrect.	Does not attempt.
Function	Correctly writes a rational function based on the participant's relative speeds for each part of the race.	Writes a rational function based on the participant's relative speeds for each part of the race, with minor errors.	Writes a rational function based on the participant's relative speeds for each part of the race, with significant errors.	Writes a function that is not rational or is not based on the participant's relative speeds.	Does not attempt.
Minimum Swim	Finds the minimum swim speed that would satisfy the swim, bike, and run times, with the given relative speed information.	Finds the minimum swim speed that would satisfy the swim, bike, and run times, with the given relative speed information, with minor errors.	Finds a swim speed that is close to but not the minimum, that mostly satisfies the swim, bike, and run times.	Finds a swim speed that is not the minimum and does not satisfy the swim, bike, and run times.	Does not attempt.
Answer to Question	Correctly answers and explains the final question.	Correctly answers the final question, with a weak explanation.	Answers and explains the final question, with minor errors.	Incorrectly answers final question, with poor or missing explanation.	Does not attempt.

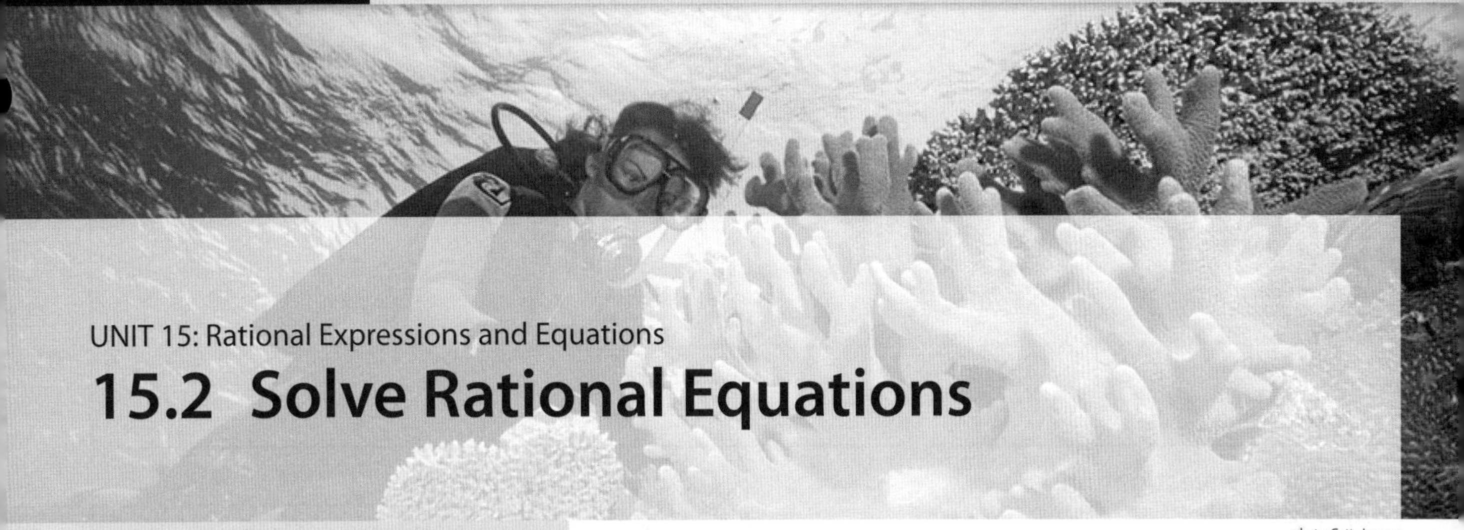

UNIT 15: Rational Expressions and Equations

15.2 Solve Rational Equations

photo: Getty Images

Lesson Objectives

- Solve rational equations and inequalities.
- Model real-world situations and solve problems with rational equations or inequalities.
- Use inverse variation to solve real-world rational functions.

Essential Question

- How does solving rational equations compare and contrast with solving polynomial equations?

photo: Corbis

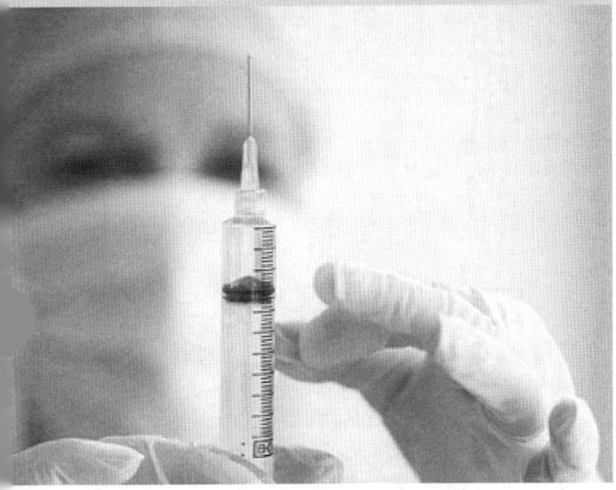

Investigations

Feeling a Little Pressure

Under the sea—a world of inverse variation.

Analyzing Work Rates

Working together lightens the load and increases the rate of work.

Explore Ways to Solve Rational Equations

Solve rational equations. Are all of the solutions defined?

Prescribing the Correct Medication

Solve rational inequalities. It's a balancing act of the numerator and denominator.

Inverse Variation

Graph inverse variations that model real-world situations. Use key features to make sense of the situation.

Key Vocabulary

extraneous (roots, solutions), inverse variation

Discover

As you complete Engage and the investigations, record the most important ideas you've learned.

Engage	
Investigation 1	**Investigation 2**
Investigation 3	**Investigation 4**

Name ______________________________ Date ______________

Check for Understanding

Solve Rational Equations: Investigation 1

1. Working alone, a landscaper could build a retaining wall in 38 hours, while his apprentice could build the wall in 62 hours. If t represents the number of hours it would take them to build the wall while working together, which equation is correct?

   A. $\frac{1}{38t} + \frac{1}{62t} = 1$ B. $\frac{1}{38t} \cdot \frac{1}{62t} = 1$ C. $\frac{1}{38} + \frac{1}{62} = \frac{1}{t}$ D. $\frac{1}{38} \cdot \frac{1}{62} = \frac{1}{t}$

2. Every night after closing, the moat in the hippopotamus exhibit at a zoo is drained and cleaned. The zoo currently has one pipe that can drain the moat in 50 minutes. It is considering installing a second pipe to speed up draining. The table below shows different times that the new pipe might take to drain the moat alone. In each row of the table, fill in the number of minutes it would take for both pipes to drain the pool simultaneously. *Round your answers to the nearest tenth of a minute, when necessary.*

Time for New Pipe to Drain Moat Alone (minutes)	Time for Both Pipes to Drain Moat Simultaneously (minutes)
25	
40	
55	
70	

3. Herb and Eleanor have part-time jobs in which they stuff envelopes with fliers that are mailed out to prospective customers. Working alone, Herb can stuff 80 envelopes in 16 minutes, and Eleanor can stuff 84 envelopes in 12 minutes. Determine how long it would take each of them to stuff 210 envelopes working alone. Then, using what you found, set up a rational equation in the variable t, where t represents the number of minutes it would take them to stuff the 210 envelopes working together. Finally, solve the equation, and explain why the solution makes sense given the information provided.

Name ______________________________ Date ____________

Check for Understanding

Solve Rational Equations: Investigation 2

1. What is the solution to the rational equation $\frac{4}{a} + \frac{a+4}{5a} = \frac{8a-9}{6a} - \frac{1}{12}$?

A. $a = 1$ **B.** $a = 5$ **C.** $a = 6$ **D.** $a = 12$

2. Which of these rational equations have a solution of $p = 3$? *Select all that apply.*

A. $\frac{12}{1-p} - \frac{p+5}{p+2} = \frac{20}{1-p}$

B. $\frac{3}{p-3} + \frac{8}{p^2+p-12} = 4$

C. $\frac{p}{6} + 1 = \frac{p}{2} - 3$

D. $\frac{1}{p-5} = \frac{9}{p^2+p-30}$

E. $\frac{15}{3-p} + \frac{3}{p+3} = \frac{18}{p+3}$

F. $\frac{8}{p+5} - \frac{10}{p^2+3p-10} = 5$

G. $\frac{p}{10} + 3 = \frac{p}{3} - 4$

H. $\frac{p+2}{p+7} - \frac{6}{p-6} = \frac{25}{p+7}$

I. $\frac{7}{10p} + \frac{2}{p} = \frac{16}{15} - \frac{1}{2p}$

3. Ken attempted to solve a rational equation by cross-multiplying and factoring as shown. He claims that $x = 8$ is the equation's solution, while $x = 2$ is an extraneous solution.

$$\frac{16}{x^2+6x-16} = \frac{10}{x^2-4}$$
$$16(x^2-4) = 10(x^2+6x-16)$$
$$16x^2-64 = 10x^2+60x-160$$
$$16x^2-64-10x^2-60x+160 = 10x^2+60x-160-10x^2-60x+160$$
$$6x^2-60x+96 = 0$$
$$6(x^2-10x+16) = 0$$
$$6(x-2)(x-8) = 0$$
$$x = 2, 8$$

Lena attempted to solve the same equation, but instead of cross multiplying, she wrote the fractions with their least common denominator and then set their numerators equal to each other. She claims that there is no extraneous solution. Who is correct? Explain.

Name ______________________ Date __________

Check for Understanding

Solve Rational Equations: Investigation 3

1. In which intervals do all values of x satisfy the rational inequality $\frac{(x-2)(x+4)}{(x+5)} \le 0$?

A. $(-\infty, -5)$ and $(-5, -4]$

B. $(-\infty, -5)$ and $[-4, 2]$

C. $(-5, -4]$ and $[2, \infty)$

D. $[-4, 2)$ and $(2, \infty)$

2. A number line can be divided into four intervals to help find the solution to the rational inequality $x - 5 < \frac{x+27}{x}$. Fill in the blanks to complete the sentence below.

Left to right, the first interval should include values less than ______, the second should include values between ______ and ______, the third should include values between ______ and ______, and the fourth should include values greater than ______.

3. To help find the values of x that satisfy the rational inequality $\frac{x^2 - 2x - 8}{x - 4} > 0$, Conrad correctly graphed the function $f(x) = \frac{x^2 - 2x - 8}{x - 4}$ on his graphing calculator as shown.

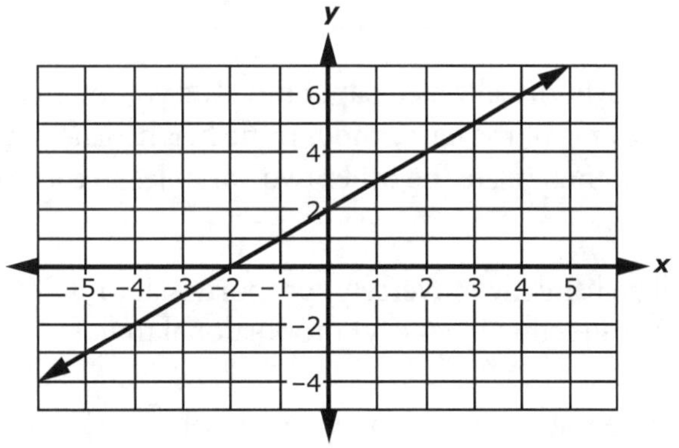

Based on the graph, he concluded that all values of x in the interval $(-2, \infty)$ satisfy the inequality. Is Conrad's conclusion correct? Why or why not? Explain your reasoning.

Name ______________________________ Date ______________

Check for Understanding

Solve Rational Equations: Investigation 4

1. A cello string's length, l, is inversely proportional to its frequency, f, when vibrating. A 27.5-inch string vibrates at a frequency of 220 cycles per second. Which equation correctly relates the length, in inches, of a cello string to its vibration frequency?

   A. $l = \frac{1}{8}f$ **B.** $l = 8f$ **C.** $l = \frac{6{,}050}{f}$ **D.** $l = \frac{f}{6{,}050}$

2. The number of hours it takes for a block of ice of a certain size to melt is inversely proportional to the surrounding air temperature in degrees Fahrenheit. At 50°F, a block of ice of this size takes 4.8 hours to melt. *Fill in the missing values in the table below.*

Temperature (°F)	Hours for Ice to Melt
60	
	3.75
75	
	3
96	

3. A manufacturer is designing a cylindrical waste paper bin. The volume, V, of the bin is given by the formula $V = Bh$, where B is the area of the base, and h is the height. The manufacturer plans to set one variable equal to a constant and then relate the other two variables to each other in an equation. *Circle the correct words in the sentence.*

   In the equation, either V will be [directly / inversely] proportional to B, V will be [directly / inversely] proportional to h, or B will be [directly / inversely] proportional to h.

4. Ingrid will bike across her home state to raise money for charity. She knows that the duration in hours of her trip will be inversely proportional to the average speed in miles per hour at which she travels. She wants to write an equation relating duration to average speed, but she isn't sure which she wants to be the independent variable, and which she wants to be the dependent variable. How will the graph of her equation change, if at all, depending on what she decides? Explain how you got your answer.

Summary

Before you attempt the Practice Exercises, review what you've learned.

In this lesson, you solved rational equations and inequalities of varying complexity by applying what you know about rational expressions in one variable.

> A **rational equation** is an equation in which one or more of the terms contain a rational expression.

Solving Rational Equations

EXAMPLE

Solve $\frac{1}{2} + \frac{x}{6} = \frac{18}{x}$.

SOLUTION:

Look for any restricted values of the domain. Since x is in the denominator of $\frac{18}{x}$, $x \neq 0$.

$\frac{1}{2} + \frac{x}{6} = \frac{18}{x}$	Find the LCD of all the terms of the equation. The LCD of the denominators 2, 6, and x is $6x$.
$(6x)\left(\frac{1}{2} + \frac{x}{6}\right) = \left(\frac{18}{x}\right)(6x)$	Multiply each side of the equation by the LCD.
$(6x)\left(\frac{1}{2}\right) + (6x)\left(\frac{x}{6}\right) = \left(\frac{18}{x}\right)(6x)$	Distribute $6x$.
$3x + x^2 = 108$	Simplify.
$x^2 + 3x - 108 = 0$	Write in standard form.
$(x + 12)(x - 9) = 0$	Factor.
$(x + 12) = 0$ or $(x - 9) = 0$	Solve.
$x = -12$ or $x = 9$	

The solutions to $\frac{1}{2} + \frac{x}{6} = \frac{18}{x}$ are $x = -12$ and $x = 9$.

Check all possible solutions by substituting each solution in the original equations.

> Any possible solution that does not result in a true statement for the original equation is an **extraneous solution.**

Substitute -12 for x in the original equation.

Substitute 9 for x in the original equation.

$\frac{1}{2} + \frac{x}{6} = \frac{18}{x}$	$\frac{1}{2} + \frac{x}{6} = \frac{18}{x}$
$\frac{1}{2} + \frac{-12}{6} \stackrel{?}{=} \frac{18}{-12}$	$\frac{1}{2} + \frac{9}{6} \stackrel{?}{=} \frac{18}{9}$
$\frac{1}{2} + -2 \stackrel{?}{=} -1\frac{1}{2}$	$\frac{1}{2} + 1\frac{1}{2} \stackrel{?}{=} 2$
$-1\frac{1}{2} = -1\frac{1}{2}$	$2 = 2$

Since -12 and 9 satisfy the conditions of the original equation, they are both solutions of the equation.

My Notes

Summary *(continued)*

EXAMPLE

Solve $\frac{2}{b-4} = \frac{16}{b^2-16}$.

SOLUTION:

Look for any restricted values of the domain. Since $b - 4$ and $b^2 - 16$ are in the denominators of the terms of the original equation, $b \neq 4$ and $b \neq -4$.

$\frac{2}{b-4} = \frac{16}{b^2-16}$	Find the LCD of all the terms of the equation. The LCD of the denominators $b - 4$ and $b^2 - 16$ is $(b - 4)(b + 4)$.
$(b-4)(b+4)\left(\frac{2}{b-4}\right) = \left(\frac{16}{b^2-16}\right)(b-4)(b+4)$	Multiply each side of the equation by the LCD.
$2(b + 4) = 16$	Simplify.
$2b + 8 = 16$	Distribute 2.
$2b = 8$	Subtract 8.
$b = 4$	Divide.

The solution $b = 4$ is extraneous, because it makes the denominators of the original equation equal to 0.

So $\frac{2}{b-4} = \frac{16}{b^2-16}$ has no solution.

My Notes

EXAMPLE

Solve $\frac{5}{b-7} - \frac{b}{2} = \frac{2b-9}{b-7}$.

SOLUTION:

Look for any restricted values of the domain. Since $b - 7$ is in the denominator of two terms in the original equation, $b \neq 7$.

$\frac{5}{b-7} - \frac{b}{2} = \frac{2b-9}{b-7}$	Find the LCD of all the terms of the equation. The LCD of the denominators is $2(b - 7)$.
$2(b-7)\left(\frac{5}{b-7} - \frac{b}{2}\right) = \left(\frac{2b-9}{b-7}\right)2(b-7)$	Multiply each side of the equation by the LCD.
$2(b-7)\left(\frac{5}{b-7}\right) - 2(b-7)\left(\frac{b}{2}\right) = \left(\frac{2b-9}{b-7}\right)2(b-7)$	Distribute $2(b - 7)$.
$10 - b(b - 7) = 2(2b - 9)$	Simplify.
$10 - b^2 + 7b = 4b - 18$	Distribute b and 2.
$b^2 - 3b - 28 = 0$	Write in standard form.
$(b - 7)(b + 4) = 0$	Factor.
$(b - 7) = 0$ or $(b + 4) = 0$	Solve.
$b = 7$ or $b = -4$	

The solution $b = 7$ is extraneous, because it makes the denominators of the original equation equal to 0.

Check the remaining possible solution by substituting the solution in the original equation.

Substitute -4 for b in the original equation.

$$\frac{5}{b-7} - \frac{b}{2} - \frac{2b-9}{b-7}$$

$$\frac{5}{-4-7} - \frac{-4}{2} \stackrel{?}{=} \frac{2(-4)-9}{-4-7}$$

$$\frac{5}{-11} + 2 \stackrel{?}{=} \frac{-17}{-11}$$

$$-\frac{5}{11} + \frac{22}{11} \stackrel{?}{=} \frac{17}{11}$$

$$\frac{17}{11} = \frac{17}{11}$$

Since -4 satisfies the conditions of the original equation, it is a solution of the equation.

Summary *(continued)*

Solving Rational Inequalities

A A **rational inequality** is an inequality that contains a rational expression.

EXAMPLE

Solve $\frac{(x-5)(x+3)}{x-2} \geq 0$.

SOLUTION:

To solve a rational inequality, determine any values that make the denominators 0. These are values that are restricted from the domain of the original problem, and therefore are eliminated if they surface in the solution set.

$$\frac{(x-5)(x+3)}{x-2} \geq 0$$

The value that makes the denominator $(x-2)$ equal 0 is 2.

Determine any values that make the rational expression equal to 0.

$$\frac{(x-5)(x+3)}{x-2} \geq 0$$

The values that make the rational expression equal to 0 result from the factors of the numerator $(x-5)$ and $(x+3)$. For this problem, they are 5 and −3.

Define intervals and select test points to examine if the rational function is positive or negative and to determine the intervals where the value of function changes signs.

$$\frac{(x-5)(x+3)}{x-2} \geq 0$$

	$x < -3$	$x = -3$	$-3 < x < 2$
$(x+3)$	−	0	+
$(x-5)$	−	−	−
$(x-2)$	−	−	−
$\frac{(x-5)(x+3)}{x-2}$	−	0	+

	$x = 2$	$2 < x < 5$	$x = 5$	$x > 5$
$(x+3)$	+	+	+	+
$(x-5)$	−	−	0	+
$(x-2)$	−	+	+	+
$\frac{(x-5)(x+3)}{x-2}$	undefined	−	0	+

Determine the intervals that satisfy the inequality. These are your solutions.

The intervals [−3, 2) and [5, ∞) include the values for x that make the inequality true.

Confirm the solutions by graphing the related function $f(x) = \frac{(x-5)(x+3)}{x-2}$. The solutions represent the intervals where the graph is above or on the x-axis.

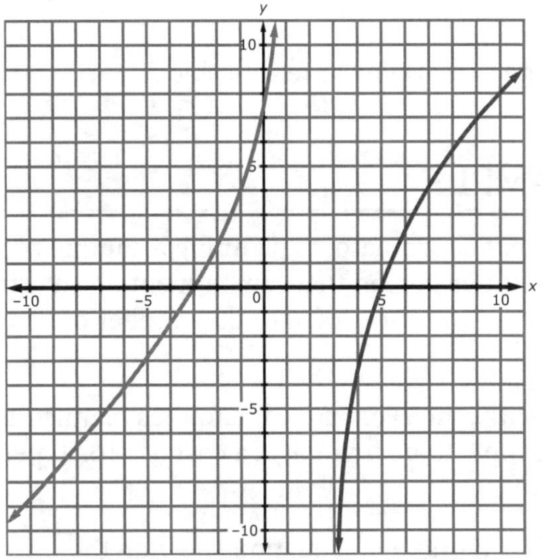

Be sure that your inequality is written so that one side of the inequality is zero.

My Notes

Summary *(continued)*

Inverse Variation

Quantities in an inverse variation are said to vary inversely with each other or to be inversely proportional to each other.

- They can be modeled by equations of the form $y = \frac{k}{x}$, where k is a nonzero constant.
- The product of x and y is a constant, $xy = k$.
- As the value of one quantity increases, the value of the other decreases.
- They are nonlinear relationships.

EXAMPLE

An event coordinator is planning a banquet. The amount per person that can be spent on catering is inversely proportional to the number of people invited.

Currently, the event coordinator has planned for 250 people to be invited, at a cost of $16.50 per person.

Write and graph an equation showing the relationship of the cost per person to the number of people invited.

My Notes

Suppose the client decides to invite 50 more people. Assuming the catering budget is fixed, how much per person can be spent on catering?

SOLUTION:

The inverse variation equation is $y = \frac{k}{x}$. Let y be the amount per person that can be spent on catering, and x be the number of people invited. The constant of variation k can be expressed as xy, so for the given information:

$$\begin{aligned} k &= xy \\ &= (250)(16.50) \\ &= (4{,}125) \end{aligned}$$

So, the constant of variation is $k = 4{,}125$, which in the context of the problem means that the total catering budget is $4,125.

The equation is $y = \frac{4{,}125}{x}$, shown on the graph. The point (250, 16.5) represents the amount, $16.50, that can be spent per person for 250 people invited.

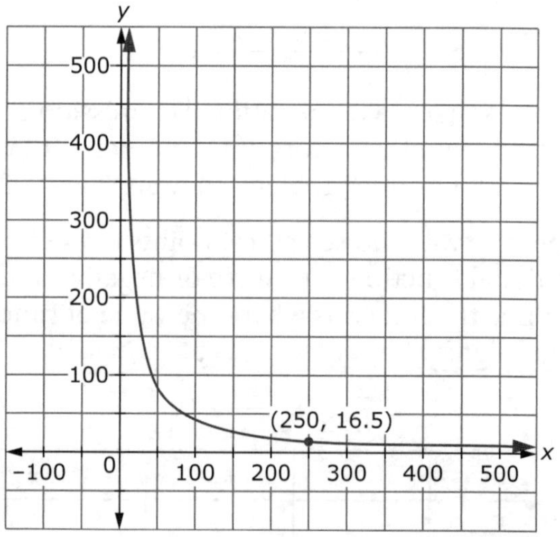

If 50 more people are invited, then $x = 300$. If you trace the graph, the y value at $x = 300$ is $y \approx 13.8$. You can also solve algebraically:

$$\begin{aligned} y &= \frac{4{,}125}{x} \\ &= \frac{4{,}125}{300} \\ &= 13.75 \end{aligned}$$

So, the event coordinator can spend $13.75 per person on catering if there are 50 more people.

Practice Exercises

Review what you've learned using these practice problems. For practice problems with feedback, try the Coach and Play items in the Practice section online.

1. Which equations are equivalent to $\frac{m}{m+3} = \frac{m+1}{m}$? *Select all that apply.*

   A. $m = -\frac{3}{4}$

   B. $m = \frac{4}{3}$

   C. $m^2 = m^2 + 4m + 3$

   D. $0 = 2m^2 + 4m + 3$

   E. $4m = -3$

   F. $m(m + 3) = m(m + 1)$

   G. $2m = (m + 3)(m + 1)$

2. Which statement best describes the process of solving rational equations?

   A. Solving rational equations is like solving quadratic equations because they always have 0, 1, or 2 solutions.

   B. Solving rational equations is like solving absolute value equations because they can both be represented by piecewise functions.

   C. Solving rational equations is like solving proportions because they can be solved by multiplying both sides of the equation by the least common denominator.

   D. Solving rational equations is like solving polynomial equations because they both have a domain of all real numbers.

3. Consider this rational equation.

   $\frac{6}{r} + \frac{4}{r-4} = 1$

   For which two values of r is the rational equation undefined?

   ____________ and ____________

   For which two values of r is the rational equation true?

   $r =$ ____________ and $r =$ ____________

4. A remote control airplane has a velocity of v miles per hour. The current wind outside is 10 miles per hour. The airplane can fly 0.5 miles with the wind in the same amount of time it takes the airplane to fly 0.3 miles against the wind.

   Which statements describe the airplane? *Select all that apply.*

   A. The airplane has a velocity of 50 miles per hour.

   B. If there is no wind, the airplane will fly 0.4 miles in the same amount of time.

   C. In 3 minutes, the airplane can fly 15 miles with the wind.

   D. In 15 minutes, the airplane can fly 10 miles with no wind.

   E. The airplane will not travel any distance forward if the speed of the wind against it is 40 miles per hour.

5. Consider the rational equation, $\frac{2x}{x-1} = 1 + \frac{5x-1}{x^2-1}$.

   One way to solve rational equations is to multiply by the least common denominator and solve the resulting polynomial equation:

   $$\frac{2x}{x-1} = 1 + \frac{5x-1}{x^2-1}$$
   $$(x^2-1)\frac{2x}{x-1} = \left(1 + \frac{5x-1}{x^2-1}\right)(x^2-1)$$
   $$(x+1)2x = (x^2-1) + (5x-1)$$
   $$2x^2 + 2x = x^2 - 1 + 5x - 1$$

   Determine solutions to this polynomial equation and to the original rational equation.

   A. -3 **B.** -1 **C.** 1

   D. 2 **E.** 3

   Solutions to $\frac{2x}{x-1} = 1 + \frac{5x-1}{x^2-1}$:

   ____

   Solutions to $2x^2 + 2x = x^2 - 1 + 5x - 1$:

   ____

Practice Exercises *(continued)*

6. A rectangular banquet room has an area of 182 square meters and a perimeter of 54 meters. Let l and w represent the dimensions of the room in meters.

   Which statements describe the dimensions of the banquet room? *Select all that apply.*

   A. The length of the banquet room is equivalent to $\frac{182}{w}$.

   B. The length of the banquet room is equivalent to $27 - w$.

   C. The equation $\frac{182}{w} = 54 - 2w$ can be used to solve for the width of the banquet room.

   D. Both the length and the width are even numbers.

   E. The length of the banquet room is 1 meter longer than the width.

7. The relationship between x and y is an inverse variation and can be modeled by the equation $y = \frac{k}{x}$.

   If $y = 50$ when $x = 20$, then $y =$ ______ when $x = 25$.

8. What are the solutions of the following inequality?

   $\frac{4z}{z-3} > z - 3$

   Complete the inequalities that represent the solutions. *Write the complete inequality.*

   $z <$ ______

   or

   ______ $< z <$ ______

9. Consider $R(x) = \frac{x^2 - 4}{x - 4}$.

   For each of the intervals below, determine if $R(x) > 0$ or $R(x) < 0$. *Some intervals may not be used.*

   A. $(-\infty, -2)$ **B.** $(-2, 2)$ **C.** $(-2, 4)$

   D. $(2, 4)$ **E.** $(2, \infty)$ **F.** $(4, \infty)$

   $R(x) > 0$: ______

   $R(x) < 0$: ______

10. The table shows the relationship between the speed, in miles per hour (mph), of a treadmill and the number of calories burned per hour.

Speed (in mph)	Calories per Hour
1	120
2	240
3	360
4	480
5	600
6	720
7	840

Use the information in the table to complete the statements.

At a treadmill speed of 2 mph, it will take ______ seconds for a person to burn one calorie.

A. 0.417 **B.** 4 **C.** 6.67

D. 10 **E.** 15

The time it takes to burn one calorie ______ as the speed of the treadmill doubles.

A. doubles

B. is halved

C. remains the same

An equation that models the relationship between the speed s, in mph, of a treadmill and the time t, in seconds, for a person to burn one calorie is ______, for some nonzero constant k.

A. $t = s + k$ **B.** $t = \frac{s}{k}$

C. $t = \frac{k}{s}$ **D.** $t = ks$

The value of k is ______

A. 120 **B.** 30 **C.** 15

D. 10 **E.** 5

Apply

How Big Should the Family Patio Be?

Will the client approve of the patio design?

A patio construction company has been hired to design and install a new poured-concrete patio. The client wants a basic rectangular-shaped patio, with a width that is 7 feet shorter than the length. The client would like the patio to be as long as possible. Concrete is sold by the cubic yard, and the client's budget allows for 1.5 cubic yards of concrete. A concrete expert for the company recommends that the depth of the patio should be at least 5 inches.

Create a design plan for the client that shows the length, width, and depth of the patio. Length and width measurements given to the client are rounded to the nearest tenth of a foot.

Show what you've learned by completing the other performance tasks in the online Apply section.

Apply *(continued)*

Your answer to Apply will be assessed on the following criteria:

1. Writing a rational inequality to represent the problem
2. Graphing the rational inequality
3. Calculating the solution for length
4. Completing the problem for all components with explanations

Criteria \ Scale	4 Exceeds Criteria	3 Meets Criteria	2 Progressing to Criteria	1 Below Expectations	0 No Expectation
Writes an Inequality	Writes a correct rational inequality with all measures converted correctly to feet or cubic feet.	Correctly converts all measures to feet or cubic feet, but makes a minor error in the rational inequality.	Correctly converts all measures to feet, but does not write a rational inequality, or writes a correct rational inequality, but does not convert to feet.	Writes an inequality or equation that has major flaws, such as not using a rational expression.	Does not write an inequality or equation.
Graphs the Inequality	Creates a clearly labeled, accurate graph that shows applicable asymptotes and constraints.	Creates an accurate graph that lacks some detail (labels, asymptotes, and/or constraints, such as using the portions of the graph where real answers do not occur).	Creates a graph that is not accurate but resembles the correct graph, such as having a wrong asymptote or wrong labels.	Creates an inaccurate graph, such as one having no asymptotes, or the graph of a polynomial.	Does not graph the inequality.
Solves for Length	Solves the inequality for length, with appropriate answers given to the nearest tenth.	Solves the inequality for length with answers given to the nearest tenth, with one calculation error.	Solves the inequality for length with answers given to the nearest tenth, but has two or more calculation errors and/or does not use a table.	Attempts to solve the inequality, but does not have a correct solution.	Does not solve for length.
Calculations	Correctly calculates each part of the problem for the given inequality, leading to the correct solution, and explains the solution.	Completes calculations for the given inequality for each part of the problem, with minor calculation errors (rounding, etc.), and explains the solution.	Completes calculations for the given inequality for each part of the problem, but has major calculation errors (such as multiplication error) and/or no explanation.	Does not complete the calculations for the given inequality.	Does not attempt calculations.

Notes

Notes